T0175623

Properties of Life

Toward a Theory of Organismic Biology

Bernd Rosslenbroich

The MIT Press
Cambridge, Massachusetts
London, England

The MIT Press would like to thank the anonymous peer reviewers who provided comments on drafts of this book. The generous work of academic experts is essential for establishing the authority and quality of our publications. We acknowledge with gratitude the contributions of these otherwise uncredited readers.

This book was set in Times New Roman by Westchester Publishing Services. Printed and bound in the United States of America.

Library of Congress Cataloging-in-Publication Data

Names: Rosslenbroich, Bernd, author.
Title: Properties of life : toward a theory of organismic biology / Bernd Rosslenbroich.
Description: Cambridge, Massachusetts : The MIT Press, [2023] | Series: Vienna series
    in theoretical biology | Includes bibliographical references and index.
Identifiers: LCCN 2022046854 (print) | LCCN 2022046855 (ebook) | ISBN 9780262546201
    (paperback) | ISBN 9780262375382 (epub) | ISBN 9780262375399 (pdf)
Subjects: LCSH: Organisms. | Life (Biology)
Classification: LCC QH307.2 .R66 2023 (print) | LCC QH307.2 (ebook) | DDC 570—dc23/
    eng/20230126
LC record available at https://lccn.loc.gov/2022046854
LC ebook record available at https://lccn.loc.gov/2022046855

10   9   8   7   6   5   4   3   2   1

Properties of Life

Vienna Series in Theoretical Biology
Gerd B. Müller, editor-in-chief
Thomas Pradeu and Katrin Schäfer, associate editors

The Major Transitions in Evolution Revisited, edited by Brett Calcott and Kim Sterelny, 2011

Transformations of Lamarckism, edited by Snait B. Gissis and Eva Jablonka, 2011

Convergent Evolution: Limited Forms Most Beautiful, by George McGhee, 2011

From Groups to Individuals, edited by Frédéric Bouchard and Philippe Huneman, 2013

Developing Scaffolds in Evolution, Culture, and Cognition, edited by Linnda R. Caporael, James Griesemer, and William C. Wimsatt, 2013

Multicellularity: Origins and Evolution, edited by Karl J. Niklas and Stuart A. Newman, 2016

Vivarium: Experimental, Quantitative, and Theoretical Biology at Vienna's Biologische Versuchsanstalt, edited by Gerd B. Müller, 2017

Landscapes of Collectivity in the Life Sciences, edited by Snait B. Gissis, Ehud Lamm, and Ayelet Shavit, 2017

Rethinking Human Evolution, edited by Jeffrey H. Schwartz, 2018

Convergent Evolution in Stone-Tool Technology, edited by Michael J. O'Brien, Briggs Buchanan, and Metin I. Erin, 2018

Evolutionary Causation: Biological and Philosophical Reflections, edited by Tobias Uller and Kevin N. Lala, 2019

Convergent Evolution on Earth: Lessons for the Search for Extraterrestrial Life, by George McGhee, 2019

Contingency and Convergence: Toward a Cosmic Biology of Body and Mind, by Russell Powell, 2020

How Molecular Forces and Rotating Planets Create Life, Jan Spitzer, 2021

Rethinking Cancer: A New Understanding for the Post-Genomics Era, edited by Bernhard Strauss, Marta Bertolaso, Ingemar Ernberg, and Mina J. Bissell, 2021

Levels of Organization in the Biological Sciences, edited by Daniel S. Brooks, James DiFrisco, and William C. Wimsatt, 2021

The Convergent Evolution of Agriculture in Humans and Insects, edited by Ted R. Schultz, Richard Gawne, and Peter N. Peregrine, 2022

Evolvability: A Unifying Concept in Evolutionary Biology?, edited by Thomas F. Hansen, David Houle, Mihaela Pavlicev, and Christophe Pélabon, 2023

Evolution "On Purpose": Teleonomy in Living Systems, edited by Peter A. Corning, Stuart A. Kauffman, Denis Noble, James A. Shapiro, and Richard I. Vane-Wright, 2023

Properties of Life: Toward a Theory of Organismic Biology, Bernd Rosslenbroich, 2023

Contents

Series Foreword

Biology is a leading science in this century. As in all other sciences, progress in biology depends on the interrelations between empirical research, theory building, modeling, and societal context. But whereas molecular and experimental biology have evolved dramatically in recent years, generating a flood of highly detailed data, the integration of these results into useful theoretical frameworks has lagged behind. Driven largely by pragmatic and technical considerations, research in biology continues to be less guided by theory than seems indicated. By promoting the formulation and discussion of new theoretical concepts in the biosciences, this series intends to help fill important gaps in our understanding of some of the major open questions of biology, such as the origin and organization of organismal form, the relationship between development and evolution, and the biological bases of cognition and mind. Theoretical biology has important roots in the experimental tradition of early-twentieth-century Vienna. Paul Weiss and Ludwig von Bertalanffy were among the first to use the term *theoretical biology* in its modern sense. In their understanding the subject was not limited to mathematical formalization, as is often the case today, but extended to the conceptual foundations of biology. It is this commitment to a comprehensive and cross-disciplinary integration of theoretical concepts that the Vienna Series intends to emphasize. Today, theoretical biology has genetic, developmental, and evolutionary components, the central connective themes in modern biology, but it also includes relevant aspects of computational or systems biology and extends to the naturalistic philosophy of sciences. The Vienna Series grew out of theory-oriented workshops organized by the KLI, an international institute for the advanced study of natural complex systems. The KLI fosters research projects, workshops, book projects, and the journal *Biological Theory*, all devoted to aspects of theoretical biology, with an emphasis on—but not restriction to—integrating the developmental, evolutionary, and cognitive sciences. The series editors welcome suggestions for book projects in these domains.

Gerd B. Müller, Thomas Pradeu, Katrin Schäfer

1 Introduction

When we observe an animal or a plant, we usually perceive immediately that it is a living being, and we experience something different from when we look at inanimate objects. As part of the living world, we naturally have an intuitive grasp on life. Such an inherent intuition usually tells us that living beings share some property, some basic dynamic attribute, which is responsible for their forms and functions, which brings forward their robustness, their responsiveness, or their capacity for growth and reproduction.

In scientific terms, however, specifying this difference turned out to be particularly difficult. Despite all modern knowledge about living organisms, it is still a challenge to answer the question of what life really is. In a certain sense, life is still an enigma.

Throughout history several possible solutions have been proposed to deal with this question. For Aristotle, life was an expression of the soul that he conceived as a fundamental and irreducible property of nature, which he described in a hierarchical structure. His concepts dominated for a long time, until in the seventeenth century Descartes formulated his dualism of mind and matter. This was quite influential for an understanding of life, because the living body was assigned to matter, which then had to be explained mechanistically. All physical processes, nonliving and living alike, had to be explained by physical principles. Only the human mind was distinguishable from this, experiencing its existence exclusively within its own thinking.

Especially during the nineteenth century, the tendency to reduce life to physical and chemical principles grew stronger. Classical physics, especially mechanics, became highly successful and advanced to the leading science. It increasingly demonstrated how to control and gain advantage of the physical world, and it developed as the basis for nascent technologies. As a consequence, organisms were increasingly described as machines, whereby the explanation in linear chains of cause and effect, as in classical physics, was regarded as the epitome of science in general. Even human beings were expected to be—at least in principle—describable as machines.

This physicalism fueled the attempts to reduce all biological and also psychological phenomena to the smallest material components, in which the driving causal processes were expected. This is the reductionistic method, which was characterized by Emil Du Bois-Reymond in his famous lecture *Über die Grenzen des Naturerkennens* ("The Limits of our Knowledge of Nature"): "Understanding nature . . . is to attribute changes in the material world to movements of atoms, which are caused by central forces independent of time, or the dissolution of natural processes into the mechanics of atoms" (Du Bois-Reymond 1872, 2, translation by author).

This attitude, however, also generated resistance from scientists questioning the worldview of physicalism. They were referred to as "vitalists," which was mainly a label proffered by their opponents attempting to distinguish and enforce the mechanistic doctrine. In different approaches, the vitalists assumed that life has its own principles. Many of those who looked for alternatives posed important questions and made major contributions to biology (Normandin and Wolfe 2013).

A common problem of many alternative theories was that they assumed forces that turned out to be untestable and thus unprovable: a "vital force," an "entelechy," an "élan vital," a "vis essentialis," and so forth. The mystery of life was therefore being explained by even more mysterious terms. By the mid-twentieth century, the vitalists became overshadowed by experimental research, which dominated science with its systematic search for causal factors.

However, beginning at the end of the nineteenth century there were several thoughtful scientists and philosophers who tried to develop a third way of thinking. They were skeptical about the often extremely mechanistic reductionism of their time, but they were also aware that the postulate of some vital forces was not an answer that could be translated into practicable and fruitful research questions. Thus, they developed an approach that aims to describe and investigate the actual properties of living organisms. This is not a uniform theory, nor is it a coordinated tendency and includes some heterogeneity, but overall it can be coined "organismic biology" or "organicism." In chapter 2, I will summarize the history of these movements against the background of main developments in the field.

During my studies, I came to the conclusion that the history of organismic thinking in biology can be divided mainly into three phases. The first phase begins around 1900 and runs well into the first decades of the twentieth century. A second phase can be described in the middle of this century, even if it existed in the shadow of mainstream thinking of that time. Now, at the beginning of the twenty-first century, we can see a growing third phase, which has been initiated by several prominent scientific authors in different disciplines.

Apart from such considerations, mainstream biology today has a strong predominance of analytic molecular and genetic research concepts, and confidence is growing that life is in principle explainable ultimately as a chemical machinery. Thus, the concept of the machine seems to have finally won recognition. Until today this research program has been working quite successfully, so that much is known about the physiological processes in organisms. However, new doubts continue to arise regarding whether the mere continuation of this analytical approach will finally generate a fundamental understanding of living entities. At the beginning of the twenty-first century the quest for a new synthesis, and a renewed Organicism, has been started even by analytical empiricists themselves. Carl Woese, for example, asked for "A New Biology for a New Century." He first refers to the attitude of many of his colleagues, who think that we will soon know how organisms work by just completing the molecular analytical program.

Look back a hundred years. Didn't a similar sense of a science coming to completion pervade physics at the 19th century's end—the big problems were all solved; from here on out it was just a matter of working out the details? Deja vu! Biology today is no more fully understood in principle than physics was a century or so ago. In both cases the guiding vision has (or had) reached its end, and in both, a new, deeper, more invigorating representation of reality is (or was) called for.

A society that permits biology to become an engineering discipline, that allows that science to slip into the role of changing the living world without trying to understand it, is a danger to itself. Modern society knows that it desperately needs to learn how to live in harmony with the biosphere. Today more than ever we are in need of a science of biology that helps us to do this, shows the way. An engineering biology might still show us how to get there; it just doesn't know where "there" is. (Woese 2004, 173)

Some pages later he demands: "Let's stop looking at the organism purely as a molecular machine" (176).

And Turner (2013, 272) formulates: "We are, arguably, presently at such a point, where the limits of the materialist approach to life are beginning to rise dimly into view."

Thus, the beginning of a new search for a more general understanding of life is now clearly on the agenda. Many recent publications point in this direction (e.g., Bateson 2005; Bock and Goode 1998; Capra and Luisi 2014; Denton et al. 2013; Dupré 2012; Fuchs 2018; Gilbert and Sarkar 2000; Grunwald et al. 2002; Henning and Scarfe 2013; Kather 2003, 2012; Kirschner et al. 2000; Lewontin et al. 1984; Nagel 2012; Nicholson and Dupré 2018; Noble 2006, 2017a; Normandin and Wolfe 2013; Nurse 2008; Penzlin 2014; Rehmann-Sutter 2000;

Rose and Rose 2013; Rose 1981, 1988, 1997; Turner 2007; Woese 2004). Looking further back in time, the list could be enlarged substantially. Nicholson (2014b, 347) sees a "return of the organism as a fundamental explanatory concept," after it had been an "undernourished orphan in biology, unwanted because it is not understood by biologists interested in levels above and below."

The following chapters will be characterized by an attempt to synthesize the modern insights of biology with an organismic conception of life. I attempt to show that by looking into the results and observations of contemporary science, the alleged contradictions between scientific facts and an organismic view of life are vanishing.

The inquiry will show that the mechanistic approach to biology is valuable from one special perspective, but that it is not sufficient to fully describe a living organism. Empirical research is developed far enough today that it reveals by itself the material and prerequisites to better understand more of the specific properties of the living. My hypothesis will be that—without recourse to mysterious forces—it is possible to generate answers to the old question of the specific properties of life, just using recent empirically generated knowledge. It does not contradict the results of reductionistic research but rather grants them meaning within the context of the whole organism and may also increase their practical usefulness.

After an extensive chapter on the history of organismic thinking (chapter 2), I will elaborate my hypothesis a bit more (chapter 3). The subsequent text in the long chapter 4 will describe fifteen specific properties of living entities, which cannot be reduced to physicochemical principles. The aim is to demonstrate that the results of empirical research show both the necessity as well as the possibility of the development of a new conception of life to build a coherent understanding of living entities. The conception is seen as a synthesis, which builds extensively on several earlier concepts and drafts of such an approach.

The synthesis of empirically describable specific properties of living entities proposes a coherent concept for an organismic understanding of life. However, an important characteristic of the concept will be that it is variable and may be replenished. In this sense it should be evolvable.

Today, biology has a tremendous amount of knowledge of biological facts. This knowledge comes from different levels of the organization of life: from the study of whole ecosystems to systematic relations of groups, from single organisms and their morphology and physiology, down to molecular details of the cell. In all these areas there is still a lot of work to do, as life continuously holds more enigmas than have been solved so far. Yet the knowledge of well observed facts about life processes is larger today than ever before.

To make progress, scientists must specify phenomena and find ways and techniques to observe them reliably on the one hand and find notions to explain them on the other hand. In doing so, they may inadvertently create a "conceptual framework," a way of thinking for their field, with associated assumptions, concepts, rules, and practice, all of which allow them to get on with their work. Conceptual frameworks are necessary in science, but they and their associated practices inevitably encourage some lines of research more readily than others. Hence, it is vital that the conceptual frameworks themselves evolve in response to new data, theories, and considerations. This is not always straightforward, as habits of thought and practice are often deeply entrenched. In this regard, alternative conceptual frameworks can be valuable because they draw attention to new constructive ways of thinking or new lines of inquiry. The following chapters try to show that, without leaving scientific methodology, a conceptual framework is possible that comes closer to the actual phenomena of living beings than thinking only in physicochemical terms.

The concept of the organism is of central importance for theory building in biology. Many theories in the subfields of biology, such as developmental biology, morphology, or physiology, depend on how an organism is perceived. Whether the organism is thought of as an aggregate of more or less independent parts, or as an integrated system with strong interconnections, makes essential differences in embryology and evolutionary biology. Whether the organism is thought of as a passive stimulus-response machine or an active agent predetermines theories about behavior. Regarding medicine, the basic notions of the human organism play a vital role in prevention, diagnosis, treatment, and research. And in agriculture, there is a fundamental difference in conceiving an ecosystem, the soil, or an animal as a randomly formed conglomerate of molecules, or considering these entities as organismic units with their respective characteristics, having the quality of being alive.

The following sentence has been attributed to Albert Einstein: "It is the theory which decides what we can observe." I want to add: "and how we will act."

2 History and Presence of Organismic Thinking in Biology

Epistemological Opening

Ludwik Fleck (1979) was one of the first philosophers of science to point out that scientific theories are not only shaped by facts but also by certain styles of thought and ways of reasoning ("thought collective"). Thomas Kuhn (1962) used the example of physics to elaborate the concept and coined the term *paradigm*. According to Kuhn the advancement of science is repeatedly marked by shifts in currently acknowledged paradigms revealing the empirical phenomena in a new light by interpreting them differently than before. He called these shifts "scientific revolutions" and claimed that the history of science was a succession of such scientific revolutions.

The philosopher of science Imre Lakatos (1980) developed the concept further during the 1980s and spoke of "scientific research programs," referring not only to individual theories or hypotheses but to entire scientific systems with a problem-solving apparatus, a "heuristic," which uses sophisticated methods not only to integrate normal observations into its concepts but to handle anomalies as well. In this manner research programs can be very efficient and stable. One possible problem, however, is that they refer to only a certain segment of reality and tend to reject new developments and incompatible observations. Research programs can be very tenacious even if they have already started to decline. In this situation alternative research programs play a significant role and may offer new solutions.

Lakatos inquired how scientific revolutions come about. He described a situation with two competing research programs, where one makes progress and is capable of new predictions while the other is failing; some scientists tend to join the advancing program, whereas others try to either initiate modifications in the diminished program or just adhere to thinking habits. According to Lakatos, Kuhn was wrong in interpreting a scientific revolution as a sudden switch in perception. In this context not only do new empirical facts play an important role but also styles of thought and ultimately worldviews.

Here a further component shall be added to these considerations: the history of science has shown again and again that alternative research programs can exist side by side, competing with each other for long periods, and that major progress is often made when a synthesis of both becomes realizable. After prolonged dispute and argumentation, it becomes clear that both sides are right from a certain perspective but there is a solution that is able to integrate aspects from both factions. This integration, however, does not occur in the sense of a political compromise to reach agreement but rather to achieve actual heuristic advances and produce a new progressive research program. The entire process may then yield innovative results that those involved would never have imagined.

Evolutionary theory may well serve as an example (Rosslenbroich 2016a). When in the eighteenth century the notion of evolution gradually emerged, two fundamentally opposed concepts for an understanding of the world confronted each other: on one side, the concept of the *scala naturae*, and on the other atomism.

The basic concept in the theory of a scala naturae was that the animate and inanimate world was perceived as ordered in a hierarchical, static sequence of steps, reaching from inanimate matter to animals of a higher order and to humans or even beyond. Materialism, however, led to an opposite worldview. It was closely connected to atomism, which perceived the world as composed of small indestructible particles combined in an unlimited number of ways and capable of building the more complex structures of nature. From this perspective nothing is predetermined, everything is undergoing continuous change, and the aim of science should be to explain processes by way of the properties of matter and physical laws. Both views date back to philosophers of antiquity, so that this contrast constitutes one of the classical conceptual antitheses of European cultural history.

The Enlightenment was essential in stirring up traditional thinking about nature. The concept of progress in relation to society played a major role in this context. At the same time, notions of advances in human conditions and abilities also served to render an idea of a changeable nature imaginable. Such deliberations then led to a synthesis of the two predominant thought patterns in which each side introduced essential elements but was also forced to abandon some of its principal assumptions.

Atomism contributed the description of a basic mutability of nature. Atoms and all components of nature may be assembled in constantly changing forms without being bound by any predetermined principle that would only allow certain combinations. The changeability of nature was therefore conceivable as almost unlimited. But atomism had some key disadvantages: it was not suitable for explaining a consistency of form, the observable order and its laws; nor did it give direction to a change of form, for example, in the meaning of an ascending development; nor could it explain the observable existence of the relatively stable species. If everything is in constant change, it remains unclear why there are definable and identifiable species of certain form and character.

The scala naturae concept on the other hand provided an explanation for a direction in evolution and for the emergence of higher organisms. But the scala naturae needed to be set in motion and temporalized to describe the changeability of nature. So scala naturae perceived nature as unfolding, that is, "in e-volution," and was thereby capable of describing the regular changeability of forms in their sequence.

Essential elements of both contrasting notions were incorporated in the evolutionary theory of the nineteenth century and contributed to the modern understanding of evolution, so that its history provides a good example of scientific advancement generated by the fusion of two originally opposing views. However, it is quite important for understanding current research to realize that these different views are more or less still present in thinking about evolution. It is possible to scan the usual interpretations of research findings for these two attitudes and to identify either a more atomistic thinking or a thinking that considers patterns, processes, and directions in evolution.

Many similar examples can be found throughout the history of biology. Will a similar synthesis, however, be possible with regard to the perpetual dichotomy between mechanistic thinking on the one hand and approaches that search for special qualities and characteristics of organisms on the other hand?

Origins of Mechanistic Thinking

The contrast between mechanistic materialism and vitalism is commonplace but also an oversimplification. The following paragraphs will summarize the main motives of the history of thinking about organisms and illustrate that since the end of the nineteenth century there is a thoughtful and carefully developed third way that proposes a synthesis of the most promising elements of the two contrasting views. History shows that this situation is not only a result of natural sciences but also of philosophical reasoning, and that even today certain forms of thought and worldviews are having a strong influence in contemporary science. They determine the questions that are asked, the projects that are planned, and the grants that are awarded. The claim of most empirical scientists today that philosophical considerations are superfluous is short-sighted and leads to a poorly reflected dealing with them. "There is no such thing as philosophy-free science; there is only science whose philosophical baggage is taken on board without examination" (Dennett 1995, 21).

The old dichotomy in the Western tradition in how to view the world and to look at living organisms goes back to early Greek philosophy (Jahn 2000; Kather 2003). For most of the ancient Greek philosophers, the world was a cosmos, an ordered and harmonious structure. They understood the order of the cosmos to be that of a living organism rather than a mechanical system. This meant that all its parts had an innate purpose to contribute to the harmonious functioning of the whole, and that objects moved naturally toward their proper places in the universe. The view of the cosmos as an organism also implied that the general properties of the universe are reflected in each of its parts. A certain relationship exists between macrocosm and microcosm as well as between the earth and the human body. The ultimate moving force and source of all life was identified with the soul, and its principal metaphor was that of the breath of life. Considerations around this general view can be found in the teachings of Plato (ca. 428–348 BC), the Pythagoreans, and other schools. Especially by means of the formulations of Aristoteles (384–322 BC), these considerations then became accepted common knowledge, and thereafter throughout the Middle Ages and the Renaissance.

Aristoteles's views had been formulated in explicit opposition to those of the pre-Socratic atomists like Democritus (ca. 460–370 BC), who took matter as the essential starting point. Democritus taught that all material objects are composed of atoms of numerous shapes and sizes, and that all observable qualities are derived from the particular combinations of the atoms inside the objects. These considerations were expanded by another atomist, Epicurus (ca. 341–271 BC), who restated that everything that occurs is the result of the recombination of atoms, and there is no purpose behind their motions nor any design of the gods. The cosmos, according to atomism, began in chaos, atoms moving randomly in the void. Some atoms chanced to encounter one another and combine. Some of these combinations were stable and persisted, and these aggregations of atoms constitute macroscopic, enduring entities, including organisms. The properties of any complex entity are simply the result of the causal mechanical interactions of its parts. The entire edifice of the world is therefore the consequence of chance encounters of randomly moving atoms and the necessary consequences of their interactions. This atomistic worldview was assimilated into Western thinking and resurfaced at various times. But predominantly it existed in the shadow of the main worldview, which later would be combined with Christian convictions.

The dominating medieval worldview, based on Aristotelian philosophy and Christian theology, attributed everything to God, his laws, and a spiritual nature. In addition, medieval thinking, particularly in folklore, was also characterized by a belief in all sorts of occult forces. This worldview began to change during the sixteenth and seventeenth centuries, a time that can be called the beginning of the modern age. Gradually the notion of an organic, living, and spiritual universe was replaced by that of the world as a machine, and the world machine became an influential metaphor within upcoming fields of science during that time. The rise of the mechanistic worldview was initiated by philosophical considerations and supported by revolutionary developments in physics and astronomy.

An essential achievement of that time was the turn to detailed observations of nature, including experimentation. Earlier, the primary observation and the direct experience of the researcher were not so important. Thinking and reasoning, mainly along the guidelines of authorities in philosophy and theology, dominated scientific work.

An important theme for many philosophers now was to develop an independent thinking, which no longer relied on theological or other authorities. It seemed to be necessary to establish an autonomous, self-controlled scientific thinking, and this was best possibly achieved with mechanistic interpretations. To what extent this would lead to a profound one-sidedness concerning living organisms was likely not foreseeable at that time.

Galileo Galilei (1564–1641) was among the first to show how important observation and experiment are (Capra and Luisi 2014; Kather 2003). With these he came to different explanations than the authorities expected and consequently ran into the famous conflict with the Catholic Church. To be effective in describing nature mathematically, he postulated that scientists should restrict themselves to studying those properties of material bodies— shapes, numbers, and movement—which could be measured and quantified. Other properties, such as color, sound, taste, or smell, were merely subjective mental projections that should be excluded from the domain of science.

Galileo's strategy of directing attention to the quantifiable properties of matter proved extremely successful in physics, but it also generated a momentous restriction. During the centuries after Galileo, the focus on quantities was extended from the study of matter to all natural and social phenomena within the framework of the mechanistic worldview of Cartesian–Newtonian science. By excluding colors, sound, taste, touch, and smell—let alone more complex qualities, such as beauty, health, or ethical sensibility—this emphasis on quantification prevented scientists for several centuries from understanding many essential properties of life. Even today this restriction is still prevalent, although successful approaches that allow scientific studies of such qualities also exist in some fields. Thus, from the start the scientific revolution introduced a certain one-sidedness, which contemporary science has not really overcome: the focus on material entities and quantities.

While Galileo devised ingenious experiments in Italy, in England Francis Bacon (1561–1626) formalized and advocated the empirical method of science explicitly. Bacon formulated a theory of inductive procedure: conduct experiments, draw conclusions from them, and test these by further experiments. This mindset profoundly changed the nature and purpose of the scientific quest (Capra and Luisi 2014; Kather 2003). Formerly, the goals of natural philosophy had been wisdom, understanding the natural order, and living in harmony with it. Science was pursued for the glory of God. Now this attitude changed dramatically. As the older view of nature was replaced by the metaphor of the world as a machine, the goal of science became knowledge that can be used to dominate and control nature.

The shift from the organic to the mechanistic worldview was initiated by one of the main figures of the seventeenth century, René Descartes (1596–1650). His philosophy marks a turning point that abandons the antique and medieval characterization of life and begins a new time of science and Enlightenment (Kather 2003; Penzlin 2014). It released European culture from outdated worldviews but at the same time introduced an estrangement of humans from the world, especially the living world. This had severe consequences during the succeeding centuries. Ernst Mayr (1997, 3) wrote:

Descartes . . . became the spokesman for the Scientific Revolution, which, with its craving for precision and objectivity, could not accept vague ideas, immersed in metaphysics and the supernatural, such as souls of animals and plants. By restricting the possession of a soul to humans and by declaring animals to be nothing but automata, Descartes cut the Gordian knot, so to speak. With the mechanization of the animal soul, Descartes completed the mechanization of the world picture. It is a little difficult to understand why the machine concept of organisms could have had such long-lasting popularity. After all, no machine has ever built itself, replicated itself, programmed itself, or been able to procure its own energy. The similarity between an organism and a machine is exceedingly superficial. Yet the concept did not die out completely until well into this century.

Descartes based his view of nature on the fundamental division between two independent and separate realms, that of mind and that of matter. The material universe, including living organisms, was a machine for him that could in principle be understood completely by analyzing it in terms of its smallest parts.

Descartes attempted to arrive at a fundamental principle that one can acknowledge as true without any doubt (Kather 2003). To achieve this, he used methodological skepticism, rejecting anything that can be doubted to acquire a firm foundation for genuine knowledge. He expanded the everyday experience—that senses can be unreliable—to the extreme. He not only doubted whether what the senses perceive is true but also whether what they discern exists at all. Consequently, Descartes even doubted the existence of his own body, at least hypothetically. Because one's own body is experienced by senses it also belongs to the outside world, similar to all other things that are perceived by senses. Therefore, everything that is somehow bound to functions of the body cannot be a starting point for secure knowledge. Descartes arrived at only one single reliable principle: thought exists. Thought cannot be separated from me, therefore, I exist (cogito, ergo sum). Descartes thus concluded: if he doubted, then something or someone must be doing the doubting, therefore the very fact that he doubted proved his existence.

Processes of the body and sensory perceptions do not belong to this mental certainty; neither do feelings or emotions. This difference between thinking and the body is grounded on the assumption that the mind cannot be divided and cannot be transformed. Mind is always identical with itself. In contrast, each body, also the human one, can be divided and can change. Because it is assembled from single parts, the body can be dissolved into its elements and vanish.

This step divides the thinking mind not only from nature but even from one's own body, which is irrelevant for the foundation of the mind's own identity. The human body belongs to the outside physical world, and because each physical entity can only be understood by its spatial expansion and its physical movement, this is also true for the human body. Changes and transitions are reduced to mechanical movements of single parts. They only differ by their mass and quality of movements, which can be represented mathematically. This means that all body functions need to be explained mechanically. The whole body is not regarded any more as an animated organism but as a "body machine."

The functions of the human organism follow the laws of mechanics in the same way as a clock works according to the position and properties of its weights and cogwheels. To understand the functions of such a machine, one only needs to know how cause and effect operate. It is possible to dissect such a machine into its parts and to replace any of the parts. A machine is indifferent to happiness and pain, aims and values. Here, essential criteria are efficiency and functionality.

For Descartes the whole material world was a machine and nothing but a machine. There was no purpose, life, or spirituality in matter. Nature worked according to mechanical laws, and everything in the material world could be explained in terms of the arrangement and movement of its parts. This mechanical picture of nature became the dominant paradigm of science after Descartes. The whole elaboration of mechanistic science in the seventeenth, eighteenth, and nineteenth centuries, including Newton's principles, was but the consequent development of the Cartesian concept.

Descartes's method is analytical and mathematical. It consists in breaking up thoughts and problems into pieces and in arranging these in their logical order. This analytical method, combined with a focus on observations that can be processed mathematically, later became an essential characteristic of science. On the one hand it has proven extremely useful; on the other hand, overemphasis of the method has led to excluding many phenomena from observation, which cannot be apprehended in that way. In addition, it led to the widespread attitude of reductionism in science, the belief that all aspects of complex phenomena can be understood by reducing them to their smallest constituent parts (Capra and Luisi 2014; Kather 2003; Penzlin 2014).

However, Cartesian dualism leaves many problems unsolved. An essential one is the question of how mind and body can interact and how the mind can control and influence the human body to make it an instrument of its decisions. Cartesian division has led to endless confusion about this relation. The problem occupies science and philosophy to the present day.

Another one is the understanding of animals. Animals do not have self-awareness, and they also lack a mind. Therefore, they too belong to the world that can simply be explained physically. Nonhuman beings are only an assemblage of elaborate but purely mechanically organized matter. Life in its vital biological functions becomes an object of physics. While animals are degraded to insensitive machines, human beings gain a privileged and outstanding position. When one denies that other beings can have their own needs and interests, they can then be used for human purposes and any compassion and sympathy with them is misplaced. Until far into the twentieth century, animals were categorized like nonliving objects by ethics and law, and science only very recently learned how to integrate subjective emotions and feelings of animals into scientific observation.

The success of this approach in many biological disciplines (but by far not all!), treating living organisms as machines, encouraged scientists to believe that organisms are *nothing but* machines, a notion that is still relevant today because a vast majority of contemporary scientists think that organisms are nothing but molecular machines. This standpoint induces a grave confusion about methodological assumptions and ontological realities.

The Frenchman Julien Offray de La Mettrie (1709–1751) tried to solve the contradictions, into which the Cartesian dualism and its psychophysical

parallelism leads, by postulating a materialistic monism. He did not differentiate the thinking mind from extended matter, and in contrast to Leibniz, he did
not consider immaterial principles within matter.

He scandalized his contemporaries with his book published in 1747 with the
title *L'Homme Machine*. Whereas Descartes believed that there were two
types of substance (mind and matter), La Mettrie argued that there was only one
(matter). Humans are not molded from any special clay than anything else, he
argued. Nature has used but one dough and has merely varied the leaven. As
Descartes before him, La Mettrie described physical processes in technological
metaphors. But now the entire human being was interpreted as a machine.

He referred to medical experiments of his time, which showed that certain
physical stimuli are able to change the mental state, and postulated that all
abilities of mind depend so much on the special construction of the brain and
the whole body that they obviously are part of the physical construction themselves. Form and direction of causality are clear: there are mechanically effective causes that induce the processes of the soul, and only these causes need
then to be studied. The fact that at the same time internal mental changes
trigger actions of the body had not been considered in such a worldview.

Increasingly the whole universe was described by means of technological
metaphors in the sense of a world machine or a clockwork. This interpretation
of nature also changed the notion of natural law. Something was regarded as
explained if one was able to account for it in terms of cause and effect. Questions concerning purpose or meaning seemed to be irrelevant to understand
the functions of the world machine. And finally, a machine is disposable and
can be manipulated, if one knows the laws that govern its functions.

Another burden inflicted by this mechanistic thinking is determinism. If every
more-or-less complex object is determined by its parts, there are no degrees of
freedom for the object. Every function is generated by physical processes on the
lowest level and thus determine the whole object. At the beginning of the twentieth century, physics had to learn that this cannot be maintained. Curiously, even
today some biologists claim a certain determination, for example, from molecular processes, genes, or neurons in the brain. This leads to controversies like
those in neurophysiology concerning the question of free will of humans.
Usually in such controversies there is hardly any consciousness of the historical
roots of these concepts and their limitations. Falkenburg (2012) analyzes this
widespread determinism in contemporary science and concludes that it is a
"modern myth" without any empirical underpinning.

A culmination of such a deterministic view was achieved by the French
mechanist and determinist Pierre-Simon de Laplace (1749–1827). According
to his viewpoint we have to regard the present state of the universe as the

effect of the past and the cause of the future. An intellect that at any given moment knew all of the forces of nature and the mutual positions of all of the beings that compose it, and if this intellect were vast enough to calculate all the data, could condense them into a single formula and know and oversee all processes of the world. For such an intellect nothing could be uncertain, and the future, just like the past, would be present before his eyes. There is nothing like chance or freedom in the world, everything is continuously determined.

Banning Agency

The extensive study by Jessica Riskin (2016) discloses an even more profound origin of mechanistic thinking. In her book, *The Restless Clock*, she examines the origins and history of banning agency from science as a property of natural entities. By *agency* she understands an intrinsic capacity to act in the world, to do things in a way that is neither predetermined nor random. Its opposite is passivity. A thing with agency is a thing whose activity originates inside itself rather than from the outside. In section 4.4 in the present book it will be argued that the banishment of agency from the study of living beings has been a severe drawback for scientific reasoning about organisms.

Riskin unravels a long-standing conflict between those who wanted to outsource agency to a divine engineer and those who reduced agency to mechanics of component parts of the perceivable organism. In both cases the organism itself is not the active entity but is driven by principles in the background. In this sense, organisms are seen as clockworks, set in motion either by divine intervention or by mechanical forces.

In this way of thinking, organisms as whole systems are relinquished from an own, immanent agency. The scientific principle, Riskin explains, banning ascriptions of agency to natural things supposes a material world that is essentially passive. This principle became dominant around the middle of the seventeenth century. Mechanism, the core paradigm of natural sciences from the mid-seventeenth century onward, describes the world as a great clock whose parts move only when set in motion either by external or internal forces.

Assuring that living beings are part of nature, according to this model, they too must be rationally explicable without appealing to intentions or desires, agency or will. This ideal of explanation is standard in the natural sciences, and even the human and social sciences frequently strive for such natural-scientific explanations whereby agency is declared absent in the system under study.

In an extensive survey, Riskin follows the history of these two attitudes toward nature, first describing the amazing interest in lifelike automata that spread across the landscape of late medieval and Renaissance Europe, a not only informative but also amusing chapter in history. These included mechanically driven animals such as a duck (which could even shit) or a group of birds, a flute player, a writing doll, a frightening devil that could roll its eyes and lift its head, or whole groups of moving figures performing a scene. This fashion inspired the sciences of life, and conversely, it gave rise to new designs of amazingly alive-looking machines. Such models of living beings were accompanied by a hypothetical figure, the man-machine, or the android of the Enlightenment thinkers. Jacques Vaucanson (1709–1782) built particularly sophisticated apparatuses in France, entertaining high society with them and also demonstrating them before the Académie des Sciences in Paris. In the nineteenth century, similar ideas led to the stories of Frankenstein's monster.

Descartes already knew such automata and developed analogies to the functions of the animal and human body. Riskin (2016) describes that he was initially interested in an epistemological way, through which organisms could be understood in the same way as a watchmaker understands a watch. An explanation should be a rational explanation of the movements of the material parts. Thus he founded a new critical rationalism for the emerging natural sciences. But this methodological or epistemological revolution (a revolution in the way people thought they should understand the world) brought with it a profound ontological revolution (a revolution in what people thought the world was). In the eyes of Descartes's followers, the whole world (with the exception of human rational thought) was just a moving piece of matter.

Riskin describes in detail and knowledgeably how then especially the philosophers of the early Enlightenment wrestled with these relations. Inspired by the living automata, they repeatedly described the human body as a machine and searched in different variations for what could then constitute one's own "I" and in what relationship it stands to the body, which, however, mostly remained in a certain indecision. "The struggle for dominion between 'me' and 'my machine' became a central drama of the age" (Riskin 2016, 165). Only La Mettrie, mentioned above, did not wrestle further with this but summarily declared everything to be a machine. Although his, as Riskin portrays, illogical and insensitive philosophy met with much resistance, his influence was lasting. Further, Riskin then shows how this problem has been literally woven into the fabric of animal and human science to this day.

Riskin's survey illustrates quite impressively how important it is to know the origin and history of current thinking, otherwise one risks entangling oneself in traditions of thinking and explaining that might be more founded on convention and habit rather than appropriateness for the object under investigation: "One major purpose of 'The Restless Clock' has been to demonstrate the importance of historical understanding to current thinking about the sciences of life and mind. Historical analysis, by revealing the now-hidden forces that shaped current scientific problems and principles, can reopen foreclosed ways of thinking. Investigating the origins and development of current scientific principles means rediscovering alternative possibilities for what it has meant, and what it can mean, to offer a scientific model of a living being" (Riskin 2016, 10).

Newtonian Mechanics

The conceptual framework created by Galileo and Descartes, the world understood as a perfect machine governed by exact mathematical laws, was realized by Isaac Newton (1642–1726), whose mechanics was the crowning achievement of seventeenth century science. In the Galileo–Descartes–Newtonian mechanistic conception of the world, all of nature works according to such laws, and everything in the material world can be explained in terms of the arrangement and movements of its parts. This implies that one should be able to understand all aspects of complex structures, including plants, animals, and humans, by reducing them to their smallest constituent parts.

In the twentieth century, developments within physics as well as the physiological analysis of organisms added new concepts for the study of the world. In part they were contradictory to the Descartes–Newtonian worldview. Thus, the notion of evolution as a random process that generates order within an ever-changing system in which complex structures developed from simpler forms, contrasts with the image of a world machine that is determined by eternal laws. Also, developments within physics itself, such as thermodynamics and electrodynamics but especially quantum mechanics, led to changes in the deterministic worldview. But somehow the mechanical models of living organisms and the essence of the Cartesian worldview survived. "Animals were still machines, although they were much more complicated than mechanical clockworks, as they involved chemical and electrical phenomena. Thus biology ceased to be Cartesian in the sense of Descartes' strictly mechanical image of living organisms, but it remained Cartesian in the wider sense of attempting to reduce all aspects of living organisms to the physical and chemical interactions of their smallest constituents" (Capra and Luisi 2014, 36).

Up to the present day, the model of humans and animals as machines has been developed further in increasingly subtle versions (Nicholson 2014a, 2018). According to the developments in chemistry and physics during the nineteenth century, organisms were considered as the sum of chemical reactions and mechanical forces. Today organisms are predominantly regarded as molecular machines, corresponding to the progress that has been made in molecular biology. Freedom of humans is declared to be an illusion, because all emotions, thoughts, and the will are just results of molecular functions in the brain. Even our self-experience as a person is an illusion generated by the neuronal machinery, which is composed of several modules that have been favorable during our history of survival. Meanwhile the term *genetic program* has been established to describe the organizational structure of living beings, and for many scientists the comparison of humans with a computer as an information processing machine is commonplace.

The advancement of medicine has gone hand in hand with that of biology. Naturally then, the mechanistic view of life also dominated the attitudes of physicians toward health and illness. The influence of the Cartesian–Newtonian paradigm on medical thought resulted in the so-called biomedical model, which constitutes the conceptual foundation of modern scientific medicine. From this perspective, a healthy person is like a well-made clock in perfect mechanical condition, and a sick person is like a clock whose parts are not functioning properly. Following this approach, medical science has largely limited itself to attempting to understand the biological mechanisms involved in disease and healing (Capra and Luisi 2014).

This point of view had two effects. On the one hand, it was possible to overcome many obscure and ineffective treatments that were common during the nineteenth century. Thinking in clear terms of cause and effect enabled, for example, Robert Koch (1843–1910) to finally identify certain bacteria as the definite cause of several plagues, and he had to struggle against many abstruse ideas and expert opinions of that time. On the other hand, the attention of physicians moved away from the patient as a whole person. By concentrating on smaller and smaller particles of the body, shifting its perspective from the study of bodily organs and their functions to that of cells and, finally, to the study of molecules, modern medicine often loses sight of the human being with emotions, thoughts, and biographical situation. This problem has often been formulated, but there is hardly any real solution in sight, especially as the connection between physical and molecular functions on the one hand and emotional and mental realities on the other hand is still not clear. An essential part of a solution, in my view, will be a more appropriate notion of life processes, which mediate between both.

On the somatic side, however, severe shortcomings also exist. Viewed from the described context it can be stated that there never will be a molecular medicine, which today some enthusiasts claim will revolutionize medicine in the near future. Of course, molecular aspects of health and disease do exist, and they might become increasingly important. But is it appropriate to concentrate such an extreme amount of research on this single level, while other levels of organismic organization and other approaches are neglected? This demonstrates that the topic discussed here is not a philosophical playground but rather has severe consequences for applied sciences.

Resistance to Mechanistic Interpretations of Organisms

Since the first formulations of a mechanistic understanding of organisms, doubts about and resistance against this view gave rise to early formulations of alternative theories. The movement enjoyed its first zenith with Georg Ernst Stahl (1659–1734) around the turn of the seventeenth to the eighteenth century. He attempted to develop a theoretical foundation for the life sciences. In the second half of the eighteenth century Caspar Friedrich Wolff (1734–1794), Johann Friedrich Blumenbach (1752–1840), and others developed theories that can be seen as alternative in connection with the debate on preformation and epigenetics in embryology.

From that time on there was always a camp claiming that living organisms were not really different from inanimate matter and should be explained by mechanical principles, and an opposing camp, asserting instead that living organisms had properties that could not be found in inert matter and that therefore biological theories and concepts could not be reduced to the laws of inorganic sciences. In many depictions of the history of biology, members of this opposing camp are collectively called "vitalists," although their concepts and thoughts were quite heterogeneous and to summarize them with this term is much too simple.

In some periods and at certain intellectual centers the physicalists seemed to be victorious, and at other times and places the opposing camp seemed to have achieved the upper hand (Mayr 1997). Ernst Mayr describes that the physicalists had been right in insisting that there is no metaphysical life component and that at the molecular level, life can be described according to the principles of physics and chemistry. At the same time, the vitalists had been right in asserting that, nevertheless, living organisms are not the same as inert matter but have numerous autonomous characteristics, particularly their historically acquired genetic programs, which are unknown in inanimate matter. Organisms are multilevel ordered systems, Mayr describes, quite unlike anything found in the inanimate world. He summarizes that the philosophy that eventually incorporated the best principles from both physicalism and vitalism (after discarding the excesses) became known as organicism.

The considerations and questions of vitalists were often quite accurate and often pointed to the actual problems life sciences are facing. However, from the beginning of such considerations special principles, which had to be added to the laws of the inorganic realm, were assumed. These vital factors or "forces" received different names throughout history: *spiritus seminalis* (William Harvey), anima (Georg Ernst Stahl), *impetus faciens* (Herman Boerhaave), *vis vitalis* (Albrecht von Haller), *nisus formativus* (Johann Friedrich Blumenbach), and so on. According to one group of vitalists, life was connected either with a special substance (called protoplasm) not found in inanimate matter or with a special state of matter (such as the colloidal state), which, it was claimed, the physicochemical sciences were not equipped to analyze. Because they usually were postulated forces or substances, but not observed ones, these considerations generated increasing difficulties for vitalistic concepts.

Johann Wolfgang von Goethe (1749–1832), most well known for his poetry, dramas, and novels, also was a committed scientist and contributed studies in botany, zoology, and the phenomena of colors (Goethe 1981). Especially in botany he made essential contributions, and many botanists of the nineteenth century and thereafter refer to his work. He was interested in studying processes and form in nature in their autonomous order, according to their own laws and principles. His main emphasis was on morphological studies, but he looked upon morphology as a functional morphology, in which structural and functional order are inseparable. His aim was to understand this order in an idealistic sense but in its immediate unity with physical phenomena. For him the direct sensory experience, including all sensorial qualities, was the starting point of his studies. So, he explicitly did not agree with Descartes's distrust of sensory perception but rather treated it as the primary source of scientific inquiry. It is remarkable that in this way he already avoided both tendencies of one-sidedness in the developing research of his time. He neither adopted mechanistic explanations nor postulated some sort of vital forces that would generate the organic phenomena, but he sought to resolve the immediate order of natural processes. In doing so, he consistently assumed their intrinsic activity, their agency, and their remodeling, such as in the description of metamorphosis, and thus he did not go along with the exclusion of organismic agency described by Riskin (2016). In his early works, the German philosopher and anthroposophist Rudolf Steiner (1861–1925) characterized the general significance of Goethe's approach for science (Steiner 1985a, 1996).

In this sense Goethe can be called the first organicist. Although he did not play a prominent role in the later foundation of the organicist movement, many organicists of later times studied his works, so that he may have been more influential than is usually recognized.

Several authors of the branch of German philosophy called *Naturphilosophie* also referred to Goethe, although this was often based on a misunderstanding of his intentions. Goethe was decidedly reserved toward some of the first authors of this branch of philosophy. The concepts were often formulated from a distance to concrete biological work, such as in physiology, zoology, or botany, and more often contained abstract metaphysical thoughts and expressions. However, again here many important contributions and formulations can be found, which demonstrate their struggling with the problems of understanding life.

The Physicalist Movement of the Nineteenth Century

Mayr (1997) describes that the nineteenth century physicalist movement in biology arrived in two waves. The first one was a reaction to the quite moderate vitalism adopted by Johannes Müller (1801–1858) and Justus von Liebig (1803–1873). It was set in motion by four former students of Müller—Hermann von Helmholtz (1821–1894), Emil du Bois-Reymond (1818–1896), Ernst von Brücke (1819–1892), and Matthias Schleiden (1804–1881). The second wave, which began around 1865, is identified with the names Carl Ludwig (1816–1895), Julius von Sachs (1832–1897), and Jacques Loeb (1859–1924). They powerfully announced that they would explain life itself and every function of an organism solely through physicochemical means. The work of Wilhelm Roux (1850–1924) also was in line with these physicalists.

These physicalists made important contributions to physiology. Von Helmholtz, along with Claude Bernard (1813–1878), deprived "animal heat" of its vitalistic connotation, and du Bois-Reymond resolved much of the mystery of nerve physiology by developing an electric and thus physical explanation of nerve activity. Schleiden advanced the fields of botany and cytology through his development of the cell theory of plants and insisted that all the diverse structural elements of plants are cells or cell products. These physiologists were quite successful in the invention of ever more sophisticated instruments to conduct precise measurements.

Du Bois-Reymond (1872) wrote that the understanding of nature "consists in explaining all changes in the world as produced by the movements of atoms," that is, "by reducing natural processes to the mechanics of atoms. By showing that the changes in all natural bodies can be explained as a constant sum . . . of potential and kinetic energy, nothing in these changes remains to be further explained" (cited in Mayr 1997, 6). Mayr comments that du Bois-Reymond's contemporaries did not notice that these assertions were only empty words, without substantial evidence and with precious little explanatory value.

It is quite obvious that such an approach is based on a certain worldview rather than on results of observation. Mayr (1997) further writes that the underlying philosophy of this physicalist school was naïve and could not help but provoke disdain among biologists with a background in natural history.

He describes that vitalism, from its emergence in the seventeenth century, was decidedly an anti-movement. It was a rebellion against the mechanistic philosophy of the scientific revolution and against physicalism from Galileo to Newton. It passionately resisted the doctrine that the animal is nothing but a machine and that all manifestations of life can be exhaustively explained as matter in motion. But as decisive and convincing as the vitalists were in their rejection of the Cartesian model, they were equally indecisive and unconvincing in their own explanatory endeavors. Indeed, there was great explanatory diversity but no cohesive theory, Mayr concludes.

Penzlin (2014) summarizes that all vitalistic theories of different provenance cannot avoid two misplaced speculations, because they are immanent within this approach: First, they need to equip their hypothetical vital factor with characteristics that in some way are able to intervene with and subjugate physical laws (the energetic-physical aspect of the critique on vitalism). Second, they need to assume that the factor is able to "decide" what is best for its purpose (the psychological-teleological aspect of the critique on vitalism), which finally leads it to psychism and mysticism. According to Penzlin, vitalistic theories have been an expression of our ignorance rather than an enrichment of our knowledge. The speculations around vital forces do not explain anything but transfer the problem unto metaphysics.

Penzlin stresses that at the same time it would be wrong to condemn vitalists all together, as has often been done in the history of biology. Vitalists need to be appreciated for drawing attention to organisms' specific abilities, which seemed to be strange within the framework of physics and chemistry at that time, and for developing important questions. But they were not able to generate answers that could resist scientific criticism. In contrast, the mechanistic physicalists have been quite successful in the analysis of causal details. While vitalists posed essential questions concerning the nature of living entities, but were not able to answer them, physicalists were able to provide many answers, but did not touch the essential properties of living organisms.

At the beginning of the twentieth century, new vitalistic views appeared, which are called *neovitalism* in distinction to the older vitalistic theories, and which initiated renewed heavy debates. Cassirer (1950) presented a remarkable philosophical analysis of the modes of thought involved. The outstanding representative of neovitalism was Hans Driesch (1867–1941). He began as an experimental embryologist and discovered that blastomeres from the sea urchin, separated from each other, could form smaller but still complete larvae. In his view, this could not be explained by a given purely material structure as the basis of the form-forming process, so he called for a natural factor that produced the ordered wholeness of the organism. In this factor he saw the decisive difference between the animate and the inanimate. He called it, based on Aristotle, *entelechy*. In many cases Driesch is also named as the one who introduced the concept of "system" into biology (Penzlin 2014). Especially his works on natural philosophy were widely read in the 1920s, but they also triggered a renewed controversy between mechanists and vitalists.

Penzlin summarizes that the controversies between vitalists and mechanists, which run throughout the whole history of biology of the last centuries, were quite fruitless and often turned polemic. Both vitalists and mechanists made the epistemological mistake of an inadmissible overstepping of boundaries. Vitalists viewed life determined according to human thinking and physicalists according to mechanistic causal relations. It seemed to be tempting to answer complex questions with a comfortable unifying principle rather than dealing slowly and carefully with the phenomena the object reveals.

During the second half of the nineteenth century, the pendulum swung quite far in the direction of mechanism, when the newly perfected microscope led to many remarkable advances in biology. This period saw the formulation of cell theory, the beginning of modern embryology, the rise of microbiology, and the discovery of the laws of heredity. These new discoveries grounded biology firmly in physics and chemistry, and scientists renewed their efforts to search for physicochemical explanations of life. When Rudolf Virchow (1821–1902) formulated cell theory in its modern form, the focus of biologists shifted from organisms to cells. Biological functions, rather than reflecting the organization of the organism as a whole, were now seen as the results of interactions at the cellular level.

Mechanistic materialism was the philosophy integral to the rise of experimental physiology. The advances in this field fostered the belief that a successful experimental biology depended on the acceptance of this doctrine and the concomitant exorcising from all biological discourse of anything implying immaterial vital forces.

Among the very influential mechanists were Adolf Fick (1829–1901) ("Fick's law of diffusion" [see Agutter et al. 2000]) and Jacques Loeb (1859–1924) (see his influential book *The Mechanistic Conception of Life* [Loeb 1912]), the latter having an aggressive mechanistic opposition to any considerations about special characteristics of life. It is illuminating that Loeb was mainly interested in ways to influence living processes for human purposes and to bring them under human control. Science represented progress for Loeb because it led to control, to the engineering of nature. He insisted, "We cannot allow any barrier to stand in the path of our complete control and thereby understanding of the life phenomena. I believe that anyone will reach the same view who considers the control of natural phenomena as the essential problem of scientific research" (Loeb 1903, cited in Allen 2005, 273).

The Downfall of the Complete Physics of Biology

However, the project of a complete "physics of biology" failed before 1900. Increasingly it was seen that the assumptions had been too naïve, and the experimental answers organic functions gave under experimentation never became as exact as in physics of nonliving matter. Moreover, the famous vitalism controversy repeatedly drew attention to the weaknesses of the mechanistic conception. Realizing that, some physiologists developed a materialist holism at the end of the nineteenth century. This also encompasses the strong belief that the study of organisms should be mechanistic, but it accepted properties of the whole organism that could be observed. Thus, although mechanistic materialism spread to the new experimental disciplines of embryology, biochemistry, and genetics over the period from 1880 to 1910, it was gradually superseded in most areas of biology by such a holistic materialism.

Claude Bernard declared that the constancy of the internal environment is a precondition for life. Bernard can reasonably be regarded as one of the nineteenth century pioneers of holistic materialism. He repeatedly emphasized, "An organism is nothing but a living machine" (quoted in Riskin 2016, 235). He insisted that such living machines work with the same forces as the rest of nature, but that they would be integrated to perform special services.

By 1932, when the popular book *The Wisdom of the Body* by Walter B. Cannon (1871–1945) was published, the implications of the new philosophy were clear. Maintaining a constant internal environment requires control mechanisms: sensors, effectors, information processing, and feedback systems. Cannon coined a single word for this overall principle: *homeostasis*. These terms were imported into the language of twentieth century physiology from control engineering in the 1940s. They seemed to contradict a classically understood mechanistic materialist view, according to which the physiological whole is nothing but the sum of its parts. Holistic materialism, on the other hand, acknowledges that a physiological whole is greater than the sum of its parts and shows a less polarized antipathy toward vitalism than its predecessor (Agutter et al. 2000).

Physiologists at the time around 1900 and at the beginning of the twentieth century developed essentially what is known today as classical physiology. It has a clear mechanistic attitude, but takes the role of organs, tissues, and cells into consideration, regarding their functions within the whole body. Homeostasis is a central notion in the context of this way of thinking. It is essentially the basis of modern medicine and of comparative animal physiology.

Neopositivism

Very influential for the further development of science in general, but especially for biology and anthropology, was the logical positivist philosophy of the
1920s, also called *neopositivism*. Members of the so-called Vienna Circle (Moritz
Schlick, Rudolf Carnap, Otto Neurath, and others) launched a philosophical
attack on the prevailing scientific method, attempting to cleanse it from metaphysics. Drawing on nineteenth century positivists, they believed that
science should be reestablished on a strictly empirical basis. Both philosophy
and ethics should be reconstituted on a scientific foundation. Moreover, the
logical positivists argued that all the sciences could be unified into a single hierarchy, in which the characteristic laws of any science would be explicable from
the laws preceding it in the hierarchy. Thus, subatomic physics would explain
atomic physics, atomic physics would explain molecular physics, molecular
physics would explain chemistry, chemistry would explain biochemistry, and
so on, right up to anthropology, sociology, and economics. However, no one
ever performed what had been conceived as a scientific ideal. Even in physics
it turned out to be impossible in most cases.

Although there was a lot of critique of the approach, and today hardly any
epistemology refers to it, these themes—the distrust of metaphysics, the belief
that ethics can be based on natural sciences, and the claim that all knowledge
is ultimately reducible to the laws of physics—regularly recur in scientific
writings up to the present day.

The influence of logical positivism led biologists to seek a more empirical
and mathematical basis for their discipline, to become *real* science equal to such
hard sciences as chemistry and physics. But while some biologists accepted that
the logic of such an outlook demanded that they view life as a purely mechanical
process, others were wary of simply reducing biology to physics. As J. B. S.
Haldane (1931, 150) put it, "biology must be regarded as an independent science
with its own guiding logical ideas, which are not those of physics."

Traces of Organicism—Phase 1

In face of the controversies between reductionistic materialists on the one hand
and vitalists on the other hand, several scientists tried to be more cautious about
their concepts and to discover a pathway between both camps. These approaches
have been summed up as "organicism" or "organismic biology" (Allen 2005;
Cassirer 1950; El-Hani and Emmeche 2000; Gilbert and Sarkar 2000; Mayr 1997;
Nicholson and Gawne 2015; Nicholson 2014b; Peterson 2016). Collectively,

those authors can be called organicists, who realized that life is not reducible to pure physicochemical processes and to small entities in the sense of atomistic interpretations but on the other hand also saw that the postulation of any vital force introduced a hypothesis that is inaccessible to scientific examination and reasoning.

The approaches of organicists were varied. One common ground, however, is that neither mechanistic-reductionistic nor vitalistic claims can alone be appropriate to study organismic or ecological functions. Thus, a third way must exist to come closer to the actual properties of organisms and living beings. A majority of such authors accept that the special organization of living beings contains complex hierarchically structured systems, which are in mutual relationships with each other. These systems generate characteristics and functions that cannot be explained by analysis of their isolated parts. They originate through integration of their components into new entities and thus give rise to new possibilities of the organism and its organs. The special properties of living beings are based not so much on their material composition but rather on their special organization, which needs to be studied and understood. Within modern-day biological sciences, organicism stresses the organization of organismic structure and function.

While reductionism is the belief that a bottom-up approach (e.g., atoms to molecules to genes to organelles to cells to tissues) is sufficient to explain all phenomena, organicism claims that this is not adequate and that top-down and bottom-up approaches must both be used to explain phenomena. For instance, reductionistic ontology would view a cell as an organized collection of organelles and a tissue as an organized collection of cells, and so forth. In contrast, organicist ontology would include those bottom-up considerations but would also take into account the functioning of the tissue within the organism and the functioning of the organism within its environment. The structure and function of a hepatocyte depend not only on the properties of the organelles composing it but also on the properties of the organ in which it resides. The properties of any level depend both on the properties of the parts beneath them and on the properties of the whole into which they are assembled (Gilbert and Sarkar 2000; Noble 2006, 2013a, 2017a).

Most organicists emphasize the importance of distinguishing between levels of organization in a complex system and investigating each level on its own. In studying organisms, this could include the atomic, molecular, cellular, tissue, organ, whole-organism, population, or ecosystem levels. This approach does not preclude starting with an analytical breakdown of a complex system into its component parts, but it does emphasize that this is not enough. For example, studying the individual cell types that make up the kidney could not be expected to provide a full picture of how the kidney functions within the intact organism. The functioning of the nephron, the main filtration site in the kidney, depends completely on its interaction with cells in different regions of the kidney. Each level of organization within the kidney—cellular, tissue, and organ—has its own properties and characteristics that cannot be understood only by examining them separately. The concept asserts that each level of organization in a complex system has its own special properties and needs to be studied by techniques appropriate for that level.

What is common to the different forms of organicism is the clear recognition that living organisms are capable of activities that have no counterpart in the mechanical world: self-replication, purposeful and ordered response to stimuli, elaborate self-regulatory capabilities, and the efficiency and active regulation of their energy transduction. To understand these complex interactive processes, it is necessary to get beyond the individual parts and to look somehow at the entire system or process.

In the period from 1900 to 1940, finding conceptual and experimental methods to accomplish these investigations was a major goal of holistically oriented but materialist biologists, including Hans Spemann (1869–1941), Richard Goldschmidt (1878–1958), John Scott Haldane (1860–1936), Oscar Hertwig (1849–1922), William Emerson Ritter (1856–1944), Jan Christiaan Smuts (1870–1950), and many others.

Also, Rudolf Steiner emphasized repeatedly that there is an own level of the living that has its characteristic regularities and processes and is not reducible to inorganic processes nor can it be grasped by psychic or mental experiences. In a series of lectures from the year 1922, *The Origins of Natural Science* (Steiner 1985b), he described how this life world—so to speak, in between—has been lost from our view during history. After humans had first been living in an imminent connectedness with nature, they developed more and more independence and increasing self-awareness. With it they confronted increasingly the world and nature as in opposition to themselves. Thereby natural science had originated, because now the external world became a riddle and gave questions that had to be worked on. But the world was increasingly described materially, physically, while the inner world of humans was a completely different one. This is the classical subject–object separation that has occupied philosophy for centuries until today. But in the process, the living itself—between the physically explained outside world and the metaphysical self-experience—fell by the wayside and became inaccessible. And then, according to Steiner, people "invented something" to be able to describe that, and that was vitalism. He thus also argued against vitalism and emphasized that the living has to be investigated according to its own laws and principles.

The organismic concepts of the beginning of the twentieth century probably describe these principles only initially and are heterogeneous among themselves. However, they are interesting as a starting point for further developments in this direction and have at least—as we will see—maintained a direction of questioning that approaches the independence of life processes.

Nicholson and Gawne (2015) recount the intentions of the generation of authors who began writing on the philosophy of biology in the 1910s and 1920s. They describe how these authors attempted to transcend the traditional conflict between mechanistic and vitalist conceptions of life by carving out a middle ground between the two positions that would retain what they deemed valuable in each position, while avoiding their respective defects.

Increasingly, the term *organismic biology* or *organicism* has been used by experimental as well as theoretically working scientists searching for appropriate nonreductionist explanations of organic functions and organization. William Emerson Ritter is acknowledged as one of the founders of organismic biology, in which the organism is conceived as a unit, its parts can only be explained with reference to this unit, and the unit must be explained through the parts (Ritter 1919). Ernst Mayr (1997) describes that according to Ritter, entireties are so related to their parts that not only does the existence of the whole depend on the orderly cooperation and interdependence of its parts, but the whole exercises a measure of determinative control of its parts (Ritter and Bailey 1928).

Oscar Hertwig is cited as another founding father of organicism. He proposed a type of materialist philosophy called wholistic organicism. This philosophy embraced the views that the properties of the whole cannot be predicted solely from the properties of its component parts, and the properties of the parts are contingent upon their relationship to the whole (Gilbert 2014). Capra and Luisi (2014) point to Ross Harrison (1870–1959) as another early exponent of the organismic school, who explored the concept of organization. He identified configuration and relationship as two important aspects of organization, which were subsequently unified in the concept of "pattern of organization" as a configuration of ordered relationships. Julius Schaxel (1887–1943) edited a series of books titled "Abhandlungen zur theoretischen Biologie" ("Treaties in Theoretical Biology") and popularized the German term *organismisch* (Laubichler 2017).

According to Capra and Luisi (2014), the biologist Joseph Woodger (1894–1981) asserted that organisms could be described completely in terms of their chemical elements, "plus organizing relations." This formulation had considerable influence on subsequent biological thought, and historians of science have stated that the publication of Woodger's "Biological Principles" in 1929 marked the end of the debate between mechanists and vitalists. Woodger and other organismic biologists also emphasized that one of the key characteristics of the organization of living organisms is their hierarchical nature (Woodger 1929).

The famous British botanist Agnes Arber (1879–1960) formulated:

The mechanist, starting from the physico-chemical standpoint, interprets the living thing by analogy with a machine. The vitalist, on the other hand, supposes a guiding entelechy, which summons order out of chaos; he thus adopts a dualistic attitude. The elements of truth in both these views are recognised, and their opposition is resolved in the organismal approach to the living creature. This approach is conditioned by the belief that the vital co-ordination of structures and processes is not due to an alien entelechy, but is an integral part of the living system itself. (Arber 1964, 100–101, quoted from Walsh 2015, 13)

Since the 1920s, the terms *holism* and *organicism* have been used interchangeably. Perhaps at first, *holism* was even more frequently used. However, during the twentieth century the term *holistic* acquired two problems: There were authors who defended holistic concepts in an extreme sense and stated that any analytic approach and any reduction to parts is useless to understand organisms. This contrasts to other organicist concepts, which just regard different levels of organization in their respective relevance and do not deny the significance of the chemical and molecular level. Another, more sociological problem is that during the twentieth century the term *holistic* has been adopted by unconventional and not seldom mystical movements in pseudoscience and medicine. Therefore, in science the word gained negative overtones. Today, however, the term is rehabilitated to some extent, provided it is clearly stated what it describes.

The less burdened term is *organicism*, although, similar to many other terms, it was misused by the Nazis. Nonetheless, it came quite undamaged out of this era, and today it is widely used throughout scientific literature to characterize the search for the actual characteristics of life.

The objection of the organicists was not so much to the mechanistic aspects of physicalism as to its reductionism, Mayr (1997) explains. The physicalists referred to their explanations as mechanistic interpretations, which indeed they were. But what characterized them far more was that they were also reductionist explanations: for reductionists, an organic function is, in principle, explained as soon as the reduction to the smallest components has been accomplished. They claim that if one has completed the inventory of these components and has determined the function of each of them, it should be an easy task to also explain everything observed at the higher levels of organization. The organicists demonstrated that this claim is not true, because reductionism is quite unable to explain characteristics of organisms that first emerge at higher levels of organization.

Mayr (1997) describes that organicists agree that no system can be exhaustively explained by the properties of its isolated components. The basis of organicism is the fact that living beings are organized. They are not just piles of characteristics or molecules, because their function depends entirely on their organization and their mutual interrelations, interactions, and interdependencies.

With these depictions Ernst Mayr (1904–2005) himself can be viewed as an organicist, which he quite explicitly summarizes as follows:

To sum up, organicism is best characterized by the dual belief in the importance of considering the organism as a whole, and at the same time the firm conviction that this wholeness is not to be considered something mysteriously closed to analysis but that it should be studied and analyzed by choosing the right level of analysis. The organicist does not reject analysis but insists that analysis should be continued downward only to the lowest level at which this approach yields relevant new information and new insights. Every system . . . loses some of its characteristics when taken apart, and many of the important interactions of components of an organism do not occur at the physicochemical level but at a higher level of integration. (Mayr 1997, 20)

Ernst Mayr was a leading and prominent figure in evolutionary theory in the second half of the twentieth century. His extensive studies and publications on the history and theory of biology were triggered by his doubts in mechanistic explanations concerning biological subjects (Mayr 1988, 1996).

During the first decades of the twentieth century the term *system* came into use by authors who wanted to express the relevance of the context within organic structures. The biochemist Lawrence Henderson (1878–1942) was influential through his early use of the term to denote both living organisms and social systems. From that time on, a *system* came to mean an integrated whole whose essential properties arise from the relationships between its parts, and *systems thinking* was the understanding of a phenomenon within the context of a larger entity. This is in fact the original meaning of the word *system* which is derived from Greek *syn+ histanai* (= "to place together"). To understand things systematically means literally to put them into a context, to establish the nature of their relationships (Capra and Luisi 2014).

The Theoretical Biology Club in Cambridge

In the 1930s Joseph Woodger, Joseph Needham (1900–1995), and Dorothy Needham (1896–1987), together with Conrad Hal Waddington (1905–1975), John Desmond Bernal (1901–1971), and Dorothy Wrinch (1894–1976), founded the Theoretical Biology Club at Cambridge University to promote the organicist approach to biology. The club opposed mechanism, reductionism, and the genecentric view of evolution, but at the same time saw the shortcomings of vitalism. Some of the members were influenced by the philosophy of Alfred North Whitehead (1861–1947). The transdisciplinary group met on a regular basis at Cambridge University with occasional prominent visitors from the European continent and North America, to clarify what exactly was meant by concepts like "organic." In that time they formed the fundament for a network of scientists who worked along principles that were identified as coming closer to what the organic processes actually show.

Peterson (2016, 7) presents a comprehensive and rich inquiry into this first phase of organicism as "an attempt to trace this sometimes murky 'third way' tradition through its derivations and renegotiations. It is also an attempt to trace a network of individuals—often outsiders for political or cultural reasons as well as scientific ones—who were advocating for the acceptance of this 'third way'. Aside from a set of scientific concepts, organicists approached the world as richer and more complex than mechanists, less mysterious and inscrutable than vitalists."

Peterson also demonstrates that organicism, though shoved into the sidelines over the last portion of the twentieth century, has been a guiding philosophy for a significant number of scholars on both sides of the Atlantic Ocean for over a century. It motivated members of multiple disciplinary subfields, and it made an impact on both their experimental practices and the reflective pieces they later wrote about their work.

In his lifetime Waddington thought along organismic guidelines. Although he is well known for his contributions to evolutionary biology and genetics, his epistemological background is hardly mentioned. In the 1950s he developed the concept of epigenetics as a way to conceptualize the organism as it develops from genotype to phenotype. Today the term is ubiquitous in biology as well as public depictions as a recent extension of genetics. But in its original context, epigenetics offered something far more complex than occasional breaches of the central dogma in genetics. Peterson describes in detail how the concept of epigenetics was stimulated by the search for a "third way" embodied in the discussions of the Theoretical Biology Club and the work that Waddington conducted with members of that club.

The Vivarium in Vienna

Another place where traces of organismic thinking can be found was the "Biologische Versuchsanstalt" ("Institute for Experimental Biology") in Vienna, the so-called Vivarium. It was an outstanding and innovative privately organized research institution between 1903 and the beginning of the 1940s (Müller 2017a). It was not primarily oriented toward an organismic concept, but the combination of theoretical work with experimental research meant that the controversies and theoretical perspectives of the time were alive and well in the place. Thus, important persons who later promoted the development of organismic views had at least some of their academic origins there.

Müller (2017a, 15) describes that the lasting legacy of this institution lies not so much in the styles of experimentation that were devised there, but primarily in the generation of an alternative approach to the study of the physiology, development, inheritance and evolution relationship, and in the contributions to the theoretical foundation of biology. In contrast to the then-prevailing concentration of evolutionary biology on assumed genetic variation and selectionist scenarios, the institution developed tools for precise measurements of the reactive plasticity of developmental and physiological processes to environmental stimuli in the generation of form and function. Yet in an increasingly reductionistic climate of evolutionary research, Müller explains, this approach came to be overwhelmed by statistical and, later, molecular studies of genetics.

Today, the very same subjects have again come to the fore (see section 4.12) in ongoing theoretical debates on the systemic perspectives introduced by evolutionary developmental biology, epigenetic inheritance, niche construction theories, and others. The institution in Vienna preceded this systemic view by a full century. The systems perspective can be traced from founders of the institute to Paul Weiss and Ludwig von Bertalanffy (who will be introduced subsequently) and, to some extent, still resonates in Rupert Riedl's systems account of evolution (Riedl 1978; Wagner and Laubichler 2004) and in recent suggestions for the revision of evolutionary theory.

Elements of New Ways of Thinking

During the first half of the twentieth century the ideas of organismic biologists contained elements of new ways of thinking—thinking in terms of connectedness, relationships, patterns, systems, and context (Capra and Luisi 2014). According to the systems viewpoint, the essential properties of an organism, or living system, are properties of the whole, which none of the parts possess. These properties are destroyed when the system is dissected into isolated elements. Although individual parts can be identified and studied within a living system, these parts are not isolated. However, during the second half of the twentieth century, with its increasing hegemony of molecular biology and genetics, these principles were nearly forgotten or at best banned to a shadowy existence.

When the first organismic biologists explored the problem of organic form and debated the relative merits of mechanism and vitalism, German psychologists contributed to that discourse from the very beginning. The German word for organic form is *Gestalt* (as distinguished from *Form*, which denotes an inanimate form), and the much-discussed question of organic form was known as the *Gestaltproblem* in those days. At the turn of the century, the Austrian philosopher Christian von Ehrenfels (1859–1932) used Gestalt in the sense of an irreducible perceptual pattern, founding the school of *Gestaltpsychologie*. In subsequent decades this specific denotation of the term Gestalt was taken up by the English language.

Emergence is another term used to describe higher-order entities or their origin. The history of the concept of emergence dates back to John Stuart Mill (1806–1873) in the nineteenth century, and to Charlie Dunbar Broad (1887–1971) at the beginning of the twentieth century (Pigliucci 2014). Today the term is undergoing some resurgence within the discussion of problems of reductionism. With this term theoreticians of emergence deny that a complete description of the world is in principle possible only from the knowledge of elementary particles and general physical laws. Emergence is seen as a phenomenon whereby larger entities and more complex properties arise through interactions among smaller or simpler entities such that the larger entities exhibit properties the smaller or simpler entities do not exhibit. This is another source of the formulation "The whole is more than the sum of its parts." It is one of the attempts to describe higher-level entities, or their origin, in complex systems. Emergence plays a role in theories of integrative levels and of complex systems.

Organicism in Ecology

The new science of ecology emerged out of organismic biology during the late nineteenth century, when biologists began to study communities of organisms. In the 1920s, ecologists introduced the concepts of food chains and food cycles. The word *Umwelt* (environment) was coined by the Baltic-German biologist and ecological pioneer Jakob von Uexküll (1864–1944) (Cassirer 1950; Uexküll 1909).

Organismic theories dominated ecology mainly up to the middle of the twentieth century. Early proponents of organicism in ecology were Karl Friederichs (1878–1969) and August Thienemann (1882–1960) in Germany (Kirchhoff and Voigt 2010; Schwarz and Jax 2011). Ecological communities are regarded in their entirety and species as their parts, which have a function for the whole. Species and groups are like organs for the whole, in which organisms stand in mutual relationships to each other and thus generate the community. Each organ fulfills functions within the community, and every individual is dependent on the community as a whole. Only by performing its special functions within the community is the self-maintenance of the individual possible. The ecological units are natural entities and exist independent from the researcher, who cannot separate them arbitrarily but rather must find them in nature. Succession, the sequence of species during changes within time, necessarily leads from lower integration in pioneer communities to more elaborated integrations and finally to organic units.

The opposite pole to these views in ecology are individualistic-reductionistic theories. They take their starting point from the individuals of a species. All species that reached the area and found suitable conditions coexist. In their existence they are not dependent on a function that they have to fulfill for a superimposed community. Communities and ecosystems are considered as nothing but accidental and, moreover, ever-changing collections of species, which can be fully explained in terms of individual species' adaptations to local environmental conditions.

Theories of the ecosystem, which are more widespread today, are less organismic but in some sense are holistic theories. Most ecologists regard the ecosystem as the central perspective, emphasizing the systems principle in ecological communities. However, Kirchhoff and Voigt (2010) describe that there is a large number of different variants of ecosystem theories. They describe that these cannot be viewed as a modernization of classical organicism in ecology but contain many elements of it. They developed the notion of the ecosystem and enriched the systemic way of thinking by introducing new concepts such as "community" and "network." With the study of ecosystems, large parts of ecology did not follow the strong reductionistic tendency of physiologic disciplines during the second half of the twentieth century. Ecology mainly studies the flow of substances and energy, while questions concerning the relationship between the individual and the community retreated into the background.

However, the dispute in ecology continues and several different positions are within the spectrum of being more or less radically holistic or reductionistic (Schwarz and Jax 2011). Using the example of ecology, Looijen (2000) developed the thesis that instead of dealing with holism and reductionism as conflicting views of nature or of relationships between fields of science, they should rather be seen as cooperating and mutually dependent research programs.

The Century of the Gene and the Dominance
of Molecular Biology

The second half of the twentieth century again saw a profound shift toward reductionistic thinking. Molecular and genetic mechanisms were set up as the primary level on which to search for explanations of organisms and human diseases. Classical physiology and morphology were belittled as merely descriptive and announced that now the real causes of organic processes and also of human diseases would be presented.

As biochemistry progressed, it established the firm belief that all properties and functions of living organisms would eventually be explained in terms of chemical laws. Cell biology made enormous progress in understanding the structures and functions of many of the cell's subunits. This was a result of new techniques that became available after World War II, such as the electron microscope and new possibilities to characterize proteins, carbohydrates, and other biochemical components. Increasingly, organisms were seen as merely a specially structured collection of molecules. Despite all the previous debates, this standpoint became mainstream in many areas of biology in the second half of the twentieth century.

There is no doubt that cell biology and biochemistry generated a large amount of biological knowledge. However, it advanced little in understanding the coordinating activities that integrate those phenomena into the functioning of a cell or an organ. Again, the awareness of this lack of understanding and a certain resistance to the atomistic attitude, which is inherent in the reductionistic concept, triggered a new opposition to this form of a mechanistic conception of life.

These oscillating movements between reductionistic and organismic approaches, including all the inherent controversies, are nevertheless part of the overall progress in biology over the decades. Woese (2004) charts a broad tableau from the nineteenth to the twentieth century: the nineteenth century is the era when biology came of age, consolidating itself, ridding itself of much of its ancient burden of mystical claptrap, and defining the great biological problems. Pasteur banished spontaneous generation, and he, along with Koch, Haeckel, Cohn, Beijerinck, and others, demonstrated that the living world consisted of far more than plants and animals. This research had significant practical consequences for improvements in human health as well as in farming. Darwin had demystified evolution and showed, at least in principle, that it can be treated by scientific considerations. The cell had emerged as the basic unit of biology. Coming from ideas of the nineteenth century, chemistry and physics were seen as the more mature, as the real sciences, so it was consequent now to direct their methods more toward biological entities.

The molecular treatment of cells and organisms developed an amazing dominance in biological sciences. But Woese describes that it would prove to be a mixed blessing. On the positive side, those problems that were amenable to a reductionist approach could benefit from the fresh outlook and experimental power of molecular biology. In addition, biology on the whole would benefit from the physicist's general mode of operation, that is, from the well-honed understanding of what science is and how it should be done: the crisp framing of problems, the clear understanding of what is and what isn't established, and the importance of hypothesis testing. On the negative side, Woese continues, biology's holistic problems, which were not commensurate with the new molecular perspective, would remain relatively or completely undeveloped. The result was a distorted growth of biology in the twentieth century.

Woese (2004, 174) describes:

The climate just referred to, of course, was the colorless, reductionist world of 19th century classical physics, which by that time had strongly affected the outlook of western society in general. The living world did not exist in any fundamental sense for classical physics: reality lay only in atoms, their interactions, and certain forces that acted at a distance. The living world, in all its complexity and beauty, was merely a secondary, highly derived and complicated manifestation of atomic reality and, like everything else in our direct experience, could (in principle) be completely explained (away) in terms of the ever-jostling sea of tiny atomic particles. The intuitive disparity between atomic reality and the "biological reality" inherent in direct experience became the dialectic that underlay the development of 20th century biology.

The new perspective brought about a significant shift in biological research. Whereas cells were regarded as the basic building blocks of living organisms during the nineteenth century, the attention shifted from cells to molecules toward the middle of the twentieth century when geneticists began to explore the molecular structure of the gene. Their research culminated in the elucidation of the physical structure of DNA—the genetic component of chromosomes—which stands as one of the greatest achievements of twentieth century science. With these discoveries geneticists believed that they had now pinned down the "atoms of heredity" and proceeded to explain the biological characteristics of living organisms in terms of their elementary units, the genes (Capra and Luisi 2014).

The unraveling of the structure of DNA by James Watson (*1928) and Francis Crick (1916–2004) made plausible that it would be possible to understand life in purely physical and chemical terms. Watson later formulated:

"There is only one science, physics: everything else is social work" (Speech at the Cheltenham Literary Festival, 1994, cited in Rose 1997, 8).

It is, however, remarkable that the field, which is expected to provide a firm chemical and physical basis for understanding life, early on introduced the notion of information to understand the function of DNA and thus already relinquished a purely materialistic explanation. This has not been realized by the majority of scientists. It became increasingly clear that information processes are essential. With information, a new quality emerged that does not exist in the physicochemical terminology, which focuses on material interactions, atoms, molecules, and crystals, as well as on forms of energy and their transformation (Eigen 1987; Fox Keller 2011; Penzlin 2014; Shapiro 2011).

Advancing to ever smaller levels in their explorations of the phenomena of life, biologists thought that the characteristics of all living organisms—from bacteria to humans—were encoded in their chromosomes. To a large extent this expectation was driven by strong commercial interests, hoping that manipulation of genetic and molecular mechanisms could provide starting points for biotechnology (Lewontin 1991; Lewontin et al. 1984). However, after a big hype in this industry at the end of the twentieth century, some disillusionment arose with the new century.

E. O. Wilson describes in his autobiography, *Naturalist*, the threat he felt from molecular biologists in the 1950s. James Watson was appointed assistant professor at Harvard, shortly after Wilson himself started teaching there. "He arrived," Wilson (1995) observes,

with a conviction that biology must be transformed into a science directed at molecules and cells and rewritten in the language of physics and chemistry. What had gone before, 'traditional biology'—my biology—was infested by stamp collectors who lacked the wit to transform their subject into a modern science. (219)

For those not studying biology at the time in the early 1950s, it is impossible to imagine the impact that the discovery of DNA had on our perception of how the world works. . . . Reaching beyond the transformation of genetics, it injected into all of biology a new faith in reductionism. The most complex of processes, the discovery implied, might be simpler than we had thought. It whispered ambition and boldness to young biologists and counselled them: Try now; strike fast and deep at the secrets of life. (223)

However today, after several decades of genetic analyses, things have become even more complex than had been thought at that time, because several levels of information processing are involved (Parrington 2015; Shapiro 2011). This is quite the opposite of what has been expected.

A crucial element in the breaking of the genetic code was the fact that physicists moved into biology (Capra and Luisi 2014). Max Delbrück (1906–1981),

Francis Crick, Maurice Wilkins (1916–2004), and several of the other protago-
nists had backgrounds in physics before they joined the biochemists and geneticists in their study of heredity. The main reason these scientists left physics and
turned to genetics was a short book titled *What is Life?*, published in 1944 by
the famous quantum physicist Erwin Schrödinger (1887–1961). The fascination
of Schrödinger's book came from the clear and compelling way in which he
treated the gene not as an abstract unit but as a concrete physical substance,
advancing definite hypotheses about its molecular structure that stimulated scientists to think about genetics in a new way. Schrödinger was the first to suggest
that the gene could be viewed as an information carrier, whose physical structure
corresponds to a succession of elements in a hereditary, coded script.

It does not lack some irony, though, that Schrödinger triggered a new phase
of physicalism in biology, while he himself claimed that physical principles
would not sufficiently explain biological entities: "From all we have learnt
about the structure of living matter, we must be prepared to find it working in
a manner that cannot be reduced to the ordinary laws of physics. And that not
on the ground that there is any 'new force' or what not, directing the behaviour
of the single atoms within a living organism, but because the construction is
different from anything we have yet tested in the physical laboratory"
(Schrödinger 1944, 76).

However, the belief in molecules became quite profound. In an extreme
version an organism is essentially seen as nothing but a collection of atoms
and molecules (Crick 1966).

Gilbert and Sarkar (2000) see in "genocentrism" an obstacle for the development of organicism. Geneticists, they write, have routinely made the case that
the gene is like the physicists' atom and that genetics took biology out of the
realm of natural history and made it a "hard" (i.e., physical and mathematical)
science. The analogy is so strong, that the two types of reductionism—one
with the tiniest particles of matter at the base, and the other with the genes as
"atoms" at the base—are often conflated.

The vast majority of recent studies carried out in the biological sciences
attempts to uncover the mechanisms that are expected to underpin the functioning of natural systems and to determine all properties of organisms. Most
of biological research has been framed on the hypothesis that biological organisms are essentially determined by genetic information and the molecular
processes through which such information is expressed.

If I talk to the molecular biologist in the neighboring institute of my university, he tells me that a human being is essentially a molecular machine, and
if we one day shall know all the molecular details, it will become clear how
a human organism functions. What we experience as consciousness will turn

out to be an effect of these molecular interactions, especially in the brain, and neurophysiology is very close to clarifying how consciousness is generated and how these mechanisms determine our behavior. Most medical research searches for "the molecular causes of diseases" and attempts to develop medicines that are able to influence these causes. If there are difficulties at present in understanding life chemically, these are merely transitional but not principal obstacles. Things are only different in their complexity, but not in principle. "Present day biology is the realization of the famous metaphor of the organism as a bête-machine elaborated by Descartes in Part V of the *Discours*, a realization far beyond what anyone in the seventeenth century could have imagined" (Lewontin 2009, v, citation from Nicholson 2014a, 162).

The position of physical reductionism often finds its expression in claims of "nothing but." They are handy but usually incorrect. Thus it is asserted that living beings are nothing but an accumulation of elementary particles, atoms, or molecules interacting with each other. Or it is maintained that organisms are nothing but a machine, or metabolism nothing but the turnover of free energy. Francis Crick (1994, 3) believes that "You, your joys and your sorrows, your memories and your ambitions, your sense of personal identity and free will, are in fact no more than the behavior of a vast assembly of nerve cells and their associated molecules. . . . You're nothing but a pack of neurons."

Assertions like these have been called "nothing-buttery" and are based, as shown in the beginning of this chapter, on an old worldview that resurfaced regularly throughout history, while there is still hardly any empirical evidence for such claims.

In theoretical biology and medicine, however, there exists a long tradition of critical examination and discussion of this research program, along with proposals for either extensions or improved versions of the approach. Today it is even asserted that there is a certain anti-reductionistic consensus among philosophers of biology. Nonetheless, all this does not carry weight with empirical scientists, who usually are suspicious about discussing such theoretical topics at all, or just do not have the time to bother themselves with such "philosophical stuff," after twelve hours in the lab each day.

Admittedly, this approach generated a vast amount of biological knowledge. However, now it has reached a fundamental limit which is even recognized by some empirical scientists themselves. For example, in *Nature* Paul Nurse (2008, 424) asserted:

Biology stands at an interesting juncture. The past decades have seen remarkable advances in our understanding of how living organisms work. . . . But comprehensive understanding of many higher-level biological phenomena remains elusive. Even at the level of the cell, phenomena such as general cellular homeostasis and the maintenance of cell integrity, the generation of spatial and temporal order, inter- and intracellular signaling, cell 'memory' and reproduction are not fully understood. This is also true for the levels of organization seen in tissues, organs and organisms, which feature more complex phenomena such as embryonic development and operation of the immune and nervous systems. These gaps in our knowledge are accompanied by a sense of unease in the biomedical community that understanding of human disease and improvements in disease management are progressing too slowly.

The same dichotomy was present around the formulation of the Modern Synthesis of evolution in the 1930s and 1940s. On the one hand, population geneticists like Theodosius Dobzhansky (1900–1975) and Ronald Fisher (1890–1962) provided the experimental and theoretical bases for a reduction of organismal biology to genetics. Focusing on single genetic mutations as the only raw material for evolutionary change is clearly an atomistic view. On the other hand, the infamous exclusion of developmental biology from the Modern Synthesis, as well as Dobzhansky's own research on reaction norms and gene–environment interactions, pointed to the limitations of straightforward reductionist approaches in biology. A literature parallel to the Modern Synthesis thus originated with the work of Conrad Waddington, Iwan Schmalhausen (1884–1963), Rupert Riedl (1925–2005), and others, and eventually led, among other factors, to the onset of "evo-devo" as a separate field of investigation, as well as to the resurgence of research on phenotypic plasticity, beginning with the twenty-first century (Pigliucci 2014; Pigliucci and Müller 2010; West-Eberhard 2003).

Reduction versus Reductionism

Basically, the question of reductionism in biology is long-standing. *Reductionism, anti-reductionism, beyond reductionism*, and *holism* are terms that come up again and again in writings and discussions, together with the accompanying arguments. This indicates that the central problem has not really been solved and that there are still different views on living beings, just as there were during the past centuries, as described earlier.

Usually ontological, epistemological, and methodological aspects of reductionism are distinguished (Brigandt and Love 2017; Looijen 2000):

1. Ontological aspects relate to the question of what entities or substances reality is assumed to be made up of, what properties are assigned to these entities and what relations are assumed to exist between them. Ontological reductionism attempts to describe all structures and functions of an organism, of organs or of

the cell in terms of physicochemical processes, because reductionists share the atomistic view, that all entities of the world are composed from certain parts. All things are only distinguished by their complexity, so that entities can be completely understood, if one knows all the conditions and relations of their parts. The complex of molecular processes builds at large the function of higher entities, and thus other disciplines in biology are increasingly superfluous.

Ontological reductionism postulates that living things are "nothing but" aggregations of atoms and molecules and the interactions that take place between them. In other words: All properties and performances of living beings can be reduced to the molecular level, that is, to the properties and interactions of the atoms and molecules composing them.

2. Epistemological aspects pertain to our knowledge of reality and to the way this knowledge is embodied in theories. Epistemological reductionism attempts to develop a unity of science through theory reduction. In its radical form, it claims that it is possible, if not in practice then in principle, to reduce all the concepts, laws and theories that have been developed for a certain higher level of organization to concepts or theories that have been developed for a lower level of organization. Thus, it should be possible to reduce biology as a whole to chemistry and chemistry to physics. Reductionists share a certain view of the relations between sciences which is known as the unity of science. This is the view that there are no qualitative differences between the various sciences, either in the objects of investigation or in the aims and methods of inquiry, and that the various sciences can be ranked according to their domains of investigation in a hierarchy going from the most general and fundamental science, physics, to the highest, most specific one, humanities. An extreme form is the dream of some physicists to find a "theory of everything," which has not yet been successful.

3. Methodological aspects relate to the ways of acquiring knowledge and to the principles, rules and strategies thereby used. Methodological reductionism attempts to explain nature through causal mechanisms in order to arrive at a "proper" knowledge and understanding of nature. The reductionistic expectation is, that the entities of higher biological levels of organization are composed of entities of lower, physicochemical levels, and that the former are causally determined by the latter. Therefore, the best strategy for obtaining knowledge of higher levels is to study interactions between their constituent parts through causal analysis. For this reason, reductionists can also be called mechanicists.

The unity of science through the reduction of all sciences to physics was one of the illusionary programs of neopositivism, which was doomed to failure.

Mahner and Bunge (1997, 197) write: "although the attempt to reduce biology to physics may no longer be pursued, we now face the claim that most, if not all, of biology is reducible to molecular biology. We trust that this program will fail just as miserably as its neopositivist forerunner."

Here, a distinction between reduction and reductionism will be introduced, because one feature among others distinguishes the situation today from its counterpart in earlier times—namely, there is no debate about the validity and use of physicochemical techniques in the laboratory. Their methodological value is unquestionable. Many of the forthcoming arguments for an organismic approach to biology will indeed be based on results from these disciplines. They are essential and necessary within modern biology. However, it will be argued, that the paradigmatic framework in which biology operates needs to be adjusted.

The term *reduction* describes a mode of operation in scientific studies. From a certain complex entity like a whole organism, a cell, or an ecosystem, only a selected part is regarded and studied in detail to get more information about this part. The complex reality is reduced to a defined field to pose exact questions and to perform experiments. To choose a certain species in an ecosystem, a certain type of cell within an organ, or a certain protein in a membrane are examples of such a reduction. Reduction is a mode of analysis, and the dissection of a biological entity permits a better understanding of the separated part. It should be clear that reduction is a fundamental and indispensable procedure in science. Reduction to molecular components, for example, is necessary to study structures and processes on this level.

The term *reductionism*, however, denotes the expectation that higher-level phenomena can be completely understood in terms of the properties of their constituent parts and that any further studies of higher-level entities are useless. Thus, reduction is a method of science, whereas reductionism is an ontological claim about reality.

Molecular studies are necessary and unavoidable in biology when they are carried out in the sense of reduction. However, the claim of many researchers in the field—which is mostly made unconsciously and without reflection—that everything can be explained from this level is an incomplete and imperfect biology. Comprehending just the parts of isolated entities is not sufficient. In spite of all the work that has been done under this program, it has by no means been successful in actually reducing the processes of life to inorganic or molecular principles. Also, nobody has ever constructed even the smallest and simplest living cell from nonliving material, despite regularly publicized claims about partial achievements. The approach of Jacques Loeb, to explain life completely from a mechanical concept, has still not been realized, and the more that becomes known about the complexity of the processes even within one single cell, the objective moves further away.

With respect to this ontological reality, the "scientific imperialism" (Walsh 2015, 29) of reductionism in many fields of science is highly problematic. It undervalues special sciences such as psychology, botany, zoology, or integrative physiology and tries to concentrate them exclusively on molecular research, a standpoint which is extremely influential in nearly all contemporary university politics and in research funding as well.

Organismic Thinking in the Second Half of the Twentieth Century—Phase 2

During the second half of the twentieth century again many authors debated the epistemological structure of biology and its problems. In addition, there were whole conferences dedicated to the issue with titles like *Beyond Reductionism: New Perspectives in the Life Sciences* (Koestler and Smythies 1969), *Toward a Man-Centered Medical Science: A New Image of Man in Medicine* (Schaefer et al. 1977), *The Problem of Reductionism in Science* (Agazzi 1991), *Wissenschaftstheorien in der Medizin. Ein Symposium* (Deppert et al. 1992), and *The Limits of Reductionism in Biology* (Bock and Goode 1998).

Stephen Rose published several books and papers combating genetic determinism. He especially opposed extreme forms of reductionism and their attempts to influence society and human self-understanding by claiming that even human functions and behaviors can be reduced to mechanistic interpretations (Lewontin et al. 1984; Rose 1981, 1988, 1997). Richard Lewontin constantly warned against one-sided interpretations of the genome and their consequences (Lewontin 1991, 1993; Lewontin et al. 1984) and also built bridges to the renewed debates in the twenty-first century (Lewontin 2000). Although genetic determinism is now considered largely overcome, it still lurks in the background (Comfort 2018; Krimsky and Gruber 2013).

The extreme reductionistic views of Richard Dawkins (*1941), as well as the mechanistic interpretations of brain functions by some neurologists, generated a lot of resistance, not only in biology but also in anthropology, philosophy, and the social sciences. Anthropology today is deeply divided on these issues (Fuentes 2004, 2009, 2017; Malik 2002; Rosslenbroich 2023). All this demonstrates that the question is not just an academic playground but rather has a deep impact on our self-understanding and the development of the society in which we live. Concepts and the type of thinking developed from them construct social realities.

At the same time, a next generation of scientists started to elaborate more details of an organicist program. By the end of the 1930s, most of the key criteria of systems thinking had been formulated by organismic biologists, Gestalt psychologists, and ecologists (Capra and Luisi 2014). The 1940s saw the formulation of actual systems theories. Systemic concepts were integrated into coherent theoretical frameworks describing the principles of organization of living systems. These first theories may be called the "classical systems theories," they have some connection and overlap to the development of the school of cybernetics during this time.

During the 1950s and 1960s, systems thinking had a strong influence on engineering and management, where systemic concepts—including those of cybernetics—were applied to practical problems. However, the influence of the systems approach in biology was almost negligible during that time and for the rest of the twentieth century. In the 1950s the decades of the triumph of genetics began with the elucidation of the physical structure of DNA and of the genetic code. This success totally eclipsed the systems view of life and any form of organicism.

Nevertheless, strong personalities still saw the necessity of the organismic view and used the systems perspective in their own concepts, forming a second phase of organismic thinking. However, they had to deal with an extraordinary supremacy of reductionistic research, so that talking of systems, organismic relationships, and hierarchical structures became quite unconventional and sometimes even was defeated by pejorative and polemic writings. In part this is still the case today. A new generation of biologists was trained without ever hearing anything about systems concepts and organismic thinking, and without ever realizing that there continuously is a difference of views and a controversy.

Two leading figures of this second phase were Ludwig von Bertalanffy (1901–1972) and Paul Alfred Weiss (1898–1989).

Bertalanffy's efforts to establish a theoretical framework proper to biological phenomena have been motivated by the conviction that biology is an independent field of research, which cannot be reduced to physics and chemistry. Although physicochemical analyses are necessary for the study of life, they are not sufficient. Investigations that only catalog the parts and processes of living systems and characterize them in isolation cannot yield truly biological explanations, because they provide no information about how these parts and processes are organized and functionally integrated into the complex coordinated whole that usually is referred to as an organism (Nicholson and Gawne 2015).

Bertalanffy (1933, 1952) regarded organicism within biology as having three major components (summarized by Gilbert and Sarkar 2000): (1) an appreciation of wholeness through regulation; (2) the notion that each entity is a dynamic, changing assemblage of interacting parts; and (3) the idea that there are laws appropriate for each level of organization (from atoms to eco-systems), which follow from emergent properties. Parts are organized into wholes, and these wholes are often components of larger wholes. Moreover, when at each biological level there are appropriate rules, one cannot necessarily reduce, for example, all the properties of body tissues to atomic phenomena. Given an entity as complex as the cell, the fact that quarks have certain spins is irrelevant. This is not to say that each level is independent of the lower one. On the contrary, laws at one level may be dependent on those at lower levels; but they may also be dependent on levels higher up.

On this basis, Bertalanffy started his considerations on an organismic con-ception of biology and summarized its "leading principles": "*The conception of the system as a whole* as opposed to the *analytical* and *summative* points of view; the *dynamic conception* as opposed to the static and *machine-theoretical* conceptions; the consideration of the organism as a *primary activity* as opposed to the conception of its *primary reactivity*" (Bertalanffy 1952, 18f).

Paul Alfred Weiss developed the most consequent and clearest organismic systems approach. He was an Austrian scientist who moved to the United States, where he became a leading figure in the science of his time. His contri-butions to neurophysiology and developmental biology became well known. However, curiously enough his systems approach has been nearly forgotten. Only occasionally is his concept cited, but there has hardly been any understanding of this fundamentally unique approach, which differs essentially from most of the usual approaches of systems biology today. Only recently have there been publications that appreciate the concept in a more profound way and argue for a revival of his ideas (Drack et al. 2007; Drack and Apftaler 2007; Drack and Wolkenhauer 2011; Rosslenbroich 2011a, 2016b).

Weiss describes different system levels such as cells, tissues, organs, and the whole organism consequently as integrative units that have their own charac-teristics. These units integrate the lower levels in a way that the whole system exhibits coherent and robust functions. These overall functions cannot be traced back to the activity of the parts involved, although the functions of the parts are crucial and necessary, so that the parts and the system are interdependent. This interdependence is the crucial element in Weiss's considerations. Weiss formulated his approach several times presenting many examples from his empirical work. He saw these theoretical considerations as an intellectual guide in the background, while his experimental work stood in the foreground.

One situation on occasion of a conference is characteristic. As a leading scientist he was to give the opening lecture of a conference in 1960. The conference had the title "The Molecular Control of Cellular Activity," a clear expression of the expectations of most scientists at that time. However, Weiss called his lecture: "From Cell to Molecule," setting a sharp contrast to these expectations (Weiss 1962, 1):

The privilege of introducing this series of lectures on "The Molecular Control of Cellular Activity" is all the more precious to me because it provides me with an opportunity to recenter the object whose "activities" are to be "controlled"—the cell—from the increasingly off-center, out-of-focus position which it has assumed in current thought. Of the twelve lectures of the series which are to follow, all twelve deal with important fragments of the molecular inventory of cells, and seven alone with nucleic acids. This is a true reflection of current hopes—or illusions—that it might be possible to pinpoint in the cell a master compound "responsible" for "life"—an obvious reversion in modern guise to animistic biology, which let animated particles under whatever name impart the property of organization to inanimate matter. Therefore, lest our necessary and highly successful preoccupation with cell fragments and fractions obscure the fact that the cell is not just an inert playground for a few almighty masterminding molecules, but is a system, a hierarchically ordered system, of mutually interdependent species of molecules, molecular groupings, and supramolecular entities; and that life, through cell life, depends on the order of their interactions, it may be well to restate at the outset the case for the cell as a unit. A unit retains its unity by virtue of the power of subordination which it exerts upon its constituent elements in such a manner that their individual activities, instead of being free and unrelated, will be restrained and directed toward a combined unitary resultant. In short, the story of "molecular control of cellular activities" is bound to remain fragmentary and incomplete unless it is matched by knowledge of what makes a cell the unit that it is, namely, the "cellular control of molecular activities".

Even then the early concepts of systems thinking in biology revealed crucial differences. It is difficult to determine whether these differences lead to the fundamental dichotomies in the present forms of understanding of the term *system*, which will be discussed later. However, Bertalanffy's formulations of a general systems theory were by far less specific than those of Paul Weiss. Bertalanffy expected to find the same principles in technical systems, organisms, and human societies, independent of the nature of their components. This is a misconception, because there is something essentially different going on in a nonliving complex, which is driven by some physical principles, than in an organism like that of a flatworm or a rabbit, or in human society which is organized by autonomous individuals who are able to decide on a more-or-less rational basis. Niklas Luhmann (1927–1998), for example, transferred a general understanding of systems from biology to sociology and neglected the individual, a fundamental flaw within his considerations. In contrast, Weiss just used the term *system* to understand how an organism organizes its components to generate coherent and robust functions, without overstating his approach. He was very specific about life functions in the narrow sense, drawing on his empirical experience, and this is what biology primarily must learn.

In an often-cited paper Michael Polanyi (1968) described "life's irreducible structure" from a strictly epistemological point of view. He argued that any construction of a machine needs information. This is also the case in organisms, which is known as the information content of DNA. However, he argued, if this information content makes use of the laws of physics and chemistry, living things transcend these laws. Thus, organisms are under dual control: the laws of physics and chemistry on the one hand, and the information content that generates boundary conditions for these laws on the other hand. Biological hierarchies consist of a series of boundary conditions, each level of which relies for its proper functioning on the principles of the levels below it, even while it itself is irreducible to these lower principles. "The recognition of a whole sequence of irreducible principles transforms the logical steps for understanding the universe of living beings. The idea, which comes to us from Galileo and Gassendi, that all manner of things must ultimately be understood in terms of matter in motion is refuted" (Polanyi 1968, 1312).

Brian Goodwin (1931–2009) was also a strong advocate of the view that genes cannot fully explain the complexity of organisms and that a systems view of life is important. From the perspective of structuralism, he suggested that nonlinear phenomena and the fundamental laws defining their behavior were essential to understand biology and its evolutionary paths. Goodwin held that many patterns in nature are a by-product of constraints imposed by complexity. The limited repertoire of motifs observed in the spatial organization of plants and animals indicates, in Goodwin's opinion, a fingerprint of the role played by such constraints. Accordingly, the role of natural selection would be secondary (Goodwin 1994, 2007).

Many embryologists have traditionally been skeptical about reductionist explanations, and this holds true far into the twentieth century. Processes in the embryo show that there is always a whole system that is changing and that parts like the cells are subordinated to developmental dynamics. Even today, after a period of hope that developmental dynamics can be traced back to molecular components and a "genetic program," many embryologists are again asking for an organicist approach in the study of embryos (Gilbert 2014; Gilbert and Sarkar 2000).

Several authors study form and function of plants and animals from the viewpoint that they have their own laws and principles and cannot be reduced to underlying forces. Studying and comparing observable phenomena in their own right is expected to unravel the immediate processual order. Morphology, as a

functional morphology, is important in this field, but it also uses results of analytical sciences, trying to integrate them into the context. It is a manner of
science that is explicitly based on Goethe's approach. In this sense it is also an
organismic view, avoiding both mechanistic reductionism as well as vitalism
(Amrine et al. 1987; Bortoft 1996, 2012; Harlan 2005; Holdrege 2005, 2013;
Lockley 1999; Riegner 1985, 2008, 2013; Schad 2020; Seamon and Zajonc
1998; Suchantke 2010).

Further interesting studies on organismic perspectives in biological thinking
during this second phase are Wuketits (1981) and Riedl (1978, 1979, 1984).
Wuketits presented a fine analysis of the problems of causal thinking in biology
and discussed aspects of circular causality, networks, and systems thinking, a
remarkable endeavor in that time. Riedl has been very influential via his scholars
to overcome some of the too one-sided interpretations in biology in that time and
developed a systems approach for evolutionary biology (Wagner and Laubichler
2004).

The Focus on DNA and Information Sciences

The discovery of the structure of DNA in the 1950s triggered an experimental
research program that was quite successful in identifying and characterizing
the molecular constituents of living systems. The general belief in the
explanatory power of this approach was coupled with the more specific conviction that biological information flows unidirectionally from the DNA to
proteins and that genes are the primary determinants of an organism's form,
function, and behavior. The invention of genome sequencing techniques led
to the Human Genome Project, which was initiated to acquire the complete
catalog of all genes in a human being. The Human Genome Project was conceived under the explicit assumption that this collection of data constitutes the
complete set of instructions for development of the human organism, that it
would allow the causes of major health problems to be identified, and that it was
the necessary starting point for the development of causal therapies.

However, toward the end of the twentieth century, essential elements of this
program had to be corrected. It had become clear that DNA sequences cannot
contain the entire information required to configure a human organism and
that there is no simple one-to-one correspondence between genes and phenotypes (Jablonka and Lamb 2005; Moss 2003; Shapiro 2011). The publication
of the sequence of the human genome in 2001 failed to keep its promise to
revolutionize biomedical research and its effectiveness in treating human diseases. Even the architects of the Human Genome Project have been forced to
acknowledge the impossibility of predicting the lives of organisms by their

genomes: "[O]ne of the most profound discoveries that I have made in all my research is that you cannot define a human life or any life based on DNA alone" (Venter 2007, 3, citation from Nicholson 2014b, 351). Such an insight is amazing regarding all the earlier warnings from more prudent authors such as Lewontin et al. (1984), Strohman (1993, 2001, 2003) or Hubbart and Wald (1993). In a fulminant volume published by Harvard University Press under the title *Genetic Explanations: Sense and Nonsense* (Krimsky and Gruber 2013) these shortcomings are discussed from several perspectives.

The most significant finding of the Human Genome Project was that humans have far fewer genes than expected and that organismic complexity does not correlate with the number of genes. The collapse of the standard genetic concept led to the idea that if one is to succeed in grasping the complexity of an organism, one must not only catalog and characterize the complete set of genes (the genome) but also the complete set of RNA transcripts (transcriptome), the complete set of proteins (proteome), and the complete set of molecules involved in metabolism (metabolome) (Nicholson 2014b). To attain this, high-throughput technologies have begun to be applied on a massive scale, resulting in the large-scale production of vast pools of data, which biologists are now trying to decipher using mathematical models and computer simulations. Together with a certain admission that reductionism is somehow insufficient, this has been called *systems biology* and is quite modern today. It is perceived to offer a way of overcoming the limits of reductionism, which have been acknowledged at least.

Kritikou et al. (2006, 801) write:

Systems biology is en vogue—the catch-phrase crops up in laboratories, grant applications and editorial offices in all manner of contexts. So what exactly is systems biology? Molecular cell biology encompasses a growing number of mechanistic but largely isolated insights and, increasingly, high-throughput "omics" data sets; both are generally semi-quantitative and specific to a particular experimental system. The challenge is to integrate this complex and highly diverse information into a conceptual framework—one that is holistic, quantitative and predictive. One day, this might result in a "virtual cell". Molecular cell biology seems to be on the verge of emancipating itself from a rather informal and purely reductionist hypothesis-driven approach by embracing high-throughput data acquisition, rigorous quantitation and mathematical modelling.

However, it is quite obvious that this form of systems biology is far from transcending reductionism and differs profoundly from the original understanding and intentions of classical integrative systems biology (Cornish-Bowden 2006; Saetzler et al. 2011). Consequently, O'Malley and Dupré (2005) identify two variants of the term *systems biology* that are presently in use: "pragmatic systems biology" and "systems-theoretical biology." The pragmatic version, which is the younger one, examines intracellular networks by characterizing each of the components in the networks individually and then modeling the interactions between them using computational tools. The expectation is that by means of this bottom-up model, one will arrive at a full understanding of the cell as a whole. Such an approach is thoroughly reductionistic given that it privileges the molecular level and then expects that the pooling of the data will calculate up to the behavior of the whole complex. Any integrative, regulative, and autonomous function of the cell as a whole is completely ignored.

This form of systems biology does not constitute a significant departure from reductionistic paradigms but rather enhances them by even abusing the term *system*. However, it means that systems biology in this sense would be subject to the same explanatory limitations as the earlier approaches. If the previous approach could not grasp biological organization, then neither can systems biology in this form of understanding (Nicholson 2014b).

As a result of these objections, an increasing awareness is emerging among some scientists that an ad hoc approach to systems is not enough. For systems biology to succeed where previous approaches failed, it needs to embrace a truly systemic perspective. It must move beyond the study of individual molecules and their interactions and study systems as systems rather than as collections of parts, which is at the heart of the systems-theoretical approach. This means that although one can analyze the various activities of a biological system in isolation, the system carries these out simultaneously by harmoniously coordinated interactions. Against this background a third phase of organicism is developing today.

Bizzarri et al. (2013) summarize in their abstract:

In the past, biological research has focused on questions that could be answered by a reductionist program of genetics. The organism (and its development) was considered an epiphenomena of its genes. However, a profound rethinking of the biological paradigm is now underway and it is likely that such a process will lead to a conceptual revolution emerging from the ashes of reductionism. This revolution implies the search for general principles on which a cogent theory of biology might rely. Because much of the logic of living systems is located at higher levels, it is imperative to focus on them. Indeed, both evolution and physiology work on these levels. Thus, by no means Systems Biology could be considered a "simple" "gradual" extension of Molecular Biology.

The assumption that modern developments in artificial life research, cybernetics, and information sciences can dissolve the boundary between living things and machines (Bongard and Levin 2021) is merely the modern version of the old machine metaphor. The view of a clockwork is replaced by aspects from software development and robotics, and is thus not one bit closer to what actually constitutes a living organism, even if some characteristics of the living can be imitated in a sophisticated way, and if some smart solutions for practical or medical tasks are possible.

A New Organicism at the Beginning of the Twenty-First Century—Phase 3

Since the beginning of the twenty-first century there is a strong and thoughtful renewal of interest in organismic perspectives in biology. It comes not so much from outsiders and revolutionaries but rather from established scientists of different provenience including quite prominent names. The question is no longer whether the physical and chemical laws apply within the realm of the organic. It is clear today that organisms deal with these laws and principles and that many processes and functions can be studied under this perspective. Likewise, it is not a question whether the analytical research program in biology is legitimate. This generated an impressive amount of knowledge, so that there is no doubt that the analytical method is an important approach. However, a different question is whether the analytic research program with its thinking in linear causal relations and sequences is sufficient to explain the phenomena of the living and its organization. Will the knowledge of physical and chemical details ever sum up beyond single and isolated knowledge, so that the overall functions and properties of the systems, for example, on the level of organs or of the whole organism, can be derived from molecular interactions? Is it really possible to find the ultimate causes of organ functioning on the molecular level?

Something like a "new frontier of biology" (Henning and Scarfe 2013) seems to be developing, emphasizing the importance of organization, integration, and regulation. This also includes such areas as systems biology, epigenetics, biosemiotics, emergence theory, niche construction, extended evolutionary synthesis, and many more. These are still not mainstream but have a growing influence.

Among many other publications around the turn of the century two papers were prominent and are being cited repeatedly. In 2004 Carl Woese, a leading microbiologist, published a fulminant paper summarizing the limitations and the one-sidedness of molecular biology under the heading: "A New Biology for a New Century." He clearly points out the necessity to place modern knowledge into the organismic context. Of course, he appreciates modern molecular knowledge, but he formulates clear doubts that it will be possible to reconstruct organismic structure, form, and function only from that level. Already cited in the introduction (chapter 1), this article continues: "The molecular cup is now empty. The time has come to replace the purely reductionist 'eyes-down' molecular perspective with a new and genuinely holistic, 'eyes-up,' view of the living world, one whose primary focus is on evolution, emergence, and biology's innate complexity" (Woese 2004, 175).

On the following pages he writes further:

Let's stop looking at the organism purely as a molecular machine. The machine metaphor certainly provides insights, but these come at the price of overlooking much of what biology is. Machines are not made of parts that continually turn over, renew. The organism is. Machines are stable and accurate because they are designed and built to be so. The stability of an organism lies in resilience, the homeostatic capacity to reestablish itself. While a machine is a mere collection of parts, some sort of "sense of the whole" inheres in the organism, a quality that becomes particularly apparent in phenomena such as regeneration in amphibians and certain invertebrates and in the homeorhesis exhibited by developing embryos. . . . Twenty-first century biology will concern itself with the great "nonreductionist" 19th century biological problems that molecular biology left untouched. (176)

Our task now is to resynthesize biology; put the organism back into its environment; connect it again to its evolutionary past; and let us feel that complex flow that is organism, evolution, and environment united. The time has come for biology to enter the nonlinear world. (179)

The other prominent paper comes from the context of embryology, which always had a strong organismic perspective among several of its exponents. It was published in 2000 by Gilbert and Sarkar with the title "Embracing Complexity: Organicism for the 21st Century." They make a strong point that a well-defined organicism is important for biology in general and for embryology in particular. They declare that top-down and bottom-up phenomena must be regarded simultaneously: "The properties of any level depend both on the properties of the parts 'beneath' them and the properties of the whole into which they are assembled. . . . Parts determine wholes, wholes determine their parts in the sense of allowing properties to be defined. In embryology, we are constantly aware of the parts being determined by their context within the whole" (Gilbert and Sarkar 2000, 2).

Another leading advocate of this renewal of interest in organismic perspectives is Denis Noble, a renowned physiologist at the University of Oxford in Great Britain. With a series of books, papers, and lectures he is arguing vehemently against the reductionist viewpoint and especially against the view of genetic reductionism. His main message is that there is no privileged level in the organism, such as molecules or genes. They are just organs within a set of different system levels, necessary but by no means determining the whole system or being more important than the other levels of that system. We know a lot about molecular mechanisms, Noble describes, and he is among the foremost to appreciate this knowledge. However, the challenge now is to extend that knowledge up the scale. How can we use it to throw light on the processes that govern entire living systems? That is not an easy question, he admits. As soon as we move from genes to proteins, for which they code, and then on to the interactions between these proteins, the problems become extremely complicated. Yet we need to understand these complexities to interpret the molecular and genetic data correctly, and on that basis to talk in a fresh and useful way about larger questions like "What is life?" (Noble 2006, 2017a, 2008b, 2011b, 2008a; Noble et al. 2014).

This, then, is the challenge that sequencing the genome has raised. Can we put Humpty-Dumpty back together again? That is where "systems biology" comes in. This is a new and important dimension of biological science, though it has strong historical roots in classical biology and physiology going back over a century. In recent decades, however, biologists have tended to focus quite narrowly on the individual components of living organisms. What properties does each component have? How does it therefore interact, over the short term, with other components of similar scale? Now, we are ready to ask some bigger questions. These are about systems. At each level of the organism, its various components are embedded in an integrated network or system. Each such system has its own logic. It is not possible to understand that logic merely by investigating the properties of the system's components. . . . Only, it requires a quite different mind-set. It is about putting together rather than taking apart, integration rather than reduction. It starts with what we have learned from the reductionist approach; and then it goes further. (Noble 2006, x)

The book of life, he argues, is life itself. It cannot be reduced to just one of its databases. The genome needs to be read through the phenotype, not the other way around. Noble presents a lot of examples from experimental research demonstrating that his bearing is not based so much on theoretical considerations but rather on empirical evidence.

Michael Joyner, a physiologist-clinician from the Mayo Clinic Rochester (USA) is not primarily an organicist but sees clearly the limits of reductionism, especially for medical research, where it was implemented with far-reaching promises. In collaboration with Noble, he published a series of papers showing that approaches of integrative physiology and integrative systems biology are much more informative than data from the genecentric view of human diseases. Although much effort and money are spent in research, which is grounded on the reductionist view on health problems, the outcome, Joyner states, is disappointing (Joyner 2011a, 2011b, 2015; Joyner et al. 2016; Joyner and Prendergast 2014; Joyner and Pedersen 2011). The volume of Krimsky and Gruber (2013) cited before presents examples of health problems, where the conventional approach and especially the focus on genes was not as helpful as once had been expected. In the same sense the strong focus on genes in so-called "personal medicine" (more recently, "precision medicine") is being heavily criticized (see the issue "The Precision Medicine Bubble" introduced by Paneth and Joyner 2018 and Joyner and Paneth 2019).

The ethologist Patrick Bateson (2005) writes that the longtime trend toward analysis at ever more lower levels is starting to reverse. The new integrative studies must make use of the resources revealed by molecular biology but should also use the characteristics of whole organisms to measure the outcomes of developmental processes. He discusses examples of ethological-physiological research, in which movements between levels of analysis are being applied with increasing interpretive power and promise. According to Bateson a renewed focus on the whole organism is also starting to change the face of evolutionary biology. The decision-making and adaptability of the organism is recognized as an important driver of evolution and is increasingly seen as an alternative to the gene-focused views.

Concerning the concentration of much of research on genetics, Bateson (2005) writes:

The ability to sequence genomes was a great scientific achievement. The much-heralded publication of the human genome does not and cannot provide the hoped for "Book of Life" that would enable us to understand all aspects of human nature. Numerous post-genomic projects are based on the assumption that, if clever enough mathematics and sufficient informatics were applied to the problem, somehow the code for the characteristics of whole human beings would be laid bare. The problem for biology in the post-genomic era is not, however, one of cryptography. Genes code for proteins not people. If we want to understand what happens in the life-long process from conception to death, we must study the process by which an embryo becomes a child and a child becomes an adult. (31) . . . In this much changed intellectual environment, the time seems right to rebuild an integrated approach to biology. With a whole array of promising new research areas and techniques emerging, integrative biologists have a lot to be excited about. (37)

In recent decades the term *organicism* has been widely used again. Within the field of philosophical biology, which regained importance within the last 20 years or so, the term is regularly in use today. Of course the approaches are diverse, but the literature on organismic aspects in this field is quite rich and inspiring (e.g., Bizzarri et al. 2013; Capra and Luisi 2014; Denton et al. 2013; Dupré 2012; Ebersbach et al. 2018; El-Hani and Emmeche 2000; Henning and Scarfe 2013; Kather 2003, 2012; Morange 2008; Nicholson and Dupré 2018; Nicholson and Gawne 2015; Nicholson 2014b; Penzlin 2014; Rehmann-Sutter 2000; Salthe 1985). Many related aspects for widening the scope to organisms are discussed in this field (see, e.g., Fuchs 2018; Nagel 2012). They can hardly be summarized here, but they will be essential elements in the following chapters within the respective topics.

One main problem is that these writings rarely, if ever, attract experimental scientists' attention. Anna Soto and Calos Sonnenschein are among the rare exceptions, combining experimental research on cancer with a strong interest in the philosophical background of their subject. They argue that a more organismic view could potentially lead to new solutions for curing cancer and state that the usual approach is heavily one-sided (Soto and Sonnenschein 2005a, 2005b, 2006, 2011a, 2011b; Soto et al. 2008; Sonnenschein and Soto 1999).

Heinz Penzlin (2014), a renowned German physiologist, published a book with the title *Das Phänomen Leben. Grundfragen der Theoretischen Biologie* ("The Phenomenon of Life: Basic Questions of Theoretical Biology"). However, the book is not as theoretical as the title might suggest, but rather presents a broad survey of physiological topics under a clear and detailed organismic perspective. He writes: "Attempts to reduce life to simple physico-chemical processes are condemned to fail because 'life' not only rests on matter and energy, but rather on matter, energy and information. Organisms are not only thermodynamic, but also communicative systems." (Penzlin 2014, 40, translation by author). In many examples he describes how life has an active component, a general activity, which deals with physicochemical conditions and prerequisites, that the organisms employ actively to cover their needs and to maintain their autonomy.

The physiologist and ecologist J. Scott Turner (2013, 2007, 2017) postulates that homeostasis is the crucial principle to develop "a coherent theory of life," and that homeostasis is generated by self-sustained active processes he summarizes as agency. In his view, agency, as the capacity to act, is where the distinction can be drawn between a living being and nonliving objects with certain resemblances to living beings.

Also, Turner (2017, xii) sees the whole topic as crucial for modern biology: "If biology claims to be a distinct science, on what grounds is the distinction built? For much of the twentieth century, we could afford to ignore this question, or at least not to engage it too critically, as we gathered up the pretty scientific baubles that lay newly strewn about us. We can ignore it no more, because leaving the question unresolved has brought biology to the brink of a philosophical and scientific crisis."

Dupré and Nicholson (2018) focus on the aspect that an organism essentially is a process and not so much a thing or an accumulation of substances. They see organisms as dynamic through and through and thus propose "that the living world is a hierarchy of processes, stabilized and actively maintained at different timescales. We can think of this hierarchy in broadly mereological terms: molecules, cells, organs, organisms, populations, and so on. Although the members of this hierarchy are usually thought of as things, we contend that they are more appropriately understood as processes" (3).

Nicholson (2014b) states that today biology is slowly witnessing the comeback of the organism as a fundamental explanatory concept.

Finally, it should be noted that the term *organismic biology* is now often used as a general term for the study of whole organisms (zoological, botanical, ecological) rather than from a molecular perspective, which says nothing about the search for independent functions and processes in living things described here. For example, there are institutes with this designation. Sometimes, however, they explicitly distinguish themselves from the dominance of molecular biology, so that overlaps may also exist.

A Revolution in Evolution

In the second half of the twentieth century, evolutionary theory tried to reduce evolution completely to a change in allelic frequencies of a population. Evolutionary biologists became concerned not with organisms, but with how populations behave over time as a result of random genetic changes. From this point of view, organisms have no autonomous agency of their own. Their distinctive features, and indeed their very existence, are to be explained by appealing to the causal capacities of genes. In the words of Dawkins (1976, xxi), organisms are mere "survival machines—robot vehicles blindly programmed to preserve the selfish molecules known as genes." It is the genes that are selected by the environment, rather than the organisms that embody them, and it is at the level of the genes where evolutionary explanations are to be expected in research.

The view of evolution that has mainly been advocated by the Modern Synthesis is being challenged today from various perspectives. The different critiques converge in rejecting the premise that genes are the primary, or even the sole, agents responsible for inheritance, development, and adaptation (Corning 2020; Jablonka and Lamb 2005; Laland, Odling-Smee, and Turner 2014; Laland et al. 2015; Pigliucci and Müller 2010; Shapiro 2011; West-Eberhard 2003). Naturally this generates controversies (Laland, Uller, et al. 2014). With regard to inheritance, there is growing evidence against the notion that genes are the only transmitters of information from parent to offspring. Jablonka and Lamb (1998, 2005) have drawn attention to a number of neo-Lamarckian epigenetic inheritance functions such as chromatin marking processes like DNA methylation, feedback loops of gene expression, micro-RNAs as regulatory units, and many more. These

processes evidently enable the acquisition and transmission of phenotypic variation across generations. With regard to development, it is becoming apparent that
genes do not constitute a special class of master molecules that direct and control
the developmental process. Rather, there are interdependent processes between
genes and the developing tissues and cells, including positional information.

The insight that genes do not determine the phenotype on their own led to
new considerations on the relationship between genotype and phenotype.
West-Eberhard (2003) has argued that phenotypic plasticity plays a substantial
role in evolution. It leads to phenotypic innovations that can later be consoli-
dated by means of genetic accommodation.

Another aspect of the traditional evolutionary picture that is challenged by
the new organism-centered view is the conceptualization of the relationship
between organism and environment. According to the twentieth century position this is an external relationship. A causal arrow runs from the environment
to the organism, or rather its genes, and this arrow explains why organisms
are the way they are. Adaptation is conceived as a process by which natural
selection forms organisms to fit preestablished environmental templates. Environments pose problems, and the organisms best equipped to deal with them
are the ones that survive and reproduce. This view, however, neglects the
active role that organisms themselves play—through their metabolism, their
activities, and their choices—in defining and partly creating their own ecological niches (Nicholson 2014b; Walsh 2015).

Odling-Smee et al. (2003), in reference to earlier work by Lewontin (1983,
2000), called this process "niche construction." If organisms not only adapt to
environments, but also construct them, they live in an environment that is a
result of the activities of past and present generations of organisms, generating
a feedback loop between niche construction and adaptations, which shape the
dynamics of evolutionary change.

In section 4.12 some aspects of these new considerations will be discussed.
The emerging view on evolution is much more organismic than the evolutionary theory of the twentieth century (Noble et al. 2014). It dislodges the gene
from its privileged position and restores the organism back to the center of the
evolutionary process (Nicholson 2014a, 2014b; Noble et al. 2014). In particular, organisms are increasingly seen as being active in evolution rather than
being passive objects of genetic (internal) or environmental (external) circumstances and prerequisites (Bateson 2017; Walsh 2015).

Conclusion and Prospect

Today, the scientific discussion on this topic is complex. On the one hand, mainstream science with the majority of financial resources and infrastructure largely

focuses on analytical studies of molecular and genetic aspects of life processes in basic research as well as in applied disciplines such as medicine and agriculture. These disciplines, with a strong focus on reduction of complex phenomena to isolated functions within these complexes, are constantly generating new knowledge and are making contemporary science quite successful. However, there still is a major lack of understanding of integrative principles in cells, tissues, organisms, or even ecosystems, which causes a lot of confusion about how the overall functions of such systems are established and maintained.

Therefore, many experimental researchers see the necessity of a course correction in life science, including Gilbert and Sarkar (2000), Joyner (2011a, 2011b), Joyner et al. (2016), Kirschner et al. (2000), Krimsky and Gruber (2013), Laubichler and Wagner (2001), Noble (2006), Shapiro (2009, 2011), Sonnenschein and Soto (2013), Soto and Sonnenschein (2012, 2005a, 2005b), Strohman (1993, 2001, 2003), Turner (2017), Woese (2004), to name just a few of the more prominent authors.

In addition to some conventional areas of science, which are now increasingly based on a more complete concept of the organism, some areas are explicitly based on an organismic concept. These include, for example, fields such as cardiology (Furst 2020), embodiment theories in neurology (Fuchs 2018), major parts of ecology and plant sociology (Looijen 2000; Vahle 2007), evolutionary biology (Rosslenbroich 2014; Walsh 2015), major parts of zoology and animal physiology (Holdrege 2021; Lindholm 2015; Penzlin 2014; Riegner 1985, 2008; Schad 2020), and of course medicine (Heusser 2016; Schaefer et al. 1977; Uexküll and Wesiak 1998).

In addition, in philosophy of biology there is a long-standing quest for more integrative organismic aspects in biological research (Bizzarri et al. 2013; Dupré 2003, 2012; Henning and Scarfe 2013; Kather 2003, 2012; Lewontin et al. 1984; Mesarovic et al. 2004; Moreno et al. 2008; Nicholson and Dupré 2018; Nicholson 2014b; Pigliucci 2014; Rose 1981, 1988, 1997; Walsh 2015).

At the beginning of the present chapter, I described that during the history of science alternative research programs often existed side by side for long periods and competed with each other, and sometimes major progress was made when a synthesis of both became realizable. After prolonged dispute and argument, it can turn out that both factions have been right from a certain perspective and that a solution is possible that integrates aspects from both sides.

My impression—and also my thesis—is that biology today develops, or should develop, toward such a synthesis concerning knowledge from analytical research on the one hand and an organismic understanding of life on the other hand. Indeed, the extensive knowledge of details in structures, functions, and genetic processes provides a new opportunity to understand integrative and systemic functions. The chance for an organismic conception of life on a scientific basis has never been as good as today. This is what I want to propose in the subsequent chapters.

At the same time, since life around us is being manipulated, abused, and destroyed on an alarming scale, it has never been so urgent or so necessary to know what we are doing there. We must learn to act in context and take into account the interactions that are crucial in the organic.

We must learn to respect the aliveness of the world around us and to take it into account in our dealings with nature. We must learn, so to speak, to fit into its aliveness in such a way that we enhance and promote it instead of bringing it to its knees with our techniques. The machine metaphor is unsuitable for this. We need a realistic conception of the organism and not defective metaphors that lead to a misguided handling of nature.

Science has put together a lot of information about environmental problems, constantly warning about the consequences of the many manipulations. However, it is equally necessary to operate further research with an appropriate awareness for living beings to understand the scope and dimension of human influence on organisms and ecosystems, because the basic understanding of nature has a significant influence on how we deal with nature. Ultimately, however, our understanding of life also has consequences for our understanding of ourselves, because we ourselves are also living beings.

We had heard that the origin of atomism was in pre-Christian Greece. Among its main representatives were Leukippos and Democritus. It was a conception of the world, which could recognize as the actual reality only smallest parts in an otherwise empty space. Everything else was composed of them. This tradition of thought continued through the centuries. In their opinion everything complex is to be traced back to the mechanics of the atoms, which was made again and again the starting point of scientific systems and ways of thinking. All other qualities are disregarded and neglected in such a concept as epiphenomena.

The central statement of Democritus in this regard was, "Only apparently a thing has a color, only apparently it is sweet or bitter. In reality there are only atoms and empty space."

When Jim Watson famously said "There are only molecules, everything else is sociology" (cited by Noble et al. 2013, 357), it is clear from which world-view this came.

I find that our world is much richer. It is full of qualities like color, smell, and taste. There is beauty, feelings, empathy. There are animals that have consciousness and whose sensitivity extends into tremendous differentiations. Animals can feel joy, sadness, and pain. These are all sensations that are of utmost importance to ourselves in a rich inner life. And finally there is the rich world of thought, in humans and perhaps in glimpses in animals. All these are not epiphenomena to be discussed away, but they are realities in need of explanation.

And the wonderful thing is that science today is increasingly able to describe these realities. We no longer must resort to subjective descriptions, but we can scientifically reconcile our subjective experiences with objective observations. And the more we succeed in bringing subject and object together in cognition in this way, the more we can overcome both Descartes's separation from the world and the bleakness and emptiness that atomists think we must base our lives on.

3 Working Hypothesis

If we want to attain a living understanding of nature, we must become as flexible and mobile as nature herself.
—Johann Wolfgang von Goethe, 1817 (from Goethe 1981, 56, translation by author).

The various efforts to develop organismic aspects in biology, some of which were listed in the previous chapter, are quite fragmented and diverse. Often, one particular aspect of organismal performance is emphasized and studied, while other aspects are neglected or are treated only within a different concept. In this respect, organismal biology is a rather heterogeneous and also confusing field. Only when one has studied a lot of literature on the subject does it become clear that there are many overlaps and commonalities. A comprehensive, common concept is largely missing.

In addition, although organismic aspects are often developed using empirical examples, the bridge to empirical research is not really built. This may be due to sociological reasons of science but also to the lack of an overarching common concept that could provide some kind of guideline. Such a concept would also potentially improve the visibility of the approach.

In the subsequent chapters of the present book, a synthesis as proposed at the end of chapter 2 will be attempted. Such a synthesis can form a basis for the further development of organismic aspects in biology.

The working hypothesis is that empirical research is developed far enough today to reveal by itself the material and prerequisites to allow us to understand more of the specific organismic properties of the living. Without recourse to mysterious forces, it is possible to generate answers to the old question of the specific properties of life by just applying recent empirically acquired knowledge. It does not contradict the results of conventional research but rather grants them a more complete meaning within the context of the whole organism and actual life processes. This will change our general idea of living organisms as well as many aspects of our dealing with organic entities.

The following chapters describe fifteen specific properties of living entities. The aim is to demonstrate that the results of empirical research show both the necessity as well as the possibility of the development of a new conception of life and to build a coherent, "living" understanding of animate entities. Most examples will come from animal and human physiology. The discussion will, however, also take into account single cells (prokaryotic and eukaryotic), while plants will play a minor role, although most, if not all, of these aspects can be extended to plants as well.

To make this orientation to organisms such as bacteria, unicellular eukaryotes, multicellular animals, as well as humans a little more concrete, a few of them are exemplified in figure 3.1. It is for illustrative purposes only, but it maps the stages of organization included in the discussion. For the most part, however, general characteristics found at all of these stages are described, while organizational and evolutionary comparisons will play a minor role.

This approach will neither try to define life nor will it try to find criteria to differentiate living from nonliving. There have been many attempts for both in the literature, but they do not enjoy a consensus among scientists, although these attempts include valuable concepts. An overview of this discussion is provided by several contributions to the special issue "Open Questions on the Origins of Life," which is introduced by Gayon (2010); further examples of this discussion are Cleland (2013), Emmeche (1997), Farias et al. (2021), Jagers op Akkerhuis (2010), Kolb (2007), Oliver and Perry (2006), Popa (2010), Ruiz-Mirazo et al. (2004), and Tsokolov (2009). Hengeveld (2011) even objects to giving definitions of life at all, because they may bias the work that follows. Cleland and Chyba (2002) state that many proposed definitions of life suffer problems, often in the form of counter examples or in the form of criteria also being valid for phenomena that are inanimate, such as fire. Van der Steen (1997) indicates that even if a fairly general definition existed, it probably would be difficult to apply it to specific situations.

Also, the specific properties that will be discussed subsequently are not intended to provide another list of necessary and sufficient conditions to characterize life, as is often done in textbooks and courses (Emmeche 1997). Questions on the origin of life, which are often the occasion for definitions, are not addressed here either.

Here, it will not be argued against such approaches, but the intention is quite different: Independent of questions regarding definition or discriminating criteria, recurring and specific properties of living beings will be described as they are typically found during empirical work and taken into focus. Thus, the approach is predominantly phenomenological. What characteristics do living organisms typically exhibit when they are studied by physiology, embryology, molecular biology, and so on?

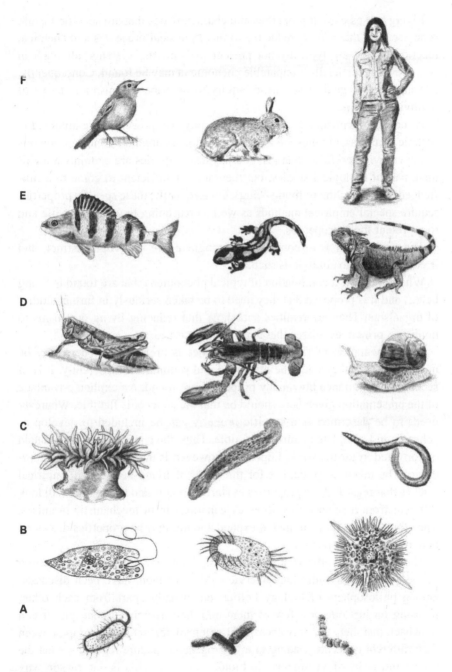

Figure 3.1
Examples for organisms of different organizational levels: (A) prokaryotes; (B) eukaryotic single cells; (C and D) invertebrates; (E and F) vertebrates. Not true to scale. *Source*: Drawings courtesy of Angela Rosslenbroich.

Living beings exhibit properties and characteristics that are specific for life processes and that are not reducible to mere physical processes and chemical reactions. They are typically not present in nonliving systems, although in some cases superficially comparable phenomena may be found. Consequently, it cannot be the goal of scientific inquiry to describe organisms in terms of nonliving machines.

Of course, chemical reactions as well as physical processes are involved in organic functions, but they are integrated within an organized and autonomous living entity. Therefore, chemical and physical inquiries are certainly a legitimate part of biological studies, but they are not sufficient to come to a full-fledged understanding of living beings. Consequently, these specific properties require special empirical methods as well as reasoning to understand life and to represent these properties.

The proposition is to consistently recognize life as living, distinct, and irreducible, and to study it as such.

What follows is a compilation of typical phenomena that are found in living beings and it is proposed that they need to be taken seriously in further studies of organisms. They are realities and show that reducing living organisms to nonliving principles misses these specific properties.

The presentation of fifteen such properties is preliminary. There may be more or less, or they could be organized and summarized differently. It is of secondary importance how many properties are found. An explicit advantage of the presentation given here should be that the concept is flexible. Whatever needs to be described as a specific property can be included to develop as coherent and complete picture as possible. Thus, the concept can be expanded or corrected in further work. Important, however, is that the specific phenomena can be taken as guidelines for the study of living beings. An empirical science that regards these properties as starting points and framework will look different from a science that seeks explanations from mechanistic premises. Therefore, the reader is invited to expand and modify the properties discussed here by further perspectives.

The envisaged synthesis at the end of the book refers to the representation of these properties within one overview. Most of them have been discussed among philosophers of biology before, but usually apart from each other, focusing on just one or a few of them and their associated concepts. It is a drawback that there are only occasional points of contact and overlap between such different concepts, leading to a heterogeneous picture regarding what the topics and tasks of an organismic biology may be. This is one reason why the theoretical consequences have had only little impact on empirical disciplines in biology so far, although the empirical work itself constantly produces examples of those properties. The philosophical work pertaining to some of these properties is usually unknown among empirical scientists, and attempts to discuss it can even produce mistrust.

However, just to present these properties within an overview may be no more than a first step. Nonetheless it can provide at least four advantages: (1) to gain a general idea about which properties need to be taken into consideration, even if they are not yet complete; (2) to understand how extensive the knowledge about organismic principles already is, clearly pointing beyond mechanistic thinking; (3) to find some relationships among these properties, and (4) to find a general principle that unifies these properties in the sense of an overall characteristic.

Presently, points 3 and 4 can only be described rudimentarily. A full synthesis in the sense of a unified notion of life appears to be unavailable thus far, and a multiperspective approach might be a preliminary solution. Multiperspectivity even might be more valuable for further empirical research than restricting the view of life to specific viewpoints, because the phenomenon of life is an extremely complex entity, which still involves a lot of enigmas.

Properties of Life

As a preliminary framework, the following fifteen properties of living beings will be discussed in chapter 4 (figure 3.2):

1. The principle of reciprocal interdependencies in organic functions

2. The principle of integrative system functions

3. The property to generate autonomy and self-determination

4. The principle of agency and self-generated activity of organisms

5. The activity to process molecules

6. The significance of information content and processing of information within organisms

7. The significance of energy content and processing of energy within organisms

8. The description of the unity of processes to generate form and shape in the sense of embodied physiology

9. The autonomy of time processes

10. The sensibility and affectability of organisms

11. Subjective experience and consciousness as a fundamental quality of organisms

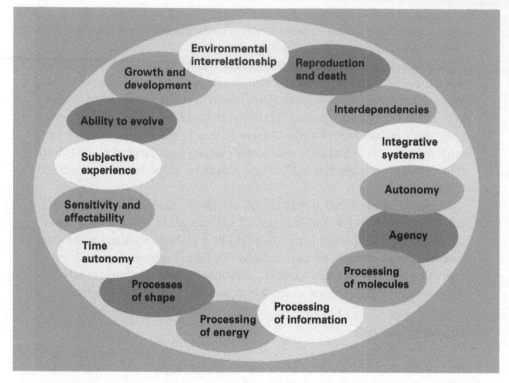

Figure 3.2
Overview of the properties of life.

12. The ability to evolve

13. The principles of growth and development

14. Processes of environmental relationships

15. Reproduction and death

We have already encountered most of these properties in one way or another in the historical overview of chapter 2. They have thus been perceived in the various sciences seeking an organismic perspective. These aspects will be brought together here in an integrated concept.

Life as an Organized Process

There is, however, an underlying general characteristic of these properties: first and foremost they involve functions, changes, and dynamics. Thus, it can be said that organisms are primarily processes. To consider organisms as processes rather than things or objects is an essential step toward an organismic view of life.

Dupré and Nicholson (2018) make a strong case for a processual concept of living beings. They propose that the living world is a hierarchy of active processes, stabilized and actively maintained at different timescales. Although the members of the hierarchy such as molecules, cells, organs, organisms, populations, and so on are usually thought of as things, they argue that they are more appropriately understood as processes. The processes in this hierarchy not only compose one another but also provide conditions for the persistence of other members, both larger and smaller.

As an example, the authors describe the liver, which provides enabling conditions for the persistence of the whole organism of which it is a part, but also for the hepatocytes that compose it. To persist, a liver requires both an organism in which it resides, and hepatocytes of which it is composed. A hepatocyte sustains a liver, and in turn a liver sustains an organism, by their respective actions. These processes themselves, which have usually been taken for things or substances, engage in processes such as metabolism, development, and evolution. "A key point is that these reciprocal dependencies are not merely structural, but are also grounded in activity. A hepatocyte sustains a liver, and a liver sustains an organism, by doing things. This ultimately underlies our insistence on seeing such seemingly substantiated entities as cells, organs, and organisms as processes" (Dupré and Nicholson 2018, 3).

This activity—in the sense of autonomous agency—will play a major role in the forthcoming chapters, especially enrolled on a more biological basis in section 4.4.

The continuous change in organisms on every level demonstrates its processual character. Subcellular structures, such as organelles within the cell, seem relatively well defined and stable. However, when they are observed for a longer period of time and within a living cell (rather than within a dead cell in the electron microscope), their continuous movements and rearrangements become observable.

Usually in student courses a general picture of a cell is used to explain its components and functions. However, such a static picture produces a static notion of the cell, which can be set in motion by microscopic films, which show the high dynamics of the various components. What we see as fixed pictures in textbooks are quasi-stationary patterns that only persist for a very tiny moment before undergoing transformations or disappearing altogether in the next moment. It is quite amazing just how dynamic a living cell is. The same is true for cells that compose a tissue. Any cell or any tissue represents a dynamic steady state, only the form of which persists, while its material constitution is constantly being exchanged.

Dupré and Nicholson distinguish the process ontology from the usual substance ontology, in which it is thought that all events in organisms can be reduced to underlying interactions of substances, especially molecules. In a historical sketch, Dupré and Nicholson (2018, 9) show that many of those authors who can be called organicists (of the first and second generation in my terms in chapter 2) argued for the relevance of activity and processes in living beings. For example, they cite Woodger (1929, 219) as saying "An organism, whatever else it may be, is an event—something happening. It is temporally as well as spatially extended." And according to Waddington (1957, 189), "The fundamental characteristics of the organism are time-extended properties."

However, here an important aspect will be added. If we regard process and change as the main principles that reign in organisms, there is—at least generally speaking—no tool to understand and to describe the order of organisms. It is self-evident that every organism at the same time has a structure, a form, and an identity, even if these features are seen within constant dynamic changes. To realize that they are in flux does not abolish the existence of these features.

Let us consider a simple example. In empirical biology it is very well possible to identify different species. There are thousands of species in the world of insects, and an entomologist is able to differentiate them by characteristic attributes such as structure, color, and many more. The chaffinch and the jay, which are presently enjoying a meal at the bird house in my garden, can reliably be categorized by their form, size, type of beak, and color of their plumage. That means the birds are not only processes, in which everything changes in any moment, but at the same time there are certain characteristics that stay the same, although there is a constant flux of substances and energy. A physiologist can reliably identify the components and structures of the heart in any mammal, because its organization is essentially the same in the respective species. Deviations can be identified as pathological. Many parts of my horse continuously change overnight or within several weeks or months, but it is still the same individual, just as I have known him for many years. He has a certain color, a shape, some typical peculiarities, and also a typical behavior and character. Thus, in a very general sense, there is not only process and change, but also identity, regularity, and order.

Thus, concurrently there are processes in which everything changes permanently on the one hand, and continuities, at least for a given time, which make up the characteristics of each respective organism on the other hand. If only the process aspect would be taken into consideration, there would be a certain dissolution (in a figurative sense).

A concept that is able to cover this reality of living beings needs to take this order into consideration. Here, this order will be called *organization*. Organization describes the typical structures, forms, and recurring features of organisms. It is the basis of morphological and physiological descriptions. Under this aspect it is neither enough to describe just the parts and molecules of an organism, nor is it enough only to describe the permanent changes of parts and components as such. An essential question is how these parts and these processes are organized—how the components have been put together and how the changes are constrained in a particular way.

How are these two perspectives interrelated? Which one is more important, the view of process or the view of organization? I think the key to understanding this relationship is to comprehend that both principles apply simultaneously at any given time: There is an organization, which is maintained by an active process. The process generates the organization, and the organization structures and directs the process. Neither of them is possible without the other. Processual turnover means that the very structure of every organism, its organization, unlike that of any machine, is wholly and continuously reconstituted as a result of its own active operation(s).

However, organization is quite an abstract term that needs to be filled out with observable biological properties: a task which is approached by the descriptions in chapter 4. As soon as some of the envisaged properties are discussed in the following chapters, a more complete view on process will be proposed in section 4.7. This view of an organized active process is a common theme in all of the following chapters, establishing a framework for describing and analyzing the specific properties of living beings.

The Principle of Concurrency

Thus, an organism is a constantly changing and processing organization. Organization enables the organism's stable identity, and the constantly running process generates and sustains the organization. Here we encounter a principle that will appear repeatedly in the forthcoming discussion of organismic properties.

It is the principle of concurrency of seemingly opposing features. The skin of many animals, for example, establishes a certain closure toward the environment, but simultaneously it is permeable for certain substances. Thus, it is relatively closed as well as relatively open. Also, for the whole organism it can be said that an individual has a specific autonomy as it has self-determination, self-regulation, and a boundary toward the environment. Simultaneously it has a continuous exchange with its environment, such as the exchange of gases and nutrients. Therefore, an organism has both relative openness as well as relative closure. There is a concurrence of both features.

In the present case it can be argued that an organism is concurrently a process and an organization. Regarding metabolism, cellular turnover, and development, an organism can be seen as a process, whereas regarding its integrity, its structures, its form, and the regularity of its physiological functions, it simultaneously has an organization, which of course is constantly in flux due to the processes.

Taking this altogether, I agree with the conclusion of Dupré and Nicholson (2018, 28), who state that the process perspective makes the ontological inadequacy of the mechanical conception of the organism explicit. If organisms are processes, structured by organization, rather than just substances, then conceiving them as machines inevitably leads to a distorted understanding of them. The reality of metabolic turnover, according to Dupré and Nicholson, means that the very structure of every organism, unlike that of any machine, is wholly and continuously reconstituted as a result of its operation. The life cycle an organism experiences is unlike anything a machine ever undergoes. Machines do not develop, nor do they reproduce; their configuration is fixed by their manufacturer, as opposed to that of organisms.[1] And, as a consequence of ecological interdependence, no organism can function, or even persist, independently of the entangled web of relationships it maintains with other organisms. In contrast, the persistence or operation of machines does not depend on their capacity to maintain relationships with other machines. By adopting an ontology based on the principle of concurrency of process and organization, we become far less likely to be tempted by the mechanical conception of the organism, which for many biologists still constitutes the standard, and often tacit, ontological understanding of living systems.

The principle of concurrency of opposite properties and functions will appear again and again in the course of the description of organismic properties. Therefore, it must be regarded as another basic principle of life.

Processing Substance, Energy, and Information

There is, however, another concurrency. The active process within an organism is at the same time dependent on a substance. There are substances, in the end molecules involved, which are moved around by the process, and the process definitely depends on that substance. Biology, and of course medicine, know well what happens if a certain substance is missing. Often the process (or whatever) is able to compensate for that deficiency in an amazingly autonomous way, recruiting another pathway or selecting another gene to generate the substance (e.g., a protein). But in many cases the lacking substance cannot be compensated for and leads either to a disaster or to illness. In general,

---

1. This is also true for modern artificial intelligence systems. For an informative discussion, see Reber (2019).

organisms are very precise with their substances, and this is of high relevance. In the sum, all the substances involved make up what is experienced as the physical existence of that organism. A horse or a rabbit have, besides being a process, a physical existence as well.

Of course, Dupré and Nicholson (2018) appreciate the existence of substances, but they regard them to be conditional on the existence of processes. They hold that things should be seen as abstractions from more or less stable processes. Processes are regarded, in some sense, as more fundamental than things, and things are seen as precipitates of processes.

A more biological answer of this relationship, although I think it is in line with Dupré and Nicholson, is this: There is an inextricably linked interdependence between the active process and the substances involved. It is legitimate to focus more on the process. This is undoubtedly of importance in a biological world of thought that is so extremely fixated on substances. Or you can focus on the substance side and make a lot of discoveries, as molecular biology demonstrates. But ultimately the relationship is irresolvable.

Or to put it another way, we look at a *substance-process* from different perspectives, sometimes from the perspective of the process, sometimes from that of the substance. But they are only the famous two sides of the same coin.

Endless debates are conceivable about this relationship, and it seems possible to find good arguments for both. But if this is so, there must be something to both perspectives. So why not dare to synthesize them? The crucial thing is *concurrency*, even if it seems an imposition for our thinking trained on the either/or principle, which is grounded in mechanical thinking. Organismic thinking needs to be thinking in simultaneity, synchrony, interdependence, and concurrency.

Even more needs to be regarded here. The process also needs energy. Energy keeps the process going, which is enabled and accomplished by a very sophisticated detailed organization of energy processing. A major part of cellular activities is dedicated to energy transformation, so that energy is available for the process in very small and handsome entities. So, there is another interdependence—that between process, substance, and energy.

Finally, all of this has a remarkable, high-level order (which science is just beginning to understand). Mechanistic biology tried hard to describe that biochemical processes just work along reaction gradients, but this proved to be impossible. All these processes in the cell are highly organized—spatially, temporally, and structurally. Such an order needs an information, an organizing principle. Much of this—but by far, not all—is provided by the DNA. Thus, at the same time the process depends on information, and information is the basis of organization.

Therefore, an essential statement of the concept developed next will be that living beings exhibit an indissoluble interdependence of process, substance, energy, and information. Or formulated somewhat differently, the living organism concurrently processes substances, energy, and information. This will gradually be developed in subsequent chapters.

Reduction and Reductionism

In chapter 2 a distinction between reduction and reductionism was made. This distinction will be of importance in the following presentation of the properties of life. Reduction has been described as an indispensable strategy for scientific investigation. It is a mode of analysis whereby the focus on a circumscribed biological entity permits a better understanding of the separated part. To choose a certain aspect, a specific perspective regarding the object of inquiry, and to pose a well-defined question belong to an appropriate armamentarium for scientists.

However, reduction—in contrast to reductionism—also involves finally putting the insights gained back into context. Especially in the organic, nothing can be considered without its context, its interdependence. It is important to understand individual processes within their context. A methodical reduction can only develop its full value when the object under investigation or the function in question is recognized in all of its functional relationships. In many cases, functions operate differently in isolated models than in the context of the whole, which exerts regulating and controlling influences on the subprocesses. In the subsequent chapters, this will be a recurring topic of consideration.

Intention of the Investigation

Some of the properties to be discussed appear in the usual lists that are drawn up when one wants to characterize living organisms. The fact that they are mentioned shows first of all that they have been perceived and that they are relevant. They are also often used to describe life, albeit in different selections and combinations, because there is no consensus on this. Thus, on the one hand, the approach presented here is not a special one.

On the other hand, however, it will be argued that the significance of these properties needs to be considered more explicitly. The point is this: If one takes these properties seriously, then this has consequences for the general way of thinking about organisms, the empirical approaches to their investigation, and for the handling and use of organisms. Biology will change if it integrates them as real entities into its empirical work as well as theory building. It will become clear that life must be understood as life and not as something else.

Today, large parts of basic research, medical research, and agricultural research are essentially performed within the machine paradigm of life and attempt to resolve processes more or less within linear causal chains. Machines can be operated, rebuilt, and manipulated to make them functional. In organisms, this turns out to be much more difficult, and the question arises whether such manipulations are even desirable within an organismic concept. If one takes the characteristics to be discussed here seriously, one comes to extended approaches to the living, which are actually not conceivable within the mechanistic doctrine but that bring us closer to the coherence, autonomy, and processuality of the living.

As described in chapter 2, developments in this direction have long since begun in the last twenty years or so. In many progressive scientific fields, this is seen as a paradigmatic reorientation of biological science with the attempt to take seriously the vitality of the living, its processual character, its intrinsic activity, and its conditionality in time. Looking at these concepts, one can have the impression that a transformation toward an organismic biology is long underway, even if it has not yet taken hold of large parts of mainstream science. However, since the literature is full of it, much of it will have to be quoted.

This paradigmatic transition still lacks a unified concept. This is what is being proposed here. Thus, it is not a question of just naming and enumerating the properties but rather of considering the consequences that result from them. It is a matter of getting a proper view on the actual meaning of these properties and their relationship to each other.

Gayon (2010) describes that the historical development of biology was led by its empirical results and driven by phenomena toward unifying concepts of the living. The cellular theory was the first step. Then came evolutionary theory, and then biochemistry and molecular biology, which showed that all known living beings are made of the same stuff (nucleic acids, proteins, etc.) and exhibit a remarkable metabolic unity (universal existence of core functions and metabolic pathways). Gayon remarks that these subdisciplines usually did not try to define life, but all of them have definitely shown that there are very strong reasons to believe that living beings share a number of properties that distinguish them from any other natural things and that justify the existence of a methodologically autonomous science. The present book attempts to focus on these properties and to prepare the next level of a unified concept.

Corning (2020) also discusses how living organisms exhibit a number of distinct properties that are often downplayed or ignored in the traditional biological paradigm and lists some of them. There are two possible techniques to do so: One is to ignore these properties all together and to conduct studies under the premise that everything is a chemical–mechanical interaction anyway. Another technique is first to mention a selection of them in a list, and perhaps even to have the textbook begin with them, and then to conduct scientific studies and their description in such a way that many of these properties are ignored. In this way they are treated immanently as epiphenomena, irrelevant in the actual scientific work and in the explanation of the real processes.

The approach proposed here will be to focus first on these phenomena and make them the starting point for investigations. In this respect, the approach is understood as a phenomenological thinking. However, this is difficult because there are many properties that must be described. We perceive them as seemingly separate from each other, and, in relation to the organism, they form a complexity that is not so easy to disentangle. Primarily, we look at a unit, a whole, when we look at an organism. Then, when we begin to describe individual characteristics, we dissect the whole according to the respective perspectives that we are focusing on.

What is described as properties are only certain perspectives on this whole, a certain point of view from which it is looked at it. Obviously it is necessary to bring them all together again at the end of the observation, which may be only partially successful. The different perspectives need to be thought "into each other" again, so to speak. Only then is there a chance to get back to the unified organism.

The intention is to work toward an understanding of organisms as best characterized by Penzlin (2014, 45):

Organisms are neither physical automatons nor automatons controlled by a mysterious force. They are holistically acting and reacting systems, which maintain their internal organization not only by a flow of matter and energy, but additionally by a flow of internally stored information. Biologists must unflinchingly devote their full attention to the peculiarities of living systems and continue in their efforts to explain them without resorting to hypothetical factors (vitalism) or useless, simplistic analogies from the inorganic (physicalism). To do this, they must develop their own systems theory, in which notions of organization, integration and specificity, and informational invariance play a central role. (translation by author)

This is not an exercise in speculative system building. It rather outlines a research agenda for theoretical as well as empirical biology with direct and wide-ranging implications for scientists and practitioners. It connects to specific areas of inquiry such as physiology, developmental biology, ecology, medicine, agriculture, and so on.

4 Properties of Life

4.1 Interdependencies

One fundamental property of functions and processes within living organisms is their mutual interaction, interdependence, and network characteristic. Most functions are characterized by interaction relationships. However, depending on the respective function and the level within the system that is being observed, a spectrum of relations may be involved, calling for different forms of scientific explanation.

Interaction and Interdependencies

Most biologists still regard the analysis of causal chains as the fundamental method of explanation for organic processes. It is expected that a cause A will, by necessity, be followed by effect B, as is the case in classical mechanics. The impression that biological activities are subserved by distinct mechanisms was in part fostered by the common strategy in scientific fields such as cell and molecular biology, in which research began by delineating a phenomenon, localizing it in a specific organelle or organ, and proceeding to decompose that unit into its parts and operations. The product of the operation of one part becomes the cause for the next part and so on. Interactions between components of different mechanisms were not denied but were largely ignored. However, in the analysis of an organic process, this form of explanation is usually limited, and the empirical results increasingly point to a much more complicated reality in living organisms.

Bechtel (2007) describes an early example: When the pioneers of biochemistry at the beginning of the twentieth century began to analyze biochemical processes, they generally assumed that metabolic reactions occurred serially and hence the pathways of reactions would be linear. Often, however, they could only arrange some intermediates in a linear sequence leading to a product with an unclear further outcome. The discovery of the most famous biochemical cycle, the citric acid cycle, resulted from such a circumstance. A number of organic acids could be arranged in a chain of reactions (figure 4.1), but then scientists faced a problem in specifying what happened next. After they dealt for a substantial time with this question, it was proposed that the last product was the new starting point for the whole process. This was at first a speculative proposal, and it took significant time to describe further intermediates. Finally, it became clear that the process was a cycle. Bechtel emphasizes that this result was not construed as theoretically significant but was just the result from trying to articulate plausible pathways of chemical reactions.

This is exactly what usually emerges in biochemistry and molecular biology: many processes turn out to be cycles. One component is dependent on the next one; there is neither a first cause nor a final effect. Rather, one process is the prerequisite for the next one, which is the prerequisite for the next one and so on. Such processes are dependent on each other (interdependent) within cycles or regulatory networks, which often is called "circular causality." Many networks, such as gene-regulatory networks, constantly grow in complexity, so

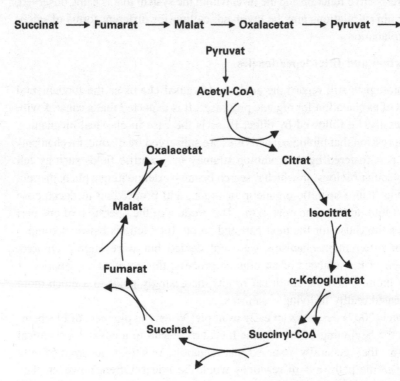

Figure 4.1
The citric acid cycle: (top) an early linear draft; (bottom) the cycle as seen today.

that diagrams of them are becoming ever more complex the more components are identified (figure 4.2). However, any linear causality obviously dissolves within such networks.

For the organism, one advantage of cyclically organized processes is that they provide means of effective regulatory feedback. In recent years this has also become obvious in genetics. It was previously thought that information always flows in just one direction, from DNA to a protein and finally to the phenotypic character, whereas today it is acknowledged that there are many feedback systems and regulatory cycles, so that it is not possible to identify DNA as a sort of primary cause (figure 4.3). Some geneticists realized this quite early (Lewontin 1991, 2000; Lewontin et al. 1984; Strohman 1993, 1997, 2002, 2003), but the conceptual and practical consequences have only recently become clearer (Dupré 2012; Krimsky and Gruber 2013; Noble 2006, 2008a, 2008b; Weiss 2018). Today, information is described that is not coded within DNA but is found within regulatory systems of the cell. This perspective derives from the new and much debated field of epigenetics (Jablonka and Lamb 2005; Jablonka and Raz 2009).

Basically it is our everyday experience that tends to understand the course of events in causal chains, often ignoring the complexity of systems. Yet in

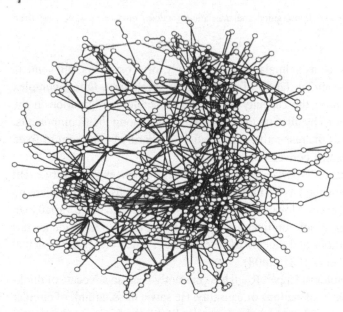

Figure 4.2
Overview of all known metabolic reactions of small molecules in a yeast cell. Reactions of glycolysis and the citric cycle are in the center of the network. *Source*: Republished with permission of the American Association for the Advancement of Science from Ravasz et al. (2002, 1552), permission conveyed through Copyright Clearance Center, Inc.

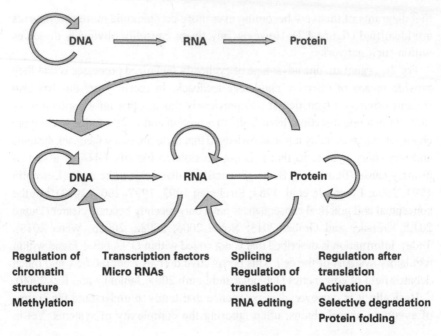

Regulation of   **Transcription factors**   **Splicing**   **Regulation after**
chromatin   **Micro RNAs**   **Regulation of**   **translation**
structure   **translation**   **Activation**
Methylation   **RNA editing**   **Selective degradation**
                                                                  Protein folding

Figure 4.3
(top) Older view of DNA transcription and translation; (bottom) modern view. *Source*: Draft
courtesy of Johannes Wirz.

organic entities such as animals, the human body, or ecosystems, phenomena
can have a very complex fabric of causes and interrelations. Often, complex
organic systems have the capacity to maintain a stable configuration in the
face of perturbations by altering the causal relations among their components.
Thus, the source of the respective organization can hardly be the sum of fixed
linear causal relations between the parts.

In general, it is an important principle of natural sciences that every event
and every phenomenon have a natural rationale, in the sense that there are no
supranatural influences. However, these reasons can be much more varied than
the simple dissection into unilinear events proposes, and modern science has
extensive possibilities and models to describe them much closer to the actual
phenomena (Wuketits 1981, 2008).

The Austrian biologist Rupert Riedl has constantly been an advocate of thinking beyond simple explanations of causality. He saw a missing link of circular
causalities between different levels in organic life. For a biologist, he formulated, it is obvious that cells and tissues, tissues and organs, and organs and
parts of the body mutually interact.

This up and down of circular causalities, he states, finally establishes a reciprocal relationship (*Wechselzusammenhang*) between the whole body and the genetic information. The whole body had increasingly been excluded from consideration due to the doctrine of classical genetics, which became even more rigorous by the so-called central dogma of molecular genetics. It describes that information exclusively flows from DNA to protein but not the other way back. This was the situation in the 1970s, and his expectation that this needed to be expanded became true when, beginning with the new century, increasingly more regulatory elements and interrelations on different levels have been described (Riedl 1978, 1979). "Functional thinking in whole systems of causes contributed to the development of the idea of an interconnectedness of these causes. Reality confirms that the idea of executive causality can only depict threads from the meshes of the net, rarely the meshes themselves and never the whole network" (Riedl 1978, 36).

Usually, the resolution of causal chains is successful on the molecular level. However, the analysis runs into difficulties when the context of further systemic levels is regarded. Even the simplest living system, a bacterial cell, is a highly complex network involving literally thousands of interacting and interdependent chemical reactions. This is increasingly recognized within present-day molecular biology, so that new ways of describing such systemic functions are being investigated (Boogerd et al. 2007a; Kaneko 2006; Noble 2006).

The Reciprocal Interrelation of Parts

Different proposals have been brought forth to describe organic functions more adequately in form of reciprocal relationships (Fuchs 2018; Haken 1983; Rosen 1991; Schad 1982; Thompson 2007). Fuchs (2018) describes the principle as "circular causality" and differentiates a horizontal from a vertical form. Horizontal circular causality is found, for example, within cellular metabolism or between cells or tissues. Vertical circular causalities are relations between different system levels such as cells, organs, and organisms. Fuchs describes them both together as integral causality.

Other researchers focus more on the network principle itself, considering living systems predominantly as self-organizing networks, the components of which are all interconnected within "nonlinear dynamics" (Capra and Luisi 2014; Stewart 2002).

The limits of causal reasoning in biology have often been discussed (Huang 2011; Mahner and Bunge 1997; Wuketits 1981). The development of more far-reaching concepts, however, is still a challenge.

The philosopher Immanuel Kant already saw the problems a mechanistic view of organisms would run into and formulated specific properties of living entities (Walsh 2015). He formulated the evident inability of mechanistic thinking to explain regularities of organismal function and development and expressed the idea that organisms are self-organizing, self-synthesizing entities. He considered Newton's laws of motion to have described the very nature of matter. However, Newton's laws claim that matter is inert, not self-moving. When it changes, it does so through the influence of external forces. Matter does not organize and replicate itself. Yet organisms do all these things. If Newton's laws lay down the rules by which matter conducts itself, organisms flout them flagrantly.

Walsh (2015) explains that Kant made much of the fact that organisms synthesize the materials out of which they are made. Consequently, an organism has the parts it has, in their particular arrangements, precisely because in its development the organism has made those parts and has put them there to serve its crucial vital functions. As a result, the relation between an organism and its parts is vastly different from the relation between a run-of-the-mill machine and its parts. Like machines, organisms are the outcomes of the interactions of their parts. However, unlike machines, an organism's parts are the consequence of the activities of the organism as a whole. The constituent parts and processes of a living thing are thus related to the organism as a whole by a kind of reciprocal causation. In this sense, an organism is the cause and effect of itself. Organisms are thus crucially unlike machines. They are self-building, self-regulating, highly integrated, functioning wholes, which exert a particularly distinctive influence on the capacities and activities of their parts. The essential definition Kant offered for organic form was that of the reciprocal interrelation of parts and, consequently, of the priority of the whole over the parts in the constitution of the entity. This leads to the property of agency, which will be discussed in section 4.4.

In a very general sense, it can be stated that the epistemic principle of causation, today mostly understood as molecular causation, is only one principle of many, which are to be found in living organisms. However, at present there is still a certain domination of explanations in terms of molecular pathways as chains of causation. A process or an event is regarded as being explained when a cause has been identified. But in most cases, it is just one component or one factor out of many that may be involved. Regarding the reality of complex relations within functions and processes, it is obviously necessary to widen the scope of explanations to come closer to what really takes place within the organism. It is definitely valuable to know the components involved, but a next—and often more difficult—step is to study the context in which these factors are integrated. Therefore, today there are many endeavors to develop approaches beyond the one-dimensional description of chains of causality.

Robert Rosen, for example, focused on the interdependence of processes within metabolism, which constitutes a closure of network processes. Metabolism, and the organism in general, is producing itself, or rather a new instance of itself, which is different from a machine that needs external agencies to construct and maintain it. Thus, Rosen is trying to overcome thinking in linear causalities, which has a limited relevance for living entities. Central to Rosen's work is the idea of a "complex system," defined as any system that cannot be fully understood by reducing it to its parts. In this sense, complexity refers to the causal impact of organization on the system as a whole (Bechtel 2010a; Cornish-Bowden and Cardenas 2005; Cornish-Bowden et al. 2007; Rosen 1991).

A related approach to the problem of causality is found in Francesco Varela's theorizing (here quoted from Bechtel 2010a) about living systems that maintain themselves "through the active compensation of deformations" (Varela 1979, 3). In his treatment of the subject, Varela builds upon Cannon's conception of homeostasis, the idea that living systems are organized to return to a target condition whenever perturbed. He does this in two steps: first, "making every reference for homeostasis internal to the system itself through mutual interconnections of processes," and second "by positing this interdependence as the very source of the system's identity as a concrete unity which we can distinguish" (Varela 1979, 12–13). Homeostatic operations in organisms are efficiently caused from within the system, and it is the continued existence of the set of causally dependent processes that constitutes the continued existence of the system. In terms of these ideas, he introduces his concept of autopoiesis: "An autopoietic system is organized (defined as a unity) as a network of processes of production (transformation and destruction) of components that produces the components that: (1) through their interactions and transformations continuously regenerate and realize the network of processes (relations) that produce them; and (2) constitute it (the machine) as a concrete unity in the space in which they exist by specifying the topological domain of its realization as such a network" (Varela 1979, 13).

For Varela, the cell is the basic autopoietic system. By focusing on how the coordinated activity within an autopoietic system enables the system to maintain its identity, Varela's framework focuses attention back on the cell itself and not just on the operations occurring within it. This is in conjunction with the idea that living systems, including cells, are autonomous systems. Varela introduced the idea of autonomy in his account of autopoiesis: "Autopoietic machines are autonomous: that is, they subordinate all changes to the maintenance of their own organization, independently of how profoundly they may be otherwise transformed in the process" (Varela 1979, 15).

Maturana and Varela (1980) applied the concept of autopoiesis beyond cells to whole organisms and especially to cognitive systems (see also the discussion in Capra and Luisi 2014).

Bechtel (2007, 293) summarizes: "Autopoiesis is important according to Varela because autopoietic systems can be autonomous, where autonomous systems are those that perform the necessary operations to maintain their own identity. This notion of autonomy provides a powerful way to conceptualize what is special about living systems. It also provides a perspective from which to view any additions to the initially conceived minimal autonomous system— they are ways of extending the autonomy of the system."

This point will be further elaborated in section 4.3.

Forms of Interdependencies

Figure 4.4 is an attempt to gain some systematic overview of possible interrelations within organic functions. It represents predominantly the different forms of explanations and interpretations that are used and thus primarily depicts the epistemological perspective. However, because these concepts should be adequate for certain organic functions, they also represent perspectives on organic reality in specific situations. Indeed, there are chains of linear causality in many cases, but as soon as functions and processes are studied within their context, different forms of interdependencies can be discovered. Already within the feedback loop it makes no sense to talk about cause and effect, because they coincide. This is even more applicable within complex networks.

Here are some possible basic forms of interrelations that can be found in organismic processes:

• *Linear causation*: One example, in which the dissection into sequences of causal events is helpful and applicable in biology, is especially on the very basic biochemical level, regarding single steps of chemical reaction within the cell or in specific tissues. Also, there are many events and conditions in body movement, limb function, orientation of the body in space (e.g., upright posture of humans) and ecosystems, in which causal mechanical analysis and descriptions are appropriate.

• *Multiple effects*: One cause may generate several effects.

• *Multiple causality*: In other situations, several causal influences may generate one reaction or one outcome within the organism.

• *Circular causation*: Cyclic organization came to the focus of biology in the twentieth century with the discovery that large numbers of biological systems

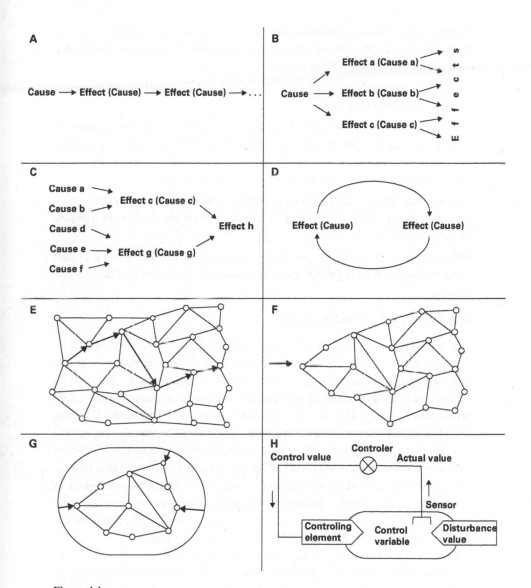

Figure 4.4
Possible forms of interrelations: (a) linear causation, (b) multiple effects, (c) multiple causation, (d) circular causality, (e) network, (f) trigger causality, (g) constraints, (h) regulatory system.

involve cycles. These include various biochemical metabolic cycles, the cell cycle, and cycles of reproduction, as well as cycles involving the biosphere such as the carbon and nitrogen cycle.

• *Networks*: Network approaches try to describe the complexity of intercon-nections between components.

• *Trigger causality*: Many influences on the organism generate a complex cascade of reactions and processes so that these influences act as a trigger. One event or a cause can set off a highly complex reaction by the organism. The effect then usually is an autonomous active reaction.

• *Constraints*: Organs and functions integrate, regulate, and organize their com-ponents and thus limit the degrees of freedom available to their elementary particles.

• *Regulatory (cybernetic) system*: A cybernetic system incorporates a closed signaling loop, which regulates specific properties of a controlled function or process.

Regulatory (Cybernetic) Systems

The development of cybernetics in the middle of the twentieth century can be seen as a step in overcoming a too one-sided causal thinking in biology and was acknowledged and incorporated into physiology at that time (Wuketits 1981). Classical physiology with all its applications, such as in medicine, would not be conceivable without thinking in regulatory units and functions. Wuketits describes that, thanks to cybernetics with its thinking in control circuits and feedback systems during the 1960s and 1970s, a new debate of self-regulation and autonomy became possible without taking recourse to vitalistic speculations. This can be understood as an organismic aspect that has been established in physiology in a pragmatic manner. "Therefore cybernetics can justifiably be seen as an important cornerstone for a systems-theoretical approach to the problem of causality" (Wuketits 1981, 79, translation by author).

Bechtel (2007) depicts that historically it took a long time for science to realize the significance of feedback processes and that it was especially Norbert Wiener and his collaborators who championed the notion of negative feedback as a fundamental principle in biological systems. Today, it can be seen as the simplest of nonlinear modes of organization to understand. Negative feedback plays a central role in nearly all physiological descriptions. Also, the organiza-tion of positive feedback systems is part of physiological descriptions although it often seemed to be more difficult. However, it has gradually been recognized that in some cases positive feedback can enable systems to self-organize.

Several points are quite remarkable in cybernetic thinking. First, it presupposes a specific information, which is the information about the set point. Thus, the control circuit is not just a material thing, an aspect that is usually neglected. A predefined target, the set point, is the information about the value to which the system needs to be adjusted. The specific quality of information in general will be discussed in section 4.6. In this context it is appropriate to point to the fact that the cybernetic model goes beyond a purely mechanistic interpretation and has long been established in biological concepts. The fact that cybernetic principles are used in technical systems does not contradict this statement, because in this case the set point always needs to be added by the constructor.

The second point, which is quite important, is that the regulatory system establishes a relative autonomy of the system toward internal and external changes and variations. In higher animals this stabilizes, in a dynamic way, properties such as temperature, gas exchange, circulation, and many components of blood composition. An earlier work demonstrates that these abilities have been enhanced during evolution (Rosslenbroich 2014).

The third point is that a regulatory system is an integrating process, which subjugates the components involved under an overall system. It establishes constraints for these components. And a fourth point, which will be discussed in its significance in section 4.4, is that it exhibits agency, because only the constant activity of the organism generates the equilibrium which is strived for. A regulatory system is an active and flexible *process*.

Although these principles were well known around the middle of the twentieth century, tendencies to establish mechanistic thinking in biology repeatedly emphasized linear causal thinking. The development of a strong focus on chemical and genetic principles in the second half of the twentieth century again contained this mindset. Whereas classical physiology had learned to think in regulatory circuits and feedback systems, molecular biology first appeared with the objective that it now will reveal the "real causes" of physiological processes, dismissing classical physiology as purely descriptive. Joyner (2011a, 2011b) discusses this critically from today's perspective.

Of course, the field had to learn rapidly that within the organismic context single processes must be seen with their integration into regulated systems, and today many molecular components of regulation have been described successfully. The pretension to deliver the real causes, however, has not been fulfilled so far. Meanwhile the knowledge about regulatory circuits and networks is constantly growing, becoming more complicated. Indeed, this has been recognized, and there have been many attempts to develop appropriate methodological extensions. The point to make here is just this: relevant empirical work, which takes the consequences from its results seriously, will encounter interdependencies, interactions, feedback processes, and context dependence of all organic functions and processes. This is characteristic of all living entities.

The organs of the whole human body, for example, influence each other, and it is hardly possible to identify a prime mover or a dominating organ or function. Throughout the body, the circulatory system and the blood system mediate between the manifold components and especially convey the regulatory components. The hormone system as well as the nervous system act as agents between the different functional contributions of the different organs. Only the regulated integrity of the whole constitutes a healthy organism.

There is no such thing as a hormone that "controls" something, as it is regularly formulated. It is the body or the system that produces and uses the hormone to generate a signal or a regulatory influence.

Repeatedly one organ has been emphasized as the central one, neglecting (more or less) its context. Thus, the brain has been seen as a central unit that directs and governs the organism. In contrast, especially embodiment theories now point to the brain's integration and mutual interconnections within the body (Fuchs 2018). In the same manner, the heart has been seen as the central cause for blood circulation, while recent theories emphasize its integrated role within the whole circulatory system (Furst 2020). It is remarkable which detours a field such as genetics first had to make before it—"surprisingly"—had to learn about the feedback systems from the organism itself (Parrington 2015; Shapiro 2009, 2011, 2017). It would have been possible to know that earlier and to avoid serious positions of one-sidedness (Krimsky and Gruber 2013; Lewontin 2000; Lewontin et al. 1984; Strohman 1993, 2003).

The Notion of Constraints

Several authors point to the principle that organs and functions integrate, regulate, and organize their components and thus limit the degrees of freedom available to their elementary particles. Thus, these functions develop constraints on the constituent parts to organize and coordinate processes and structures. This type of organization affords macroscopic objects to behave in their specific ways (Moreno and Mossio 2015; Winning and Bechtel 2018).

Winning and Bechtel describe that by regarding constraints, scientists can develop generalized accounts for the interactions of macroscale objects that restrict the degrees of freedom of their constituents, which are incorporated into the macroscale objects. The framework of constraints can be applied iteratively—a macroscale object can be further constrained by incorporating it into a yet larger-scale object. For example, the ways in which a macromolecule might move are constrained when it is embedded within a membrane of a cell, and that membrane is further constrained within the cell and when the cell is incorporated within a multicellular organism. Winning and Bechtel argue that the causal organization of a system consists in its spatiotemporal organization combined with the operative constraints.

Constraints in organisms may change and variate throughout altering time and life situations rather than being fixed and statically determined. Organisms can also respond to perturbations in a flexible way. Both properties distinguish them from mechanical machines. Organisms must be responsive to changing conditions, for example, routing free energy in different ways on different occasions and seeking a different form of free energy when the current one is no longer available. This requires exercising control over time-dependent, flexible constraints so that the components perform the work required to maintain the whole set of functions, which constitutes the organism in a far-from-equilibrium condition. Should this fail, the organism dies.

Noble (2017a, 50) points to the membrane itself as a key controller. He writes that what characterizes living organisms is not the atoms of which they are made but how these are controlled, and that the cell membrane with its proteins is one of the key controllers because the channels are selective. He also points to the principle of multicellularity. Compared to single cells, which first appeared in evolution, the aggregation produced remarkable changes in the properties of the individual cells. The whole ensemble then determines how the individuals behave (figure 4.13 in section 4.3). These are some out of many examples for constraints.

If, Noble further writes (172), we focus on observing the motions of the smallest particles, such as molecular compounds, we will find that they do indeed obey the laws of physics and chemistry in their interactions with other molecules. In that sense, and in that sense alone, biology can in principle be reduced to physics and chemistry. It is when we focus on more global, but nonetheless real, properties that we can see the limits of that reduction. These interactions are always constrained by the requirements and the activity of the cell and the tissues, for example. Even constraining some gas molecules by putting them in a container constrains their motions so that there will be an overall property, the pressure of the gas, within the container. In any equations we use to describe what is happening, that constraint will appear in the boundary conditions that must be inserted into our model to enable a solution to be found. All models, even entirely determinate ones, require initial and boundary conditions to be inserted into the equations before we can make any predictions on what may happen. In organisms these constraints are established by an activity, not only by a spatial condition.

Gene-Regulatory Networks

Bizzarri et al. (2019) make a strong case for an extension of explanations in the study of gene-regulatory networks. They describe that genes that control the development of specific tissues and are organized into pathways have been identified. This evidence has increased the understanding of how organs are formed and what goes wrong in disorders and diseases. However, as a consequence, biological processes tend to be explained as a series of genes. The result is a view of genes assigned fixed and specific functions within hierarchical mechanisms, in which master regulators drive developmental processes. Yet, to provide conceptual insight into how and why processes occur, they propose that a shift of attention is required from genes to patterns and dynamics of the connections between components. This approach, they describe further, reveals more sophistication and subtlety than is implied by simple hierarchical genetic wiring diagrams. Nonlinearity and feedback in even small systems can have unexpected consequences. Linear logic becomes inadequate because the distinction between cause and effect is lost, and the explanation of how a process occurs will require an understanding of how relationships change over time. Describing and understanding these dynamics is the challenge facing modern developmental biology. "The billiard ball model of causality has lured biologists towards linear, unidirectional, unilevel models of biological systems that are not actually built this way. Linearity is broken by branching pathways and unidirectionality is broken by feedback. Models of single-level interactions are broken by biological hierarchy. . . . Despite biologists' general awareness of redundancy and homeostatic control circuits, we still largely do not understand the corrective, self-organizing processes that reliably reach complex, systems-level patterning goals" (Bizzarri et al. 2019, 262).

One of the most striking results of the examination of these networks is that genes do not act as units of phenotypic control in the manner supposed by classical and early genetics. The relation between individual genes and phenotypes is not a 1:1 mapping; it is exceedingly complex (Noble 2017a; Wagner 2012; Walsh 2015). In many cases, networks are able to compensate a missing genetic component, for example, if it is missing by experimental knockout technique. If one such gene fails to work appropriately, the network often alters its dynamics in a way that compensates, ensuring that the appropriate output is preserved. In other cases, if an essential protein is missing within the respective function and cannot be produced otherwise, the function is disturbed. However, such systems are hardly linearizable.

Also, Walsh (2015, 39) points to the description of gene-regulatory networks and describes that genes typically do not work as isolated agents, but as parts of complex self-regulating systems:

The discovery of gene regulatory networks arose in part out of the failure of traditional mechanist approaches to the study of gene action. The usual way to understand what a gene does is to remove or disable it and see what happens. It turns out that "knocking out" a gene in a gene-regulatory network often has no overall effect. But this doesn't mean that the target gene makes no causal contribution to the activities of the system in normal circumstances. The phenomenon is a reflection of the fact that the relations among the components of a gene regulatory system are nonlinear, cyclical and self-regulating. An interference to one part of the network ramifies, causing changes to all the other parts. These changes are compensatory; they permit the system as a whole to maintain its function, despite the perturbation. The capacity of regulatory networks to compensate in this way is called "distributional robustness" and it is no mere exotic quirk. It is integral to the stability, growth and function of organisms.

Walsh cites Wagner (2007, 176), who presented a lot of examples for such a robustness in recent years: "Living things are unimaginably complex, yet also highly robust to genetic change on all levels of organization. Proteins can tolerate thousands of amino acid changes, metabolic networks can continue to sustain life even after removal of important chemical reactions, gene regulation networks continue to function after alteration of key gene interactions, and radical genetic change in embryonic development can lead to an essentially unchanged adult organism."

Walsh further describes that distributional robustness consists in the capacity of a system to regulate the activities of its parts, and that it calls for an alternative methodology, in which the dynamics of the entire system are integral to an explanation of the activities of its parts. The component genes in a gene-regulatory system behave in the way they do precisely because their behavior is regulated by the system. Each gene in each regulatory system has a repertoire, a range of activities in which it could partake. Which element of its repertoire is activated in a given context depends on this very context. "This sort of adaptive dynamics seems to be the rule in biology; gene regulatory networks, genomes, cells, tissues and entire organisms exhibit this kind of complex, adaptive, nonlinear dynamics. Scientists and philosophers are beginning to realise that new conceptual tools are required. Increasingly they are looking toward emergence as an alternative to deductive mechanism" (Walsh 2015, 40).

Bizzarri et al. (2013, 34) write: "Genomic functions are inherently interactive . . . , and biological processes flow along complex circuits, involving RNA, proteins and context-dependent factors (extracellular matrix, stroma, chemical gradients, biophysical forces) within which vital processes occur. Indeed, no simple, one-to-one correspondence between genes and phenotypes can be made. . . . The concept of 'gene' inherited by molecular biology has therefore been broadly revised, taking into consideration that gene function is in fact 'distributed' along a connection of corporate bodies that interact among them according to a not-linear dynamics" (see also Fox Keller 2000; Moss 2006; Noble 2008b; Shapiro 2009).

Describing Networks

Realizing the complexity in interconnections between components, parts, or functions within living organisms, network-based approaches for modeling and explaining complex biological systems have increasingly been developed in diverse fields of biology. Examples describe and analyze the organization, function, and stability of ecological communities, interactions of proteins and metabolites, brain circuits, gene regulation, or evolving organisms (Bechtel and Richardson 2010; Green et al. 2018; Kostić et al. 2020).

In such approaches, network modeling is utilized as a research strategy for the organization and interpretation of data. Unlike mechanism diagrams, network models typically do not contain details about the molecular properties of the components. By representing interactions among a vast number of molecular species, they enable an analysis of the organizational structure of larger systems, sometimes involving automated pattern detection. In some cases, this allows the study of the complex structure and dynamic operation of large-scale networks (Bechtel 2020; Green et al. 2018; Huang 2011). One example is explained in figure 4.5.

One approach to study networks has been developed by Barabási and colleagues (Barabási and Oltvai 2004; Jeong et al. 2000; Santolini and Barabási 2018). They describe the aim of recent postgenomic biomedical research to systematically catalog all molecules and their interactions within a living cell. The emerging results are forcing the acknowledgment that, notwithstanding the importance of individual molecules, cellular function is a contextual attribute of patterns of interactions between the myriad cellular constituents. These authors declare that uncovering the generic organizing principle of cellular networks is fundamental to our understanding of the cell as a system.

Although they study networks from an abstract point of view, they describe some general organizing principles, which are interesting for the topic being discussed here (Barabási and Oltvai 2004). They assert that the components within a cell can be reduced to a series of nodes that are connected to each other by links, with each link representing the interactions between two components. The nodes and links together form a network. In this way, interactions between molecules such as protein–protein, protein–nucleic acid, and protein–metabolite interactions have been conceptualized using a general node-link nomenclature.

They depict how older network models assumed a "democratic" or uniform character of connectivity of the individual nodes within the network. In such models the links are placed randomly among the nodes, and it is expected that

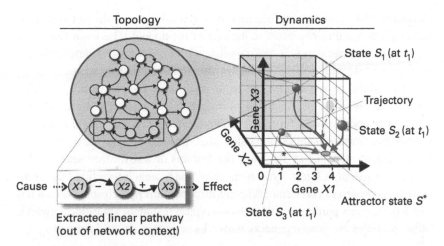

Figure 4.5
Schematic representation used by Huang (2011) to illustrate the difference between the analysis of linear pathways and the investigation of dynamics within the state space of a system with networks. The conventional identification of a linear pathway as a chain of causation consisting of genes *X1*, *X2*, *X3* extracted out of the network context is shown as a contrast underneath a network topology map. On the right, a three-dimensional state space capturing the dynamics of a hypothetical three-gene network (genes *X1*, *X2*, and *X3*) is shown. Any point in this space represents a theoretical network state *S* at time *t*. Three arbitrary states (balls), S_1, S_2, and S_3 are shown. Because, as most states, they do not represent stable network states, they are driven by the network interactions to seek a stable state; hence, they move in state space along trajectories (solid lines) that lead to the stable attractor state. An attractor is a stable solution to a set of conditions to which a system will tend to move. The dashed trajectory represents an example of a trajectory that has been perturbed (e.g., by drugs that affect expression of genes *X1*, *X2*, and *X3*) away from its natural course defined by the network interactions into regions of the state space that are even less stable, and hence quickly returns to the trajectory that leads to the attractor. In summary, the states, S_1, S_2, S_3 and the perturbed trajectory all lie within the state space region that "drain" to the particular attractor S^*; hence, they all lie within its basin of attraction. *Source*: Republished with permission of The Royal Society (UK) from Huang (2011, 2248), permission conveyed through Copyright Clearance Center, Inc.

some nodes collect only a few links whereas others collect more. In a random network, the nodes' degrees follow a Gaussian distribution, which indicates that most nodes have roughly the same number of links, approximately equal to the network's average degree. More recent findings, however, indicate that the random network model cannot explain the topological properties of real networks. The deviations from the random model have several key signatures, the most striking being the finding that real networks usually are highly nonuniform. Most of the nodes have only a few links, while a few nodes have a large or very large number of links. Such nodes with a high number of connections are called hubs. Networks with such a disparate distribution of connections are called *scale-free* (a name that is rooted in statistical physics literature). This indicates the absence of a typical node in the network (one that could be used to characterize the rest of the nodes). The main feature of such scale-free networks is the coexistence of nodes of widely different degrees (scales) with nodes with one or two links to major hubs.

Most cellular networks turn out to be such scale-free networks. The first evidence came from the analysis of metabolism, in which the nodes are metabolites and the links represent enzyme-catalyzed biochemical reactions. Meanwhile, further studies indicated that protein–protein interactions in diverse eukaryotic species also have the features of a scale-free network.

This is apparent in figure 4.6, which shows the protein interaction map of the yeast *Saccharomyces cerevisiae*. Whereas most proteins participate in only a few interactions, a few participate in dozens—a typical feature of scale-free networks. Other examples are gene-regulatory networks or protein domain networks.

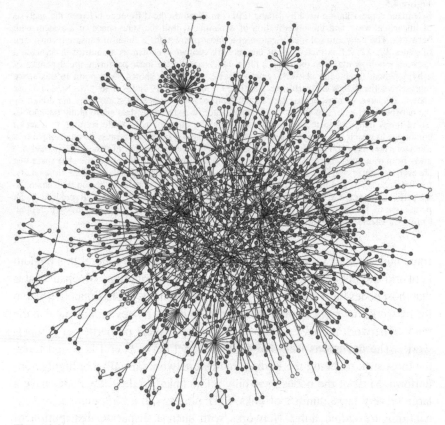

Figure 4.6
Yeast protein interaction network. A map of protein–protein interactions. *Source*: Reprinted by permission from Springer *Nature* from Jeong et al. (2001, 41).

Barabási and Oltvai further explain that one key feature of scale-free networks is their robustness, which refers to the system's ability to respond to changes in the external conditions or internal organization while maintaining relatively normal behavior. This again points to the principle of autonomy discussed before. Intuition tells us that disabling a substantial number of nodes will result in a functional disintegration of a network. This is true for a random network. Scale-free networks, however, can be very resilient against component failure, withstanding even the incapacitation of many of their individual components and many changes in external conditions. This is because random failure affects mainly the numerous small degree nodes, the absence of which does not disrupt the network's integrity. Conversely, this reliance on hubs induces a so-called attack vulnerability—the removal of a few key hubs splinters the system into small clusters of isolated nodes.

This feature of scale-free networks indicates that there is a strong relationship between the hub status of a molecule (e.g., its number of links) and its role in maintaining the viability and/or growth of a cell. Deletion analyses indicate that in *S. cerevisiae* only ~10 percent of the proteins with less than five links are essential, but this fraction increases to more than 60 percent for proteins with more than fifteen interactions, which indicates that the protein's degree of connectedness has an important role in determining its deletion phenotype (Jeong et al. 2000). Furthermore, only ~18 percent of *S. cerevisiae* genes (~14 percent in *E. coli*) are lethal when deleted individually, and the simultaneous deletion of many *E. coli* genes is without substantial phenotypic effect whatsoever (Gerdes et al. 2003; Giaever et al. 2002; Winzeler et al. 1999; Yu 2002).

However, the structural topology of networks can only be one component of the robustness of the cell, as the cell's activity and its ability to alter pathways and functions are involved during the reaction to perturbation. In principle, cells as well as multicellular organisms always attempt to maintain their overall integrity, which is a function of the system as a whole, and also involves some active flexibility in pathway changes. Therefore, a more complete description of cellular networks requires that the cell's activity and the temporal aspects of the interactions are considered as well. The activity in time and the linkage of components within networks establish the cell as a functional unit.

Green et al. (2018) present an overview of different network approaches. Besides the approach of Barabási and colleagues, the survey also takes the following models into account: a hierarchical modular network approach, which focuses on the hierarchical organization of several modules in larger networks; approaches that describe the analysis of temporal expression data of proteins, thus representing the dimension of time; approaches that are able to regard context-sensitivity of processes; and approaches that are able to regard specific dynamic states of the respective system. Consequently, they propose a shift of focus from linear molecular pathways to dynamic states of whole networks. Green et al. (2018, 1775) conclude: "Biologists have been drawn to these tools as they confronted the challenge of coping with the complexity of highly interconnected and non-linear biological systems."

Conclusion

Overall, the former ideas, that cellular and molecular mechanisms are complex chains of causally reacting components that together result in the generation of the phenomenon of interest, are being increasingly replaced by understanding functions by means of nonlinear circular causation, interdependencies, as well as system- and network-characteristics. In every respect, the components of an organism are much more interdependent in manifold ways than has been previously assumed. Organismic functions are bound together in networks of interaction relationships. Analysis of single molecular steps and their description in reaction chains has been a necessary methodological prerequisite. However, developing models and explanations, which are able to describe the real characteristics of organic processes, is an essential part of organismic thinking. These challenges are being increasingly acknowledged, so that in recent years systems biology has become an important research field. This will be discussed in more detail in section 4.2.

4.2   Integrative Biological Systems

Within an organism, whether as a single cell or as a multicellular organism, all parts and processes are integrated into a coherent entity so that they contribute to the overall function, survival, and activity of this entity. All processes are regulated in such a way that the organism can achieve a certain autonomy. The complex of interacting components, together with the coherent entity itself, can be called an integrative system. Hence, key features of a biological system are active integration, regulation, interaction of components, and the autonomy of the respective system as a whole. Altogether, they typically exhibit a structure in space and time.

Systems Biology

Melham et al. (2013) describe that around the turn of the twenty-first century, the course of molecular biology began to change as researchers started to realize that approaching the molecular activity in a cell or organism in a piecewise way, one molecule at a time, was not going to explain biological activity, because too much was going on and too many proteins were involved in even the simplest of phenomena. Earlier concerns that too little was known about the mass of proteins had been met around that time by the availability of high-throughput technologies with attempts to describe the "proteome." Such technologies generated large amounts of molecular data on a wide range of physiological, developmental, and evolutionary phenomena, but earlier hopes that this plethora of information could be integrated on the basis of bottom-up approaches had soon been dashed, Melham describes. The new subject of systems biology reflects the wish to make sense of all this data in a much broader context.

There is much to be discussed about systems biology, which became quite acknowledged in recent years. Topics include definitions, self-understanding, methods, and concepts (Bizzarri et al. 2013; Boi 2017; Boogerd et al. 2007a; Greene 2017; Kitano 2001; Noble 2006, 2017a). However, the present consideration will only be about the properties of integration and regulation as key features of organismic entities, which are the focus of some sections of systems biology. Methodological or modeling questions of the field will not be discussed here.

From the viewpoint of classical physiology, it is hardly surprising that all functions and processes within an organism are regulated and meticulously integrated into the overall functionality of the organism regarding the challenges of survival and action. All of physiology, including medical physiology, is full of knowledge and descriptions of how an organism manages to coordinate and regulate its organs, its blood circulation, and its exchange of nutrients and gases, for example. Just remember how exactly the human body regulates the heart rate and the amount of blood being transported in every moment and how the heart itself is part of this regulation. The heart is regulated and is a regulator at the same time (Furst 2020). Several components are involved and can be studied for their respective contributions. They are, however, always aligned to the overall performance of the entire system.

In a more analytical sense, it is still a challenge to say what this integration really is. How does a multicellular organism, or even a single cell, really manage to integrate and to organize the molecular functions and particles? The fact that systems biology has become so popular in recent times highlights the relevance of this question.

Two Different Meanings

In a rough distinction there are two schools of thought in current systems biology, as already mentioned in chapter 2. O'Malley and Dupré (2005) call them "pragmatic systems biology" and "systems-theoretic biology." The majority of today's systems biologists belong to the pragmatic school, which studies large sets of molecular data to reconstruct systemic properties. Here, the term *system* covers a range of molecular interactions ("omic" approach), which generate the whole function in question. It relies principally on high-throughput technologies and on massive data analysis by means of mathematical modeling. This approach still relies on a molecular level rationale as the privileged level of explanation.

In contrast, theoretical systems biologists recognize that complex physiological phenomena also take place at biological levels of organization higher than the subcellular one. For this approach it is crucial to analyze systems as proper systems, not as mere collections of parts. Systems are taken to constitute a fundamental ontological category, which integrates the components and processes involved. A living system acquires only a limited number of configurations (forms, processes) as a consequence of the constraints exerted on its parts by the system as a whole (Bizzarri et al. 2013). Therefore, this understanding will be called here *integrative biological system*. This feature unravels the existence of different hierarchical levels of causality in living matter and outlines the relevance of the supramolecular order.

The principle of integration allows the system to have a history: this means that the present behavior of the system is in part determined by its past behavior. Such a system displays both sensitivity and resilience (robustness, autonomy) with respect to the perturbations exerted by internal and/or external stimuli. Integration also allows a specific *organization* of the system to be displayed, including a form, a shape, and a time structure.

Superficially, these two lines of thought can be characterized as "bottom-up systems biology" and "top-down systems biology" (= "downward causation"). The bottom-up approach expects that all processes and functions can be explained purely through the interaction of components. The top-down approach takes into account distinct properties of the system with constraining and regulating influences on components at the lower levels. Thus, the overall system needs to be regarded as a real entity.

Here the point will be made that basically both aspects are important. There is something special in the living organism, that there are complex molecular interactions that are the prerequisite for all functions and structures, and that simultaneously the coherent overall unit generates integration and control of the constituents and their functional interaction. To use a pointed formulation, bottom up and top down are simultaneously involved at any time within an integrated overall process. This again is the *principle of concurrency*.

Hence the conceptual differences of both approaches might not be so substantial. Both concepts finally try to describe the superordinated organization, looking from two perspectives. Here again a synthesis might provide conceptual progress.

If one of these perspectives would be taken to an extreme (which presumably hardly anybody does today), each of them would generate epistemological shortcomings: a purely pragmatic view would overlook and neglect the integrative aspect of higher-level organization. A purely holistic view on the other hand, which would refuse to look at anything different from a whole organism or a whole organ, would ignore the components that make the system possible.

The mutual interdependence between the parts and the overall system, between molecules and cellular and organismal organization, within its permanent processsuality, is the principle of a living system. In this sense a system is not a thing at all, but rather a process (Dupré and Nicholson 2018). However, the process also generates structures and forms. If we are able to describe this inextricable interrelation, we will have an organismic theory of living systems. The task, then, is to describe the dynamic behavior of these systems.

This might approximately be the same point of view that Boogerd et al. (2007b, 12) describe: "Such a systems biological approach, an integrative interlevel approach, that combines both molecular and systems properties and that attempts to explain the systemic behaviour of organisms in terms of their functional organization appears promising."

Importantly, Boogerd et al. (2007b, 15) point to the cell as the smallest unit of life. Although molecules constitute living systems, they are by themselves not alive. No matter how thin the dividing line is, there is a qualitative jump between a living and a nonliving system (Mahner and Bunge 1997). Therefore, much of biology and systems biology starts with cell biology.

Hofmeyr (2017) describes from his research, how he stepwise found elements of the regulation of metabolism and how metabolism is integrated into the physiological requirements of the cell. Then he summarizes the quintessence of his empirical biochemical work: "Our theory explicitly recognises that the evolved properties of any part of a system can only be understood in relation to the whole—it is, therefore, a systems theory. While it is perfectly possible to describe any part of a system fully in terms of its constituents only, it is impossible to understand why that part of the system has the properties it has without considering it in the context of the intact, whole system" (121). Then, referring to the famous sentence of Dobzhansky about biology and evolution, he formulates: "Nothing in an organism makes sense except in the light of functional context" (121). Systems biology "situates the phenomenon of life somewhere between molecule and autonomous organism. The focus throughout is on organism: life emerges from a system of material components that are functionally organised in such a way that the system can autonomously produce and repair itself, can distinguish itself from the rest of the world, and can adaptively restructure itself within its genomic constraints in the face of

environment fluctuations. Systems biology therefore goes beyond the properties of individual biomolecules, taking seriously their organisation into a living
whole" (122–123).

In the same sense Walsh (2015, 39) summarizes:

Many systems are organized in such a way that the various parts have highly context-
sensitive "nonlinear" and "cyclical" causal relations. Contrary to the presumptions of
the analytic method, they do not "behave in new arrangements as they have tried to
behave in others". The activities of one component of the system alters the activities
of other components, which frequently, in turn, redound upon the first component,
through a complex feedback loop. In systems like this there is a sort of causal reciprocity
between part and whole. The way that a part affects the whole can only be understood
in the context of the way that the entire system affects, controls and regulates the
activities of the parts. The strategy of causal decomposition fails because the causal
contribution of a component to the dynamics of the system cannot be understood in
isolation from the system as a whole. Indeed, this sort of cyclical, nonlinear, causal
structure is the norm in biological systems.

Integrative Biological Systems Theory

Attempts to describe integrative system levels are among the older concepts
and were introduced in the 1920s by Ludwig von Bertalanffy and Paul Weiss,
who are regarded as the founders of biological systems theory. Especially Paul
Weiss presented a concept with a consequent integrative character, and many
results of modern research are in accordance with his concept (Drack et al.
2007; Drack and Apftaler 2007; Drack and Wolkenhauer 2011; Rosslenbroich
2011a, 2011b). Therefore, his concept will be presented in greater detail, and
it will be proposed that it is of major importance for the development of an
organismic understanding of systems.

Weiss (1963, 1968, 1969, 1971, 1973, 1977) characterized the relation of
analysis and synthesis and described how we first recognize nature as an
immense cohesive continuum. Then we start to identify discrete fragments in
it and isolate single entities to learn more about their exact properties. Subsequently, we find out that modifications of such an entity, that may be called
entity A, are regularly associated with a series of modifications in another entity
called B. By studying this regularity, a rule can be established from which all
future correlations between A and B can be extrapolated. We then proceed to
study A in its relation to C, and C in its relation to B, and so on, to learn how
different parts of nature, erstwhile mentally dissected and separated, are actually interdependent. At this stage it is expected that it should be possible to
turn the process around—either physically or mentally in our imagination—
linking by way of consecutive synthesis such coupled pairs into complex
chains, reconstructing the whole system in a quasi-mechanical way.

As Weiss explained, in practically all of biological thinking the opinion dominates that by application of this synthetic method, science will eventually succeed in describing and comprehending all entities and processes in nature. Weiss stated that physics had already begun to depart from such a micro-mechanistic attempt whereas biology had not. However, in an organism the mere reversal of the analytic dissection can yield no complete explanation of its behavior as a living system.

What is overlooked, Weiss continued, is that during isolation of A, B, and so forth, already a lot of information has been neglected to characterize these entities. However, especially in an organism, each entity depends on the inter-actions with others. This means that in the absence of C, neither A nor B can exist. The coexistence and cooperation of all three is indispensable for the existence and operation of any one of them. Only by artificially neglecting the so-called boundary conditions can A and B be studied in an isolated manner. Experimentally, this procedure may often be adequate. Nonetheless, it is overseen that the information, which is neglected during this process, cannot be reconstructed through a synthesis from the knowledge of the proper-ties of these parts. The analytical procedure has been very successful in science, but obviously it must be complemented by a scientific method that also regards the system's properties on the whole.

Weiss points out that in contrast to the infinite number of possible interac-tions and combinations among the parts, in the living system only an extremely restricted selection from the opportunities for chemical processes is being real-ized at any one moment—a selection that can be understood solely in its bearing on the concerted harmonious performance of a task by the complex as a whole (Weiss 1969). This is the feature that distinguishes a living system from a dead body or a functional process from a mere list of parts involved. "The systems concept is the embodiment of the experience that there are patterned processes which owe their typical configuration not to a prearranged, absolutely stereo-typed, mosaic of single-tracked component performances, but on the contrary, to the fact that the component activities have many degrees of freedom, but submit to the ordering restraints exerted upon them by the integral activity of the 'whole' in its patterned systems dynamics" (Weiss 1969, 9).

Weiss defines a system as a relatively independent and stable entity. Accord-ingly, a system generates restricting and regulative functions and imposes them upon its component parts, so that the functionality of the whole system is maintained. Thus, the system itself contains constituting properties and also information, which does not necessarily derive from the components. At the same time each system depends on its components. The central illustration of Weiss is shown in figure 4.7 (left).

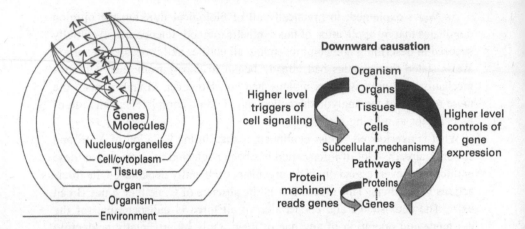

Figure 4.7
Schematic representations of the systems concept. (left) *Source*: Reprinted with permission from Weiss (1959, 18). Copyright 1959 by the American Physical Society. (right) *Source*: Noble (2006, 51), reproduced with permission of the Licensor through PLSclear.

The cell is an example of such a system. It contains many components, but the system integrates them into a functional whole. The cell depends on the components but is only viable by regulating them within the system. "[T]he basic characteristic of a system is its essential invariance beyond the much more variant flux and fluctuations of its elements or constituents" (Weiss 1969, 12).

This means that the component processes are not bound within determined processes but rather can be variable, as is well known from cell biology today. Whether and when information is transcribed from DNA, whether certain proteins are generated and how long they stay in function, or which components are integrated into the membranes to keep them within an optimal state of fluidity, is permanently regulated by the cell, according to its respective situation. "This is exactly the opposite of a machine, in which the structure of the product depends crucially on strictly predefined operations of the parts. In the system, the structure of the whole determines the operation of the parts; in the machine, the operation of the parts determines the outcome" (Weiss 1969, 12).

Weiss further describes that a cell does not always work directly with its molecules but rather has subsystems, the organelles, which perform partial functions. Also, the whole organism can be regarded as a system with several subsystems, the organs. Hierarchic stepwise delegation of tasks to subsystems is nature's efficient instrument to let an organism maintain order without having to deal with all its molecules directly.

Weiss sees systems as not being closed inherently. They have a relative stability and thus an organizational closure, but at the same time they are open to influences from their surroundings. Weiss demonstrates this using the cell as an example: a cell is a well-characterized entity and can be regarded as a system. However, a multicellular organism needs to be regulated and thus must have a certain openness to regulative influences. To guarantee this, the cells of multicellular animals have a multitude of membrane receptors, which mediate signals from the surroundings. They also need to have a regulated exchange of substances with the environment to maintain their basic functions as well.

This concept of Weiss is compatible with the more recent writings of Denis Noble, who draws new consequences from the knowledge available today (Noble 2006, 2008a, 2008b, 2011a, 2011b, 2012, 2017a). Noble (2006) describes how within recent decades much progress has been made in dissecting systems into their smallest components. Now the challenge is to extend that knowledge up the scale. At each level of the organism, its various components are embedded in an integrated network of systems. Each such system has its own logic, so that it is not possible to understand that logic merely by investigating the properties of the system's components. The challenge is to learn more about the properties and conditions of each of these systems. In this sense the genome is not privileged, but is rather just one level within this hierarchy, and there is no reason to assume that the whole complex of integrated networks is determined by just that level.

Noble uses the representation of figure 4.7 (right), which also points out the interdependencies between the different systemic levels. This concept has clear parallels to the ideas of Weiss. The object, life itself, seems to imply such concepts.

Noble describes ten principles of systems biology, which are represented in table 4.1 (Noble 2008a). His first principle describes that biological functionality is multilevel. He bases this on the simple physiological observation that functionality of any organism is attributable to the organism as a whole, and that it controls all the other levels. Claude Bernard, who was very influential in the development of classical physiology, detected this first with his principle of the milieu interieur. For this reason Noble calls Bernard the first systems biologist (Noble 2008a).

Two further principles indicate that the transmission of information is not one-way, and DNA is not the sole transmitter of inheritance. Today the former view of information flow from DNA to proteins is replaced by the knowledge of multilevel interactions between the many components involved, as mentioned in the previous chapter. Noble gives several examples including the feedback control of gene expression during individual development. All nucleated cells in the somatic body contain exactly the same genome. But the expression pattern of a cardiac cell is completely different from that of a

Table 4.1
The ten principles of systems biology

First principle: biological functionality is multilevel.

Second principle: transmission of information is not one-way.

Third principle: DNA is not the sole transmitter of inheritance.

Fourth principle: the theory of biological relativity—there is no privileged level of causality.

Fifth principle: gene ontology will fail without higher-level insight.

Sixth principle: there is no genetic program.

Seventh principle: there are no programs at any other level.

Eighth principle: there are no programs in the brain.

Ninth principle: the self is not an object.

Tenth principle: there are many more to be discovered—a genuine "theory of biology" does not yet exist.

Source: Compiled from specifications by Noble (2008a).

hepatic or bone cell. Whatever is determining those expression levels is accurately inherited during cell division. This cellular inheritance process is robust, it depends on some form of gene marking. It is this information on relative gene expression levels that is critical in determining each cell type. However, this is relevant information. In the processes of differentiation and growth, it is just as relevant as the raw DNA sequences. Yet, it is clear that this information does travel the other way too. The genes are told by the cells and tissues what to do, how frequently they should be transcribed, and when to stop. This, Noble states, is "downward causation" from those higher levels that determines how the genome is "played" in each cell. In general, the DNA code is marked by the organism, modulating gene expression, which is the focus of the rapidly growing field of epigenetics. This concept also takes into account that inheritance through at least some generations can take place on additional levels beyond the DNA level (Jablonka and Lamb 2005; Jablonka and Raz 2009; Shapiro 2011, 2014).

The central principle in Noble's understanding of a system is the theory of biological relativity, which he further elaborates in Noble (2017a). It states that there is no privileged level of causality within the organism: "The principle of Biological Relativity is simply that there is no privileged level of causation in biology: living organisms are multi-level open stochastic systems in which the behavior at any level depends on higher and lower level and cannot be fully understood in isolation" (Noble 2017a, 160).

Noble (2013a) formulates that each level or scale can be seen as setting the boundary conditions for the smaller-scale events. This results, then, in a nested sequence of scales, each subject to conditions imposed by higher scales and in turn determining those at smaller scales. The idea of biological relativity can then be stated as a natural consequence of the fact that the system is open at each scale and therefore subject to the boundary conditions determined at higher scales. He states that it is a matter for empirical discovery to determine at which scale(s) a particular process is integrated and describes that in his own research on the heart, it is clear that the scale, at which the electrical rhythm is integrated, is that of the cell. The scale at which some forms of fatal arrhythmia are integrated is that of the whole organ. Noble (2012, 60) writes: "Determining the level at which a function is integrated is an empirical question. Cardiac rhythm is clearly integrated at the level of the pacemaker sinus node cell, and does not even exist below that level. The principle can be restated in a more precise way by saying that the level at which each function is integrated is at least partly a matter of experimental discovery."

Noble presents a further example from his studies on heart physiology. The cardiac pacemaker depends on ionic currents generated by a number of protein channels carrying sodium, calcium, potassium, and other ions. The activation, deactivation, and inactivation of these channels proceed in a rhythmic fashion in synchrony with the pacemaker frequency. We might, Noble describes, therefore be tempted to say that their oscillations generate that electrical potential of the overall cell, that is, the higher-level functionality. But this is not the case. The kinetics of these channels vary with the electrical potential. There is, therefore, feedback between the higher-level property, the cell potential, and the lower-level property, the channel kinetics. This form of feedback was originally identified by Alan Hodgkin working on the nerve impulse, so it is sometimes called the Hodgkin cycle. If we remove the feedback, for example, by holding the potential constant, as in a voltage clamp experiment, the channels no longer oscillate (figure 4.8). The oscillation is therefore a property of the system as a whole and not of the individual channels or even of a set of channels, unless they are arranged in a particular way in the right kind of cell. Nor can we establish any priority in causality by asking which comes first, the channel kinetics or the cell potential. This fact is also evident in the differential equations we use to model such a process, Noble describes. The physical laws represented in the equations themselves, as well as the initial and boundary conditions, operate simultaneously (i.e., during every integration step, however infinitesimal), not sequentially. "It is simply a prejudice that inclines us to give some causal priority to lower-level, molecular events," Noble states (2008a, 21).

This is also a good example for the principle of interaction and interdependence, which has been discussed in section 4.1. Again, there is no primary cause at all that generates the whole process or that in some form is leading or causing the process.

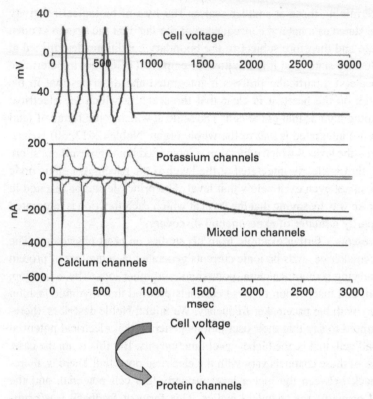

Figure 4.8
Computer model of pacemaker rhythm in the heart. For the first four beats, the model is allowed to run normally and generates rhythm closely similar to a real heart. Then the feedback from cell voltage to protein channels is interrupted. All the protein channel oscillations then cease. They slowly change to steady constant values. The diagram below shows the causal loop involved. Protein channels carry current that changes the cell voltage (upward arrow), while the cell voltage changes the protein channels (downward arrow). *Source*: From Noble (2006, 64); Reproduced with permission of Oxford Publishing Limited through PLSclear.

Multicellularity

Many examples underpin the concept of integrative systems, four of which will be presented here: the principle of multicellularity, aspects of neurophysiology, aspects of epigenetics, and an example from embryology.

Multicellular organisms originated during evolution independently at least twenty-five times from unicellular ancestors (Grosberg and Strathmann 2007). The richest level of elaboration is found in the metazoa, the global term for multicellular animals. More than 600 million years ago, single-celled ancestors of animals gained the capacity to form cellular clusters and eventually specialized to perform specific functions within a sort of integrated organism. One challenge (out of many) to understanding this transition is how the cells involved are coordinated and subordinated within the newly established association.

The principle of regulation in multicellular organisms includes at least two elements: one is the reaction to perturbations, which attempts to maintain homeostasis, and the other is the coordination and integration of specific parts and components of the organism. Free running functions of cell groups or organs would lead to a dissociation of processes involved, which would not be compatible with the survival of the organism. Therefore, all physiological descriptions of regulation and control are examples of integrative system functions.

Remarkably, many of the developmental requirements for multicellular organization, including cell adhesion, cell–cell communication and coordination, and programmed cell death, likely existed in ancestral unicellular organisms (Gerhart and Kirschner 1997; King 2004). In addition, there were some functions necessary for multicellular life, which presumably were an invention of metazoans: the initiator and effector caspases involved in programmed cell death, transcription factors related to body patterning, critical components of the immune system, and many genes affecting the nervous system function (Richter and King 2013).

Gerhart and Kirschner (1997) discuss the origin and conditions of multicellularity in some detail from the molecular perspective. They describe that the most important feature of metazoan cells for complex multicellular development may have been their ability to seal off the outside world and generate extracellular space internally in the organism (see also Rosslenbroich 2014). Within this space the cells can be regulated and integrated into the overall system, and the internal compartment is mostly under the organism's control. They use the term *contingency* to describe the cell's responsiveness to the conditions of the system (*control* and *regulation* are alternative terms, they state). They describe that most processes in the metazoan cell are more highly contingent than those in single-celled eukaryotes. For example, single-celled organisms divide primarily in response to nutritional signals, whereas the division of metazoan cells is contingent on cell–cell contacts, cell–matrix contacts, nutrition, numerous growth factors, and hormonal signals from distant tissues. In these metazoan cells, core reactions that have been conserved throughout evolution, such as transcription, RNA splicing, and translation, are contingent on a variety of conditions. The most common strategy for creating contingency is to impose an inhibition on a conserved reaction, only allowing it to become activated under certain conditions. Inhibition and conditional activation integrate conserved reactions into highly regulated networks of the system, which Gerhart and Kirschner describe in some detail.

They also describe that the main changes during the transition to multicellularity must have appeared in the supracellular organization. Molecular processes, as well as the cellular features of the eukaryotic cell, all were in place when the large transition took place:

As we look at emergent morphologies, we are struck by their number and diversity. Yet, ... there is no evidence of a corresponding molecular explosion at that time. Before the divergence of the major phyla most of the gene families were already established. Most of the signaling systems, transcriptional circuits, and principles of cell differentiation and tissue formation had already developed in the earlier and still paleontologically obscure period of the founding of the first metazoans. Most of the important steps in cellular evolution can be traced beyond the Cambrian to the origin of the eukaryotes and many to our prokaryotic ancestors. . . . Although it was an evolution of animal anatomies, it was not a time of great innovation in cellular mechanisms. (Gerhart and Kirschner 1997, 2)

Gerhart and Kirschner conclude:

Evolutionary biologists . . . need to explain why organisms have changed. Evidence of conservation, to a fist approximation, suggests that they are looking in the wrong place. The difference between birds and mammals is not going to be found in the structures of their muscles or nerves, their types of collagen or their microtubules. It will be found in their wings, their feathers, their sweat glands, their hair, and the organization of their brain cortex. Change has occurred principally in the organization of tissues and in the evolution of novel physiological and embryological mechanisms. (140)

If this is the case, then the organization within the system, the rearrangement on a supracellular and supramolecular level of the system, was important. There is no logical reason why the changes should first have come from the genetic or the molecular level. However, it surely needed the underlying support from the genome as well as molecular pathways.

In recent times, the ancestry of gene families has been confirmed by detailed genetic analysis on special components in metazoans. One element of a multicellular animal is the generation of an extracellular matrix (ECM), which holds the cells together and provides a matrix, in which the cells can migrate or interact among themselves and with the matrix. The ECM is seen as a general cellular scaffold, promoting cell adhesion and communication. The basement membrane is a specialized sheetlike form of the ECM that provides structural support to epithelial cells and tissues while influencing multiple biological functions and was essential in the transition to multicellularity. Rodríguez-Pascual (2019) describes that in recent genomic analysis a common theme emerges: unicellular ancestors of metazoans had a complex gene repertoire involved in multiple functions, however incapable of building ECM. Fueled by clonal- or aggregationlike processes, preexisting components of basement membranes gained novel capabilities (or redirected their previous roles) to start elaborating ECM scaffolds at the emergence of metazoa (figure 4.9). "Said colloquially, unicellular ancestors hosted a party on multicellularity, with the secret password co-option to get past the bouncer, where these folks joined together to make a tasty matrix punch" (Rodríguez-Pascual 2019, 771).

Co-option is a term that has been used to describe the function of ancient genes in a new context (Boto et al. 2009; West-Eberhard 2003). Other possibilities being discussed for genetic underpinnings of the transition are calling upon the modularity of the genome and its propensity for recombination, domain shuffling, gene duplication, and the evolution of gene regulation (King 2004; Richter and King 2013).

Volvocine green algae have been used for studying the transitions to multicellularity because they range in complexity from unicellular to multicellular genera (Kirk 1998, 2000; Ueki et al. 2010; figure 4.10). The question Ueki et al. pose is: How do 5,000 rowers, who previously have been independent on an evolutionary scale, coordinate their strokes to row the whole system into the sunlight? The somatic cells (as differentiated from gonidia, the reproductive

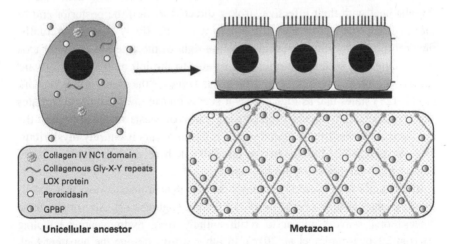

Figure 4.9
Model for the evolution of basal membrane (BM) constituents, according to Rodríguez-Pascual (2019, 771). Unicellular ancestors of multicellular animals expressed numerous genes encoding for BM components in the absence of any recognizable extracellular matrix structure. By co-option, these preexisting components gave rise to the elements required to build sophisticated BM scaffolds in the metazoan stem lineage.

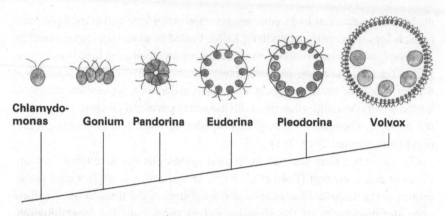

Figure 4.10
Representation of the volvocine lineage hypothesis according to Kirk (1998). *Chlamydomonas* and several genera of volvocaceans can be arranged in a conceptual sequence in which there are progressive increases in the number of cells, the amount of extracellular matrix, and the extent to which labor is divided between reproductive cells (gonidia) and sterile somatic cells with their flagella. Present considerations, however, assume that the spheroidal colony evolved twice independently within this group (Yamashita et al. 2016).

cells) are equipped with the same structures that single-celled members of the group have, especially with the two flagella and their basal apparatus. In *Volvox* these flagella are oriented in such a way within the system that all flagella beat with their effective strokes directed toward the posterior end of the spheroid, that is, that they are able to work effectively together. Actually, the beat direction is directed slightly to the right of the anterior–posterior axis of the spheroid, so that the organism rotates to the left as it swims forward (hence, the name given to it by Linnaeus: *Volvox*, "the fierce roller."). Kirk (1998, 121) states that as logically as it seems that in multicellular flagellates such an orientation should occur during embryogenesis of the spheroid, the cytological mechanism by which such a rotation occurs is entirely mysterious. The transparent ECM seems to play a major role here, as the somatic cells are held in fixed relationships to one another.

The flagella of a *V. carteri* colony beat in a synchronized wave that travels from the anterior to the posterior of the colony (figure 4.11). Amazingly, these "metachronal waves" appear to result entirely from hydrodynamic coupling (Herron 2016; Brumley et al. 2015). In other words, despite the apparent high degree of coordination among the flagella of separate cells within a colony, no actual communication between cells is necessary. Synchronization emerges from indirect interactions mediated by the liquid medium. Brumley et al. (2014) describe experiments in which somatic cells were physically separated from a colony and held at various distances from each other. Despite there being no

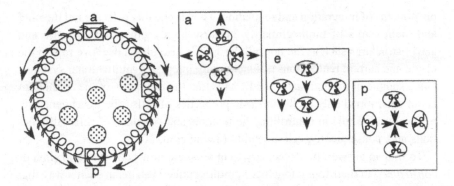

Figure 4.11
Orientation of cells within a *Volvox carteri* spheroid. (left) Somatic cells are oriented so that all flagella beat with their effective strokes directed toward the posterior of the spheroid (the end containing the gonidia). (right) En face representations of the orientations of the basal apparatus with the flagella seen in four neighboring cells at the anterior (a), equatorial (e), and posterior (p) regions of the spheroid (boxed in the left diagram). The arrows indicate the directions of the effective flagellar strokes in those cells. In the anterior, the cells are oriented so that flagella beat away from one another; along the sides, the cells are oriented so that their flagella beat in parallel, and at the posterior pole the cells are oriented so that their flagella beat toward one another. *Source*: From Kirk (1998, 122), reproduced with permission of Cambridge University Press through PLSclear.

direct physical connection between the cells, they beat synchronously when close together, with a phase shift that increased with increasing cell-to-cell distance. Other factors for synchronization seem to involve a sensibility of differential light intensities for the respective cells when the colony rotates as well as some impact of gravity (Ueki et al. 2010). Whatever the functional details are, the coordination of flagellar beats is generated by the context. There is neither a master coordinator, nor is it conceivable that every cell within the colony has a specific genetic or molecular compound that determines its role at its very position.

Another interesting point is that with the establishment of multicellularity in volvocine algae, a programmed cell death is generated. When gonidia mature to a new organism, the somatic cells of the mother entity die in a way that is induced and coordinated by the system. Such a coordinated cell death is a quite impressive example of a constraint imposed by the overall system.

Bich et al. (2019) assume that the study of general multicellularity can provide an understanding of how biological systems are functionally integrated into coherent wholes: "Multicellularity . . . presents . . . challenges to our understanding of biological systems, related to how cells are capable to live together in higher-order entities, in such a way that some of their features and behaviors are constrained and controlled by the system they realize" (1).

Bich et al. propose that besides the cells the intercellular space and its structures and functional organization need to be considered to understand the phenomena of integration and regulation. They argue that cells are not the only and main actors of multicellularity, because highly organized dynamic and active structures such as the noncellular ECMs also play a decisive role in the origin and current realizations of functionally integrated multicellular systems. One example of the activity of the ECM is that the formation of the epithelium is not determined only by the intrinsic properties of cells. The ECM generally plays an active role in controlling the features and positions of cell–cell junctions and, consequently, cell assembly (Tseng et al. 2012).

To find an answer, it is necessary to understand how cells are *controlled* or *constrained* in their living together in multicellular systems in such a way that they realize and maintain viable organized entities. When these forms of control fail, or their properties change in certain ways, this change may give rise to different (transient or stable) forms of multicellular organization or regressions (Bissell and Radisky 2001; Sonnenschein and Soto 1999; Soto and Sonnenschein 2011a, 2011b). Soto and Sonnenschein propose that the "default state" of cells is characterized by proliferation and motility, and the challenge for multicellular systems is to exert a differential and dynamic control upon these properties in a way that is functional for the whole. Therefore, the system activates the division of certain cells in specific moments in time and inhibits it in others. Moreover, depending on the state of the system, the capability of motility is also inhibited in most cells. When those constraints that act on proliferation, motility, and other capabilities, fail, or their properties are modified, these changes may give rise to different forms of disruption of the system such as in human cancer. Therefore, they propose that the occurrence of cancer might be more related to such integrative functions rather than genetic disorders of the cells involved, as the conventional paradigm supposes. They advanced the thesis that alterations of ECM may anticipate cell transformation in cancer (Sonnenschein and Soto 1999; Soto and Sonnenschein 2011a, 2011b).

In many cases the means by which the integration of parts into the system is performed are well known today. For example, cell signaling describes in detail how communication processes govern basic activities of cells and how multiple cell actions are coordinated. Methods of systems biology study the underlying structure of cell-signaling networks and how changes in these networks may affect the transmission of information. Such networks are complex systems in their organization and obviously exhibit overall integrative properties.

Cells communicate with each other via direct contact (juxtracrine signaling) over short distances (paracrine signaling, using the immediate extracellular environment) or over large distances and scales (endocrine signaling, = hormonal signaling, traveling via the circulatory system). Specificity of signaling can be controlled if only certain cells can respond to a particular hormone, and glands again are controlled by the situation of the overall system.

It is especially interesting that although paracrine signaling, for example, elicits a diverse array of responses in the induced cells, most paracrine factors utilize a relatively small set of pathways (fibroblast growth factor, Hedgehog family, Wnt family and TGF-β superfamily). Different organs in the body, even between different species, are known to utilize similar sets of paracrine factors. That means the pathways and the receptors involved are highly conserved throughout evolution, although their usage can be quite diverse (Richter and King 2013).

Neurophysiology

Another system that coordinates and integrates the components within metazoans is the nervous system. Recent considerations, however, show that the brain is not a master coordinator as has previously been thought. Discussed especially in reference to the humans, the brain rather needs to be seen as embodied within the overall organism, so that again a systems approach is much more appropriate to describe the situation of a mutual interaction between body and brain. The brain is only understandable in the context of the overall system (Fuchs 2018).

This principle has also been found in more detailed neurophysiological studies, described by Liljenström (2016). The following summary is taken from Boi (2017). Studying the causal pathways in brain dynamics, Liljenström remarks that downward causation from larger to smaller scales could be regarded as evidence that "multi-level both-way" causation occurs. He investigated, on the one hand, how cortical neurodynamics may depend on structural properties, such as connectivity and neuronal types, and on intrinsic and external signals and fluctuations. On the other hand, he examined to what extent the complex neurodynamics of cortical networks can influence the neural activity of single neurons. Liljenström attempted to show that the neural activity at the microscopic level of single neurons is the basis for the neurodynamics at the mesoscopic network level, and fluctuations may sometimes trigger coherent spatiotemporal patterns of activity at this higher level. Irregular chaotic-like behavior can be generated by the interplay of neural excitatory and inhibitory activity at the network level. These complex network dynamics, in turn, may influence the activity of single neurons, causing them to fire coherently or synchronously. Thus, Liljenström (2016) concludes "this downward causation is complementary to the upward causation" (181).

It is apparent that the intricate web of inter-relationships between different levels of neural organization, with inhibitory and excitatory feed-forward and feed-back loops, with nonlinearities and thresholds, noise and chaos, makes any attempt to trace the causality of events and processes futile. In line with the ideas of [Noble 2012] it seems obvious that there is, in general, both upward and downward causation in biological systems, including the nervous system. This also makes it impossible to say that mental processes are simply caused by neural processes, without any influence from the mental on the neural. (Liljenström 2016, 184)

Genetics and Epigenetics

In section 4.1 it was mentioned that there are many feedback systems and regulatory cycles in functions of heredity, so that it is not possible to identify DNA as a sort of primary cause (see figure 4.3). Davies (2012) makes the case that the complex epigenetic inheritance system, as it has been described in recent years, introduces a multilevel control and organization upon the given DNA template. The discovery of epigenetic factors is an interesting case of how incrementally integrative top-down functions are being detected, causing much surprise or even confusion in a field, which was certain to have found *the* determining principle of life in the DNA as the master controller of living beings. Today, the picture of heredity is becoming much more multilevel and with this much more organismic.

Davies also describes that although genes store heritable information, actual gene expression often depends on epigenetic factors external to DNA. Epigenetic changes can come about due to chemical signals received from other genes or from both chemical and physical signals originating in other cells, organs, or the external environment. These signals may serve to switch genes on or off using molecular markers or the physical rearrangement of DNA. Changes in this system can be both reversible and heritable. Whereas the genome is associated with a physical object (DNA) and with a specific location, the epigenome is a global, systemic entity. Epigenomic information can be associated with certain non-DNA molecular sequences, but often there does not seem to be a stored "epigenetic program." Instead, epigenomic control, Davies explains, is an emergent self-organizing phenomenon, and the real-time operation of the "epigenetic project" lies in the realm of nonlinear bifurcations, interlocking feedback loops, distributed networks, top-down causation, and other concepts familiar from the complex systems theory. These principles seem to provide striking examples of how bottom-up genetic and top-down epigenetic causation intermingle.

Epigenetics can seem hard to understand, Davies outlines further, "or even mysterious, because so much of our scientific intuition is based on linear quasi-isolated causal chains and reductionist descriptions." He concludes:

The necessity to consider organisms as systems subject to both upward and downward causation complicates causal reasoning and presents a challenge to physical scientists used to thinking of step-by-step cause and effect. The subject of systems biology attempts to get to grips with this fundamental and unavoidable characteristic of life. (43)

Then Davies sets forth:

[L]ife can be understood only by seeing the big picture, that is, the way in which an individual post-translational modification event conforms to an overall plan or strategy for the cell as a whole, a strategy involving countless other molecules. . . . In the case of epigenetics, there is no physical headquarters, no localized commanding officers issuing orders, no geographical nerve centre where the epigenomic "programme" is stored and from where epigenomic instructions emanate to help run the cell. The epigenome is not to be found at a place and the ultimate information source of epigenetics cannot be located anywhere specifically; rather, it is distributed throughout the cell. To be sure, the epigenome is manifested in particular structures (histone tails, nucleosome patterns, methylation patterns, chromatin packing . . .), but it does not originate there. The epigenome is everywhere and nowhere; it is a global, systemic entity. . . . Undeniably the genome provides the words, but the epigenome writes the play!

Hence, the conception of DNA itself is changing dramatically too. First, the genome is not some sort of blueprint for an organism, as sometimes has been claimed. Many authors have argued against such a view, including Lewontin et al. (1984), Moss (2003), Neumann-Held (2002), Noble (2012), Oyama (2000, 2002), and Strohman (1993).

Second, the cell is the active part interpreting the genome, "deciding" which proteins result from processes such as alternative splicing and protein regulation. And third, in many cases cells are even able to rearrange the genome. Shapiro (2011, 2013a, 2013b, 2014, 2017) argues convincingly that organisms deal much more actively with their genome than has been previously thought. They also play a much more active role in genomic changes during evolution, adding an impact of biological agency on evolutionary innovation. Shapiro sees the genome more as a "read–write genome" rather than a "read-only genome."

Positional Information and Pattern Formation in the Embryo

A particularly clear example of the interaction of different systemic levels and the importance of context in the organism is the initial predisposition of patterns in the egg and early embryo (Gilbert 2014; Müller and Hassel 2006; figure 4.12). In a developing embryo, cells must behave in a manner appropriate to their location. In one place they are supposed to build up nerve tissue, in another to form a muscle or a skeletal element. However, the DNA of the nucleus does not contain any information about where a cell is currently located. Basically, there are two processes by which the right cells are found in the right place at the end of development:

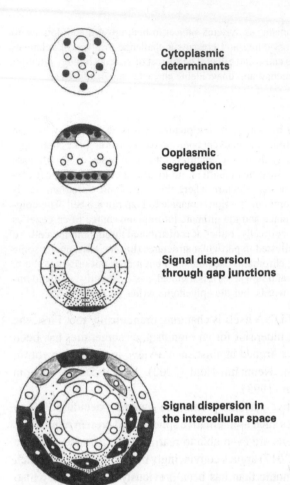

Cytoplasmic determinants

Ooplasmic segregation

Signal dispersion through gap junctions

Signal dispersion in the intercellular space

Figure 4.12
Sequence of determination in a generalized animal embryo. An initial site-dependent specification of future fate occurs through cytoplasmic determinants, which are ordered in the course of egg-internal pattern formation (ooplasmic segregation) and then assigned to different blastomeres. Later, determination and pattern formation are based on signal exchange between cells. *Source*: Republished with permission of Springer, Berlin, Heidelberg, from Müller and Hassel (2006, 323), permission conveyed through Copyright Clearance Center, Inc.

1. Cells are committed (determined) to a differentiation program. While they are differentiating or after they have completed their differentiation, they actively seek out the place where they will remain. This is found, for example, in the chromatophores of the skin, in the cells of the immune system, and in germ cells.

2. More often, however, cells find out where they are before they are determined, and then choose an "occupation" according to the location and differentiate accordingly. Or viewed differently: the environment tells a cell what

it has to do. Positionally appropriate cell differentiation is triggered from the outside. From the outside means here: from activating or permissive (allowing) influences that originate outside the cell nucleus. Such external orienting cues may come from the environment, such as gravity, light, or physical stresses and forces, or there may be cell interactions and concerted behavior through signals offered by neighbors (e.g., induction signals).

In *Drosophila*, maternal gene products are significantly involved in this (e.g., the products of *bicoid*, *oskar*, and *nanos*). They are already distributed in the cytoplasm of the egg according to specific spatial ratios. This spatial distribution contains the information for where certain subsequent steps take place and where body parts and organs develop. These cytoplasmic determinants are fixed stationarily to the cytoskeleton. The gene products contain the information for what should happen at the particular location, but the spatial arrangement is organized by the positional information. That is, positional information and genetic information work together to eventually develop the correct anatomical pattern.

In other eggs, such as those of amphibians, fish, or birds, no definitive pattern in the distribution of such determinants is yet present in the freshly laid egg. Rather, cytoplasmic currents and transport processes begin after fertilization, which redistribute maternal gene products. Physical forces may contribute to the final location of one or the other cytoplasmic component (Forgacs and Newman 2005; Newman 2014). A segregation (ooplasmic segregation) occurs in the egg, resulting in an egg-internal pattern formation, which, however, is not genetically fixed.

However, the pattern of cytoplasmic determinants can only provide a first orientation. During the further construction of the body architecture, new patterns are then created epigenetically, which are not yet preformed in the egg. The term *epigenetic* means "from processes above the level of genetic information." The processes that are crucial for pattern formation are then interactions of the cells with other cells or with extracellular substrates (extracellular matrix).

Genetic information is needed to produce the proteins that mediate these interactions. These are, for example, the molecules that are distributed in the oocyte or molecules that are exposed on the cell surface or secreted into the environment and serve as signals. But the site of interaction is determined by nongenetic patterns. There is still much debate about what positional information consists of in detail (Levin 2012; Minelli and Pradeu 2014). However, it is becoming clear that it involves interactions between different levels of systems that ultimately build the necessary organization of cells, tissues, and organs.

Integrative Interaction

As argued before, classical physiology shows that in a multicellular system there is no master coordinator. The components (molecular compounds, cells, organs) generate the system, and the system integrates the components. Physiology describes, however, that special organs or certain tissues are responsible for special regulatory functions. However, these organs are again controlled and regulated within the system itself.

As stated earlier, it would be one-sided to declare certain components to be more important (neglecting the integrative system functions), and it is equally one-sided to declare the whole system as being more important (neglecting the components). The actual situation in living systems can best be described by the term *integrative interaction*, which expresses the *concurrency* of both.

The system is not a superordinated entity, something "above" the cells or the tissues. Rather, the system is established by the cluster of cells, tissues, and organs, which overall generate the integrative system. In this sense, multicellularity, for example, establishes a top-down system (Morange 2008, 107). The cells generate the system and the system coordinates, integrates, and constrains the cells. Interaction and context dependence of the cells generate the system itself, and the system coordinates the interaction. The mutual interdependence of the process as a whole, together with the parts and detailed functions involved, is the organization of the system.

4.3 Autonomy

Living systems establish a relative autonomy in the sense that they maintain themselves in form and function within time and achieve a self-determined flexibility. These living systems generate, maintain, and regulate an internal network of interdependent, energy-consuming processes, which in turn generate and maintain the system. They establish a boundary and actively regulate their interaction and exchange with the environment. They specify their own rules of behavior and react to external stimuli in a self-determined way, according to their internal disposition and condition, and they maintain phenotypic stability (robustness) in the face of diverse perturbations arising from environmental changes, internal variability, and genetic variations (Maturana and Varela 1987; Moreno and Mossio 2015; Moreno et al. 2008; Rosslenbroich 2014; Ruiz-Mirazo and Moreno 2012; Varela et al. 1974).

General Principles of Organismic Autonomy

This matches with the systems approach discussed before. Hofmeyr (2007) suggests that "for systems biology, the defining difference between a living organism and any nonliving object should be that an organism is a system of material components that are organised in such a way that the system can autonomously and continuously fabricate itself, i.e. it can live longer than the lifetimes of all its individual components. Systems biology, therefore, goes beyond the properties of individual biomolecules, taking seriously their organisation into a living whole" (217).

Essentially, a single-celled organism distinguishes itself from the surrounding fluid medium by actively creating its *boundaries*. Simultaneously, it regulates its interactions with the environment. The existence of a boundary is a central element of a living system (Luisi 2003; Maturana and Varela 1987; Varela et al. 1974). Inside the boundary of a cell, many reactions and chemical transformations occur; the cellular membrane encloses a defined reaction room, thus contributing to the maintenance of the cell's identity.

Metabolic processes within the cell construct the boundaries, but the metabolic processes themselves are made possible by those boundaries. In this way, the cell emerges as a figure out of a chemical background. Metabolism establishes *dynamic disequilibrium*, a dynamic stage of order within a network that characterizes life. A cell that drifts toward equilibrium is dying.

Autonomy is achieved by using *energy-rich molecules*. In face of the hydrolyzing and oxidizing influences from the environment, energy-rich bonds are maintained in a relatively stable state. Organic molecules are always reduced compounds, and thus they are rich in energy. Energy from the environment is accumulated within these complex molecules so that an energetic gradient is being maintained.

In addition, each organism establishes a genetic system for building an order to avert the tendency of its material components toward entropy. Genetic and epigenetic instructions make it possible to build temporary islands of self-determined order over and over again. These principles enable the identity of the individual as well as that of the species. Information is the source for building up a higher degree of order than exists within the environment. This *informational self-determination* is part of autonomy and has also been called organizational closure (Moreno and Mossio 2015).

Taken together these are some features of autonomy: boundaries, metabolic processes, self-regulation, dynamic disequilibrium, energetic gradient, and informational self-determination (table 4.2).

Autopoietic Systems

Thompson (2007) qualifies the necessity of a strict physical boundary for an autonomous system. He states that a system can be autonomous without having this sort of material boundary; the members of an insect colony, for example, form an autonomous social network, but the boundary is social and

Table 4.2
Definition for general autonomy

Living systems are autonomous in the sense that they maintain themselves in form and function within time and achieve a self-determined flexibility.

These living systems

I.   Generate, maintain, and regulate an inner network of interdependent, energy-consuming processes, which in turn generate and maintain the system;

II. Establish a boundary and actively regulate their interaction and exchange with the environment;

III. Specify their own rules of behavior and react to external stimuli in a self-determined way, according to their internal disposition and condition;

IV. Establish an interdependence between the system and its parts within the organism, which includes a differentiation in subsystems;

V.   Establish a time autonomy; and

VI. Maintain a phenotypic stability (robustness) in the face of diverse perturbations arising from environmental changes, internal variability, and genetic variations.

Source: For details, see Rosslenbroich (2014).

territorial, not material. Autonomous systems are organizationally closed in the sense that their organization is characterized by their internal network processes, which recursively depend on each other and thus constitute the system as a unit. These processes generate a far-from-equilibrium situation as long as the system is living. Equilibrium with the processes in the environment arises when the system is dead. At the same time, living systems are materially and energetically open to their environment. They receive energy and nutrients from the environment and excrete products and waste. Luisi (2003) emphasizes that there is an interesting contradiction between biological autonomy and dependence on the external medium and that all living organisms must operate within this contradiction.

Moreno and coworkers work toward an understanding of a most basic form of autonomy of living organisms (Moreno et al. 2008; Moreno and Mossio 2015; Ruiz-Mirazo and Moreno 2012). They see autonomy as a fundamental characteristic of life and stress explicitly the significance of the principle for understanding the origin of early life on Earth. A motivation for their search for a basic autonomy is to provide a link between this fundamental principle of life and physics and chemistry, so that the idea of autonomy itself is naturalized and can serve as a bridge from the nonliving to the living domain. Because they are crucial for the generation of simple self-maintaining and self-constructing systems, they understand that these systems must engage in an interactive loop with their respective environment across some boundary condition (gradients, influx/outflux of different compounds, energy transduction, etc.) to sustain the processes of generation of internal order in accordance with the generalized second law of thermodynamics.

Moreno et al. describe that, unlike physical or chemical dissipative structures, in which patterns of dynamic order form spontaneously but whose stability relies almost completely on externally imposed boundary conditions, autonomous systems build and actively maintain most of their own boundary conditions, making possible a robust far-from-equilibrium dynamic behavior. Thus, a central question is how a system develops the capacity to channel the flow of matter and energy through itself to achieve robust self-construction (i.e., self-construction that includes regulation loops with its immediate environment).

Thompson (2007) introduces the distinction between heteronymous and autonomous systems. Whereas heteronomy literally means other-governed, autonomy means self-governed. A heteronymous system is one whose organization is defined by input-output information flow and external mechanisms of control. Traditional computational systems and many network views, for example, are heteronymous: They have an input layer and an output layer; the inputs are initially assigned by the observer outside the system, and output performance is evaluated in relation to an externally imposed task. An autonomous living system, however, is defined by its endogenous, self-organizing, and self-controlling dynamics and determines the domain in which it operates. It has input and output; however, these do not alone determine the system. It is the internal self-production process that controls and regulates the system's interaction with the outside environment. For Thompson, the principle of autonomy is essential for understanding principles such as intentionality and subjectivity of living entities, which in complex forms generate a continuity of life and mind. He attempts to understand the relation between these entities by his "enactive approach," focusing on the conditions of this continuity.

Philosophical Description of Organismic Autonomy

Fuchs (2018) gives a description of the concept of organismic autonomy to prepare a view of the human neurophysiologic functions that is more integrative. He draws on results from ecological and philosophical biology with its main exponents J. von Uexküll (1973), Plessner (1975), and Jonas (1966) and those of system theories such as those of Bertalanffy (1973) and Maturana and Varela (1987).

Fuchs describes living beings as complex entities or systems that maintain themselves in form and structure within time, although there is a continuing exchange of substances with the environment. This maintenance is an active self-organization as the organism subordinates the substances under its own principles and transforms and integrates them. They gain new properties, which they only have within the systemic context of the organism. Fuchs points to an example: The ferrous ion in hemoglobin behaves differently from iron in the outside world—it does not oxidize irreversibly but is able to bind oxygen reversibly, which is a crucial prerequisite for the turnover of energy in animals.

Beyond this, metabolism leads to a transformation of substances during decomposing digestion and resynthesis. The nutritional components are transformed into substances with the characteristics of the organism and integrated into its processes. By means of these dynamic processes, the living being encloses itself from the environment and gains—in different degrees—self-determination or autonomy. This means that its processes and its behavior are not primarily determined from the outside but rather depend on its internal disposition and condition. External influences predominantly are stimuli, which are answered by reactions of the whole organism, rather than causal effects as in mechanical cause-and-effect relations, as long as they are not destructive.

The basis for autonomy, Fuchs describes further, is the special interdependence between the whole and its parts within the organism, which include a differentiation in subsystems and organs. Although the organism consists of the sum of its macromolecules, cells, organs, and circulatory and nervous systems, it has a different relation to these component parts than a crystal to its components. The organism is itself the condition of its parts because it enables their existence. It produces and reproduces them while consisting of them. Self-maintenance is continuing self-generation. At the same time, the parts fulfill their respective functions within the organism and contribute to its overall functioning.

Of course, Fuchs also describes that the autonomy of living beings is not possible in autarky. The organism only gains its sovereignty for the price of certain requirements. The changing substances need to be available and incorporated to maintain homeostasis. Thus, organisms are always in need of factors from their environment (Jonas 1966).

According to Plessner (1975), Fuchs further describes that plants exhibit a predominantly open relation to their environment, whereas animals have a more closed form of organization. In animals, the exchange surfaces for metabolism are turned to the inside. Special internal organs and internal cavities appear, while exchange surfaces on the outside are reduced. Thus, animal life steps to a certain extent out from the direct environmental relation. The enclosure from the environment requires—on the other hand—a sensorimotor interzone, which restores the contact with the environment, however, on a new level. This condition shows separate organs for sensory and motor activity and their central nervous connections. The principle of a closed body organization enables the independent movement of the animal.

According to Fuchs, the loss of a direct environmental relation corresponds to a gain in degrees of freedom. Whereas the mimosa reacts directly to touch, the stimulus-response relationships in animals tend to be less tightly connected. Animals tend to modulate a reaction so that the probability of a certain behavior can be modified. Signals can internally be enforced, compared to other signals, and memorized. Thus, not a rigid, but rather a flexible relation between organism and environment emerges.

Robustness

In recent years, a somewhat new term developed in some areas of molecular biology. It was increasingly comprehended that many structures and functions as well as proteins and genes have certain stability in the face of environmental variations and genetic changes. Many physiological and developmental systems are resistant or "robust" to such perturbations. That is, despite these natural perturbations, the systems produce relatively invariant outputs (Bertolaso et al. 2018; Gerhart and Kirschner 1997; Kitano 2004, 2007; Larhlimi et al. 2011; Masel and Siegal 2009; Masel and Trotter 2010; Stelling et al. 2004; Wagner 2007, 2012). Robustness is understood as a property that allows a system to maintain its functions against internal and external perturbations and uncertainties. It encompasses a broad range of traits, from macroscopic, visible traits to molecular traits, such as the expression level of a gene or the three-dimensional conformation of a protein. "Biological systems maintain phenotypic stability in the face of diverse perturbations arising from environmental changes, stochastic events (or intracellular noise), and genetic variation. It has long been recognized that this robustness is an inherent property of all biological systems and is strongly favored by evolution" (Stelling et al. 2004, 675).

Masel and Siegal (2009) see it as impossible to understand whole biological systems without understanding their robustness. Stelling et al. (2004) note that robustness encompasses a relative, not an absolute, property because no system can maintain stability for all its functions when encountering any kind of perturbation.

Robustness is concerned with maintaining the possibility of a system to function rather than maintaining an actual state of a system. Thus, Kitano (2007) differentiates it from stability and homeostasis, which predominantly describe a function that keeps a condition relatively constant. A system is robust as long as it maintains functionality, even if it transits to a new steady state or if instability actually helps the system cope with perturbations. Such transitions between states are often observed in organisms when facing stress conditions. One such condition can be extreme dehydration, to which some organisms can react with a dormant state, becoming active again on rehydration. These examples of extreme robustness under harsh stress conditions show that organisms can attain an impressive degree of robustness by switching from one steady state to another rather than trying to maintain a given state.

Wagner (2012) divides the perturbations that can affect a phenotype into two broad categories. The first consists of environmental perturbations. These include changes in an organism's exterior environment, such as changes in temperature, in available nutrients, or in the abundance of other organisms, such as potential prey. They also include changes in an organism's internal environment, such as temporal fluctuations in gene expression levels, which are caused by ubiquitous intracellular noise. The second kind of perturbations is mutations, changes in an organism's genotype. Mutations affect an organism more permanently than environmental change because the changes they cause are readily inherited from generation to generation. For this reason, Wagner states that they are especially an important object of study for students of evolution.

Because the term *autonomy* describes living systems as actively distinguishing themselves from their surroundings, it overlaps to some extent with the term *robustness*. But it is not congruent with it. Robustness can be seen as a prerequisite for autonomy. Self-determination and self-maintenance need robust functions to defy perturbations from the nonbiological and biological surroundings as well as from the internal variability.

Robustness can also be regarded as a part of autonomy itself. Robustness, also in different actual states of a system, maintains basically that the system is kept in a far-from-equilibrium state. Even dormant forms are different from their immediate surroundings in a self-organized manner, including when the metabolism is completely reduced. If the system becomes like the surroundings, this results in an equilibrium state and death.

Stelling et al. (2004) mention the important point that the primary function of a system may usually be robust to a wide range of perturbations, whereas the system can show extreme fragility toward other, even seemingly smaller, perturbations. They think that the coexistence of extremes in robustness and fragility ("robust yet fragile") perhaps constitutes the most salient feature of highly evolved complexity. Making one feature robust to a class of perturbations can make the same or other features fragile to that or other perturbations. In this sense, they expect a necessary connection between complexity and robustness.

In this discussion, several principles are seen as relevant for maintaining and establishing robustness (Kitano 2004, 2007; Stelling et al. 2004). One strategy to protect against failure of a specific component is to provide for alternative ways to carry out the function the component performs. This can be called "redundancy of components." At the genetic level, this backup strategy or "genetic buffering" (Hartman et al. 2001) might be brought about by duplicate genes with identical roles or by different genes that constitute alternative but functionally overlapping pathways. In contrast to redundant systems in engineering, however, identical genes that do not diverge in functionality or regulation would not survive in evolution. Instead, structurally different entities perform overlapping functions, which seems to be a common principle in organisms, on other levels in addition to the genetic.

A further principle discussed in this regard is that of "feedback circuits" (Bechtel 2007; Stelling et al. 2004). Control circuits play a decisive role in maintaining cellular functions in the face of internal or external uncertainties. By using the output of a function to be controlled to determine appropriate input signals, feedback enables a system to regulate the output by monitoring it. Negative feedback can reduce the difference between actual output and a given set point, thereby dampening noise and rejecting perturbations. Positive feedback can enhance sensitivity. This is primarily required for robust cellular decisions that need to be derived from noisy and graded input signals and to be maintained. Well-balanced positive and negative feedback can lead to a blend of sensitivity and stability. Another possibility for achieving higher robustness consists of combining multiple levels of regulation, for instance, controlled transcription, translation, post-translational modification, and degradation. Often, when highly precise and reliable behavior is indispensable for overall cellular functionality, multiple intertwined feedback loops operate. The different levels of control for circadian clocks (Bechtel 2010b; Hogenesch and Herzog 2011; Mohawk et al. 2012) and developmental control circuits (Carroll 2005) provide good examples of these aspects.

The principle of modularity might also contribute to the robustness of organisms. The composition of cells and of organisms from "functional units" or "modules" is under increasing discussion in the literature (Stelling et al. 2004). Modules constitute semi-independent entities that show dense internal functional connections but looser connections with their environment. Modularity, as the encapsulation of functions, can contribute to both robustness of the entire system (by confining damage to separable parts) and evolvability (by rewiring of modules or by modifications in modules that are not noticeable from the outside).

Finally, the integration of cellular functionality across hierarchies seems to be important. Stelling et al. (2004) describe that cells, which under normal operation provide a certain robustness of their behavior, can collectively reduce the impact of environmental perturbations when they are components of an organism network. Thus, the "collective of cells" inherits some of the cells' robustness, augmenting it by synergistic network-level interactions. An efficient means for coordination in such networks and in complex systems is to organize the system hierarchically—namely, to establish different layers of integration. This not only might reduce the costs of information transmission but also might further enhance robustness by different level regulations, multiplying each other.

Open or Closed?

It has often been formulated that the organism is an open system. For example, Capra and Luisi (2014) describe organisms as materially and energetically open systems. And, of course, they are right in that every organism must have a constant exchange with its environment: substances are absorbed with food and with respiration, other substances are released. The same is true for energy. At the same time, however, one could formulate a certain closedness, since with the help of material components (and many more) an autonomy is permanently maintained, which has a certain stability against the turnover with the environment. By far not all material components are permanently exchanged, otherwise the integrity of the organism would not be guaranteed. The largest part of the substances in the organism is not exchanged with the environment.

So it is only certain substances that are exchanged with the environment and only a certain, limited amount of energy, and both are regulated and controlled by the organism. What is valid then: Is the organism a materially and genetically open or a closed system? I now refrain from stricter physical definitions, because we have to operate here with an organismic view. And there the decisive thing is that the phenomena show that the organism is both at the same time. It is both open and closed. The realization of this simultaneity is a fundamental principle of organisms, which cannot be resolved into an either-or, unless one describes only certain sections of it (= concurrency).

The same applies to the recent discussion about whether organisms can be regarded as individuals at all. Organisms, including humans, are so extensively associated with other organisms that they can hardly be regarded as a section of them. In this context, the extensive association with microorganisms is especially pointed out. Our intestine, our skin, and also many things in our environment have large amounts of microorganisms, which are essential to our body. Without them, this is not viable. So, are we individuals, or are we associations of several organisms? Also, this problem cannot be solved in an either-or way. One can fill many papers with it and argue one view or the other. Only the recognition of the simultaneity of both leads to an organismic understanding in this question.

Changes in the Capacity for Autonomy

An extensive study demonstrated that the ability for autonomy changed considerably during evolution (Rosslenbroich 2011b, 2014). The ability for regulation and stabilization of endogenous functions, as well as the flexibility within the environment, have been enhanced throughout evolution. In this sense, organisms not only adapted to the environment but also expanded their own individual autonomy (see definition in table 4.3).

During the major transitions, the capacity for individual autonomy was gradually enhanced and extended. Enhanced autonomy is understood as a property that increasingly allows a system to maintain its functions in face of internal and external perturbations and uncertainties, and at the same time extends the scope of self-generated flexible reactions and active behaviors. Physiological stability and versatility lead to enhanced capacities for self-control of life functions and activities. Generally formulated, it is a widening of possibilities. Thus, there are organisms that are more subject to the direct physical, chemical, and biological conditions of their surroundings, and others that can act more on their own behalf because they are more active, flexible, and selective in their interaction with the environment. At the same time, this results in extended possibilities for the regulation

These processes are described as changes in relative autonomy because numerous interconnections with the environment and dependencies upon it were retained simultaneously. Also, it is not a linear trend, but rather an outcome of all the diverse processes that are involved during evolutionary changes (for discussions of the concept, see Arnellos 2016; Lockley 2014; McShea 2015).

Several biological elements can contribute in different degrees to changes of autonomy. They are not general rules or some sort of continuous trends. Rather, they function as a set of resources that can—singly or in combination with each other—change the capacity for autonomy. One such element is *spatial separation from the environment*, such as with cell membranes, cell

Table 4.3
Definition of increasing autonomy in evolution

Increasing autonomy is defined as an evolutionary shift in the system-environment relationship, such that the direct influences of the environment on the respective individual systems are gradually reduced and stability and flexibility of self-referential, intrinsic functions within the systems are generated and enhanced. This is described as relative autonomy, while, at the same time, numerous interconnections with and dependencies on the environment are retained. Thus, organisms can undergo relative emancipation from environmental fluctuations, gaining self-determination and flexibility of behavior.

A set of resources can be involved to change autonomous capacities:

I.   Changes in spatial separations from the environment,
II. Changes in homeostatic capacities and robustness,
III. Internalization of structures or functions,
IV. Increase in body size, and
V.   Changes in the flexibility within the environment, including behavioral flexibility.

Source: For details, see Rosslenbroich (2014).

walls, or integuments. To different degrees, they all serve to keep the environment outside the organism and to regulate and direct the exchange with it. *Homeostatic functions* are means to establish and enhance internal functional stability. Another element is the displacement of morphological structures or functions from an external position into an internal position within the organism, here summarized as *internalization*. Multiple processes of internalization are involved in building up the inner anatomy of organisms, ontogenetically as well as phylogenetically. A *gain in size* during many transitions leads to a reduction of the surface-to-volume ratio. This means that in larger animals there is less direct contact with the immediate environment relative to the existing body mass. The smallest known cells, bacteria, have a large surface for environmental exchange. In larger bodies, this direct exchange capacity is reduced relative to the body mass.

These elements are prerequisites for establishing a certain amount of physiological *flexibility* within a given environment, that is, a capability of organisms to generate flexible functional answers to conditions and changes in their environment. Finally, this principle can be widened to include all forms of behavioral flexibility, emancipating organisms from mere short-term reactions to environmental factors. These changes in autonomy can be described with morphological, physiological, and ethological criteria. We followed this pattern on different levels of evolution, from the evolution of the eukaryotic cells, through the origin of multicellularity, the generation of animal phyla, to the evolution of vertebrates including birds, mammals, and humans.

In a discussion of the concept of autonomy McShea (2015, 440–441) described that at the heart of the discourse

is a notion of an organism as an entity, a whole, that arises from its parts, and that simultaneously governs the behavior of those parts. Parts and whole are both cause and consequence of each other. This back and forth between parts and whole sets the entity apart from its environment, to some extent, a separation that creates a kind of autonomy for the entity, a self that is distinct from its environment and that has the robustness to persist in the face of changes in that environment. In organisms, typically, this autonomy is achieved in part by boundaries, which insulate it, and further by the entity's ability to respond to environmental changes with internal changes that compensate. . . . Autonomy is the emergence of an entity, a local peak in "entity-ness", that accompanies a local increase in internally driven homeostasis.

Here, only one empirical example (out of many) will be presented briefly (figure 4.13): During the generation of multicellularity, an epithelial boundary is formed, which is organized as integuments. The cells, together with the occluding junctions between them, seal the internal space from the

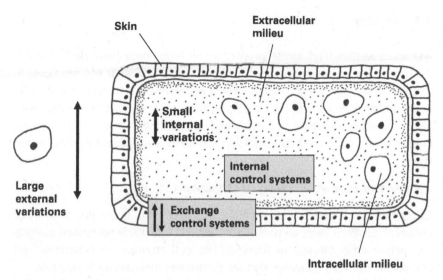

Figure 4.13
Schematic representation of the physiological principle of extracellular homeostasis.

environment so that the passage of substances can be controlled. The composition of fluids in the inner spaces can be regulated, and concentration gradients compared to the environment can be created. The extracellular matrix within the boundary effectively allows cells to create their own intercellular conditions, which regulate and protect them from the external milieu. However, there are large differences in the degrees of buffering of physical and chemical changes in this intercellular space. Some organisms have only basic regulative functions for the extracellular matrix; others develop organs such as nephridia, which regulate this composition. In summary, it can be stated that the general characteristics of multicellularity, as they evolved in early metazoans, essentially enhanced the possibilities of physiological regulation of internal compartments and tissues. These (and some further) phenomena demonstrate that the transition from single cells to metazoans included different degrees of a relative emancipation from direct influences of the environment and a relative stabilization against its fluctuations.

When the physicist Erwin Schrödinger (1944) first pointed out that living organisms are far-from-equilibrium systems relative to their physical or abiotic surrounding—a quote which has often been repeated in subsequent writings—he basically formulated the principle of autonomy, again showing that the principle is unavoidable when one wishes to describe organisms.

4.4   Agency

Agency is another fundamental property of all living organisms. Basically, it is
the overall *autonomous activity* of the organism to maintain life functions, to
establish and defend its processual relative autonomy, and to operate within the
environment. It consists in the capacity of the system to perform the processes
of its immediate existence as a living organism within a certain self-organized
time structure and to respond actively to internal and external conditions and
signals.

Agency as a Primary Property of Organisms

Every living cell performs a metabolic turnover. This is not just a series of
chemical reactions along an energy gradient, but rather it is performed actively
and permanently throughout lifetime. The cell arranges all conditions and
components in such a manner that the permanent metabolism is possible.

In animals, circulatory systems are kept running—besides many other
functions—and in many there is a heart that beats lifelong. The earthworm
digs actively through the soil, fundamentally changing it, birds move through
the air in an often acrobatic manner, and many animals appear to choose their
actions from a repertoire of internally generated behaviors.

Also, the growth of a plant is a general form of agency. In a regulated
manner some of its cells perform mitosis and thus add new tissue to the plant
again and again. Mitosis is an active process, highly regulated within a
sequence of events of several hours within the dividing cells. Plants actively
grow contrarily to gravity, roots actively penetrate the soil, and they are even
able to activate nutrients from it, as is well known today. Nowhere in the mere
chemical or physical, inanimate world are there processes, which are continuously performed, maintained, and regulated in such a manner.

For the moment this statement seems to be simple and self-evident, but it
has far-reaching consequences. The conventional view of the organism local-
izes the ability that something can be active into the physicochemical processes. According to this view events within a cell or an organism are driven
by molecular energetic reactions, being the motor of all functions. Thus, the
activity is, more or less explicitly, seen in the metabolic machinery.

However, a consequent phenomenological view shows that agency can only
come from a whole and intact organism. Minimally, agency needs a functioning cell and its integrity. Only an intact organism can arrange all conditions
and components in such a manner that a metabolism and the regulated usage
of energy within a surrounded unit becomes possible. An organism keeps its
metabolism going, obtains the necessary energy, encloses itself from the

environment, and constructs substances such as proteins and nucleic acids. Some are also able to move or to perform even more complex activities within their environment.

Nothing comparable is known from the inorganic realm. Or, as Peterson (2016, 86) asks: "Is a stone falling down a waterfall really analogous to a salmon leaping up a waterfall?"

As described in the previous chapter, organisms establish an energetic gradient toward their environment. The maintenance of this gradient and its defense against the influences from the environment need to be actively performed. Thus, agency is a primary property of an organism. At least it needs a functioning cell (procaryotic or eucaryotic), which is not reducible to subcellular functions, structures, or substances, although it uses and needs the metabolic overturn to fuel these activities. There is no subcellular unit that is able to maintain activity outside the context of the cell.

The same is true for genetic information: genes cannot be active, and they are not able to "control" or even to "determine" something within the organism, as is often said. Rather by the activity of the cell the information content (genetic, epigenetic) is processed within highly regulated functions. Information can be used in a meaningful way only within the context of the cell's agency. When the cell loses its integrity and functionality, this agency collapses, so that the substances now follow the physicochemical laws, not being integrated any longer into the organic processes.

In multicellular organisms the processes of the cells are coordinated and integrated, as described in section 4.2, so that the whole system can develop its agency.

Different Forms of Agency

Agency can appear in three different, though related forms:

1. nondirected agency

2. directed agency

3. agency with a preconceived goal

Nondirected agency (form 1) is the mere process of life and its functions involved. The maintenance of a metabolism, the production of proteins and of components of the membrane, as well as keeping the DNA intact and reading it (transcription), are basic activities of an organism. To this also belongs the generation of an external seclusion, either by a membrane in single cells or by epithelial arrangements of tissues in multicellular organisms. I call this form of agency nondirected because the functions are not directed toward a goal that is external to the organism but rather maintain the integrity and processual functionality of the organism as such. It is an unintentional agency that derives merely from metabolic self-maintenance.

It may be possible to say that self-maintenance also includes a kind of goal. The intrinsic goal is simply *to be*. And this has a value, as Jonas (1966) points out, as *to be* is better than *not to be*, which is the reason why every living entity keeps being active to maintain its existence. In some sense this is identical with the description of "minimal agency" of Barandiaran et al. (2009). However, here I want to accentuate that agency is conceivable as genuine without an external directedness, without a major goal, and of course without an intentionality directed into the future.

A *directed agency* (form 2) can be observed, when organisms move along a gradient or toward a goal. Already bacteria can direct their movements according to chemical gradients (positive or negative chemotaxis) or other factors of their environment. Such a movement is an activity which has a direction. For example, it tries to reach nutrients or to avoid harmful substances and much more. However, such activity is generated quite directly along a concentration gradient, which is perceived through receptors. There is no indication of an anticipated end state, as humans can have it in the sense of representations that might plan the action toward a goal. However, different levels of complexity might trigger a directed agency. Whereas the physiological processes involved in single cells are comparatively direct, they are extended in higher animals and the overlapping to form 3 may be broad.

Agency with a preconceived goal (form 3) is only possible within a complex organism with a highly developed central nervous system. Especially humans, in initial stages also some higher animals, can act according to an envisioned goal. In this case a future goal, which is mentally represented, becomes the reason for an activity.

This is the form of agency that is predominantly used in philosophical discussions on this topic (Schlosser 2015). It includes a background of desires, beliefs, intentions, and mental representations as we are used to have them in human actions.

In form 2, some forms of purpose or intention are involved. When a lion starts hunting, it certainly has the intention to catch prey. A marmot whistles with the intention to warn buddies about approaching danger. In his introduction, Walsh (2015) describes from his own experiences as a student in an ethology course how he had to describe behavior just by a catalog of movements, postures, and sounds. The students and scientists were pushed to free the observation of behavior from any taint of purpose or intention. He describes how the students were to identify behaviors with manifestations rather than motives. They were told that the observed Columbian ground squirrels were standing on their hind legs at full height rather than surveying the scene for predators. They were emitting a high-pitched bark rather than alerting their fellow colony members to imminent dangers. However, the lesson he took with him was that organisms are fundamentally purposive entities, and that biologists have an aversion to purpose. So he posed the question: Why should the phenomena that demarcate the domain of biology be off-limits to biology? In today's ethology this restriction has been overcome, especially in cognitive ethology. But the field had to struggle for a long time against conventional scientific paradigms.

"This is more than just an idiosyncratic bemusement on my part," Walsh then describes.

It is a leitmotif that runs through the entire history of biology. Since its inception, biology seems to have been torn by the evidently incompatible demands of treating organisms as natural entities, like everything else, and as singularly peculiar (naturally purposive) things, unlike anything else. One of the common strategies has been to attempt to minimise the distinctiveness of organisms, to show that the problematic nature of organisms is incidental to a comprehensive understanding of biology, and that the principles by which we account for nonliving phenomena are wholly adequate to the explanation of living phenomena. (Walsh 2015, ix)

However, the distinction between form 2 and form 3 may help to avoid the expectation of a general purposiveness in nature, which introduces further theoretical problems. It is possible to describe forms of agency without anticipated final states of the activity. Animals, for example, even more complex ones, mainly live in the present. If they are hungry, they look for food, but they don't imagine the necessity for caring about food tomorrow. For plants this seems to be even more true. To free the phenomenon of agency from an anticipated goal makes this essential property of living organisms negotiable for science.

Beyond this, it also frees the topic from the misunderstanding that the introduction of agency necessarily entails anthropomorphism. The human version of agency needs not to be introduced into the behavior of other organisms (Hoffmeyer 2013).

The distinction of three forms of agency touches the debate about the problem of teleology, the study of functions, properties, and behaviors of organisms under the premise that they are directed toward goals, a debate that has been conducted quite extensively throughout the history of biology (Brandon 1981; Di Paolo 2005). This debate is burdened too much by the assumption that the description of a behavior or a function as being teleological would in any case include the expectation that a goal in the sense of form

3 is expected. However, this is not necessarily the case. If one differentiates according to the three forms of agency, exact descriptions are possible without assuming such goals in each case. Many functions and behaviors of organisms are based on agency without an imagined future state. In many cases it "just happens," it is just the activity within the present situation. As humans we are used to our ability to envision a future goal, but this is not the case in most other forms of nature.

Biology never got rid of teleological descriptions and formulations, although at times the field tried hard to rely solely on explanations of cause and effect, in which there is no room for any teleological explanation. This illustrates that this is a misunderstood property of organisms, which revolves around the problem of agency and teleology in living organisms.

Minimal Agency, Autopoiesis Theories, and Self-Organization

Many contemporary authors are looking at the principle of agency from the perspective of autopoiesis theories, in which the principle of autonomy is tightly linked to agency. Although agency has usually been understood and discussed by cognitive science in connection with high-level human cognitive abilities, researchers increasingly apply the concept of agency to phenomena throughout all of biology, at least in the sense of agency form 1. Most of these authors call it "minimal agency" (Arnellos et al. 2010; Barandiaran et al. 2009; Hoffmeyer 2013; Moreno and Mossio 2015). Together with a minimal form of autonomy, agency appears within single cells like bacteria. A cell is seen as an autopoietic system based on a self-organizing network of biochemical reactions that produces the described features of autonomy (Maturana and Varela 1987; Varela et al. 1974). Such a cell actively relates to its environment to maintain its viability. Its sensory responses serve to trigger motor and other behavior subject to the maintenance of its autonomy and regulated by internal needs, thus existing within an interactive relationship to its environment. This mode of coupling with the environment is recapitulated in a more complex form by more elaborate organisms, including their nervous systems. The nervous system establishes and maintains a sensorimotor cycle, whereby what the organism senses depends directly on how it moves and behaves, and how it behaves depends directly on what it senses.

In her analysis of the concept of directed behavior as a property of complex systems, Juarrero (2002) reminds us that the modern idea of self-organizing systems runs counter to the scientific tradition of seeking explanations exclusively in causal relations, being based on the assumption that causes are external to their effects. It is general scientific belief that nothing can move, cause, or act on itself. That a chicken develops from an egg is not due to immanent activities in the egg as a substantial thing, it is due rather to some determination that has to be found either in parts of the egg or in external determining factors. Even Kant saw this problem, as Juarrero (2002) describes: "Organisms' purposive behavior resists explanation in terms of Newtonian mechanics and is likewise a major impediment to unifying science under one set of principles. These considerations convinced Kant that natural organisms cannot be understood according to mechanism in general or its version of causality in particular. Since only external forces can cause bodies to change, and since no 'external forces' are involved in the self-organization of organisms, Kant reasoned that the self-organization of nature 'has nothing analogous to any causality known to us'" (47).

Recently there have been many attempts to view complex systems as having dynamic properties that allow self-organization to occur (Capra and Luisi 2014; Haken 1983; Kauffman 1993). However, these concepts are still not mainstream in science.

Juarrero attempts to establish a different logic of explanation, one more suitable to complex living systems from the viewpoint of action. Such systems are typically characterized by positive feedback processes in which the product of the process is necessary for the process itself, as she formulates. This circular type of causality is a form of self-cause, she continues. When parts interact to produce wholes, and the resulting distributed wholes in turn affect the behavior of their parts, interlevel causality is at work. Interactions among certain dynamical processes can create a systems-level organization with new properties that are not the simple sum of the components that constitute the higher level. In turn, overall dynamics of the emergent distributed system not only determine which parts will be allowed into the system, the global dynamics also regulate and constrain the behavior of the lower-level components. Far from being the inert epiphenomenon modern science claims all wholes are, complex dynamic wholes clearly exert active power on their parts such that the overall system is maintained and enhanced. This is mainly identical to what has been described as "integrative systems" before.

Hoffmeyer (2008, 2013) states that the field of biosemiotics is based on an understanding of agency as a real property of organismic life. Agency, he explains, is rooted in the capacity of cells and organisms to interpret events or states as referring to something other than themselves or, in other words, the capacity to interpret signs. These signs need not necessarily be emitted with a purpose of communication; in fact, by far most signs are not part of a sender-receiver interaction at all but simply important cues (internal or external) that organisms use to guide their activities. This semiotic perspective introduces a further aspect of the properties of life, exhibiting a strong relation to information processing in organisms, which will be discussed below (for further details, see Sharov and Tønnessen 2021).

Functions Are Actively Performed

Many processes, which an earlier mechanistic physiology claimed to function passively along physical gradients or chemical reactions, turned out to be actively performed by the cell or the organism. Agutter et al. (2000) for example describe the history of thinking about diffusion of substances and gases through tissues or throughout the cell. Physiologists of the nineteenth century tried to explain it just from physicochemical gradients that would force substances to move. According to physical (i.e., inorganic) laws, the net movement of matter is attributed to the random movements of molecules taking place from a region of high concentration to one of lower concentration. This would for example generate gas exchange in the lung alveoli or through cell membranes. Step-by-step physiology and molecular biology had to learn that transport processes are active or at least that the conditions are actively arranged by the organism. Therefore, it is quite clear today that physicochemical models and laws of diffusion do not apply to the complicated situations in tissues or cells. Against this background Agutter et al. wonder why even today diffusion theory continues to be accepted uncritically and taught as a basis of both biology and medicine.

Physicochemical data can be obtained from living systems, but the whole corpus of modern cell and molecular biology makes it clear that such data can seldom, if ever, be interpreted in the same way as in non-living systems. Yet the temptation of reductionism persists. The dream of the mechanistic materialists [of the nineteenth century], that physicochemical data can and should be interpreted in precisely the same way irrespective of whether the system is living or non-living, remains seductive. And it is in the context of that dream that diffusion theory still survives in biology. (Agutter et al. 2000, 104)

Agutter et al. especially emphasize movements of molecules that occur without the involvement of any specific mechanism, and they describe that even these movements are not wholly passive as diffusion theory assumes. Much clearer is the activity of cells and tissues in specific transport processes like membrane transport systems; the movement of materials by the cytoskeleton, exocytosis, and endocytosis (the process of bulk export of material out of and into the cell); the partitioning of proteins to different subcellular compartments; events at the nuclear pore, and so on; all of which are currently productive fields of research.

In a similar sense early biochemistry presumed that the interior of the cell was largely unstructured. Researchers focused their interest on chemical reaction sequences, and they were only conceivable within homogenous solutions.

The interior of the cell was seen as a "colloidal solution" of large molecules in which physicochemical concepts could be applied to cellular activities. Today it is well known that the interior of the cell, the cytoplasm, is highly structured. The network of the cytoskeleton and of organelles provide a sophisticated system of spaces and surfaces, along which all the molecular constituents and reactants are actively transported and sorted by the cell.

However, physiology in general is a science about activities. The physiologist and ecologist J. Scott Turner (2007, 2013, 2017) postulates that homeostasis is the crucial principle to develop "a coherent theory of life." Homeostasis, he holds, is generated by self-sustained, active processes, and the capacity to act is where the distinction can be drawn between a living being and nonliving objects of certain resemblances with living beings.

Concerning the activity of the genes Hoffmeyer (2013, 155) formulates that agency is present with the cells, the tissues, or the organism, not with the genetic system. He cites Richard Lewontin, who formulated that "genes do nothing" and that the concept of "selfish genes," as Richard Dawkins formulated it, is outright nonsense. Language use in modern biology, Hoffmeyer continues, is abundant with hidden homunculi, which is not based on observation.

In recent years several authors have stressed this point extensively (Capra and Luisi 2014; Noble 2006, 2008b, 2011a; Shapiro 2011, 2014, 2017). DNA alone cannot be active by itself. It must be read by proteins to produce other proteins. There are several dozen proteins to do this reading, and many others to express a gene. The relevance of DNA, they hold, needs to be understood in terms of systems biology, according to which it is the organism itself that determines which genes will be read and expressed. The extensive findings of epigenetics presently are building a bridge to a functional understanding of such a view.

Agency in Evolution

Denis Walsh (2015) has contributed the most thoughtful work on the topic of agency so far. He is very explicit about the significance of agency, not only to understand organisms in general but also as the central element within the evolutionary process. He describes that the category "organism" plays only a very little role in evolutionary theory. Conventional theory today principally deals with the dynamics of supraorganismal assemblages (populations) of suborganismal entities (genes). The distinctive properties that make organisms organisms play virtually no part in the explanation of evolutionary phenomena within the synthetic theory of evolution that has grown to such prominence throughout the twentieth century. This has not gone unnoticed, he describes, and cites Brian Goodwin: "Something very interesting has happened to biology in recent years. Organisms have disappeared as the fundamental unit of life.

In their place we now have genes, which have taken over all the basic properties that used to characterize living organisms" (Goodwin 1994, 1). And then, of course, he cites Richard Dawkins, who took this view to the extreme: "Evolution is the external and visible manifestation of the survival of alternative replicators. . . . Genes are replicators; organisms . . . are best not regarded as replicators; they are vehicles in which replicators travel about. Replicator selection is the process by which some replicators survive at the expense of others" (Dawkins 1976, 82). Walsh argues for an organism-centered alternative to this view, since organisms participate in evolution as agents, and proposes an "agent theory" as an option to the conventional "object theory."

This is a significant departure from the place accorded to organisms in orthodox synthetic theory, in which organisms are treated as objects of evolution. The objective is to explain organismal form as the result of assumed evolutionary processes such as selection, drift, mutation, and migration, impinging on populations of genes. Very little attention is paid to the way that organisms themselves participate in the process of evolution.

However, Walsh does not just transfer the active role solely unto the organism but rather describes the relation of the active organism within the conditions and prerequisites of the respective environment. In this context he formulates an "organismic perspective by arguing that the unit of greatest theoretical significance for evolution is not the gene, or for that matter even the organism per se. It is the organism situated in a system of affordances. Affordances are emergent entities; they are properties of a system, in this case, a system comprising an organism and its conditions of life. Affordances are constituted in large measure by the ways that organisms can exploit or ameliorate these conditions" (Walsh 2015, 209).

Thus, Walsh hypothesizes that organisms actively participate in the process of evolution. Through their activities they not only create evolutionary change, but they also modify the conditions to which evolutionary change is a response. This again includes the principle of interdependence, as described in section 4.1.

Walsh (2015) distinguishes between object theories and agent theories. Objects underlie forces, laws, and initial conditions, according to the Newtonian paradigm. In the Newtonian paradigm we describe the dynamics of a system by answering two simple questions: What are the possible configurations of the system? And what are the forces that the system is subject to in each configuration? Crucially, in the Newtonian paradigm the forces, laws, and initial conditions are extraneous to the objects and exist independently of them. Agent theories, however, are characterized by "immanence" and "explanatory reciprocity." In an agent theory the entities in the domain include both agents and the principles we use to explain their dynamics. The agent's activities are generated endogenously; agents cause their own changes in state in response to the conditions they encounter. These conditions, in turn, are largely of the agent's making. So, as agents implement their responses to their conditions, they not only alter their own state, they also change the conditions to which their activities are a response.

Walsh describes that there is a dialectical relation between the activities of the entities in the domain and the principles we call upon to explain them. The activities of the agent can be explained as a response to its conditions, and reciprocally, the change in conditions can be explained as a consequence of the activities of the agent. The objective of an agent theory is to account for the interplay between the entities whose dynamics we want to explain and the principles we use to explain them.

This is strongly related to the principle of organismal autonomy, which is described in section 4.3. In the words of Walsh (217): "In constituting their affordances through self-maintaining, self-regulating activities, agents forge for themselves a degree of freedom from the vicissitudes of their environments. In doing so they determine which of the conditions is salient, and what they afford. And they set themselves up to exploit the opportunities those conditions have to offer. This is the sense in which agents are 'autonomous'."

Walsh also points to the principle of integrative systems, which was discussed in section 4.2. He claims that organisms have the capacity to construct their own parts, and to enlist and regulate the causal capacities of their parts in such a way as to ensure the pursuit of their goals. Furthermore, he stresses the importance of this top-down explanation for understanding the process of evolution. The activities of individual genes, genomes, cells, tissues, organs, and immune, thermoregulatory, metabolic, behavioral, and cognitive systems can be explained for the most part by adverting to the purposive dynamics of the organisms of which they are a part. Organisms must make their parts work in ways that are conducive to the fulfillment of their needs.

Contemporary mechanistic theorists insist that organisms inherit their causal capacities from their parts. And so the capacities of whole organisms can be accounted for from the bottom up. However, that is only half of the explanatory picture, as Walsh further explains. The parts have their particular capacities, in these instances, because of the systems in which they are embedded. There is a reciprocal relation of regulation between the activities of an organism's parts and the activities of the organism as a whole. Organisms are truly, as Kant formulated, both causes and effects of themselves. An overreliance on the bottom-up strategy of mechanism obscures the explanatory reciprocity between parts and the whole. It gives us organisms as effects of their parts but not organisms as causes of the existence and the activities of their parts. Mechanism can tell us why, given the causal capacities of an organism's parts, organisms behave in the way they do. However, it cannot answer the counterquestion: Why do the parts have those particular causal powers rather than other ones? The parts of organisms get their particular causal powers from their contexts—the activities of the organism as a whole (Walsh 2015, 219, mainly in his words). This relation, importantly, again contains the principle of concurrency.

Prominent statements in this direction have also been made in the field of molecular biology: "On the side of generating phenotypic variation, we believe the organism indeed participates in its own evolution, and does so with a bias related to its long history of variation and selection" (Kirschner and Gerhart 2005, 252).

4.5 Processing of Molecules

One characteristic of a living organism is that it actively processes molecular compounds, which in turn enable the activity of the organism. Composition, organization, transport, and distribution of molecules of very different kinds are exactly regulated and thus underlie the activity of the organism and especially its cells. Within this organization the molecular compounds unfold their qualities and functional capacities, which fuel the activity of the cell. This interdependence makes functions and processes of living entities possible. Thus, the active part is neither an activity of the cell as such, which processes the molecules, nor is it the activity of molecules that generate the activity of the cell. It is the interdependence between molecular potentials and the agency of the whole organism that enables the overall processuality.

The Activity of the Cell

As mentioned repeatedly now, the second half of the twentieth century has been characterized to a large extent by approaches that assumed that main processes and functions within the cell can be explained by physicochemical principles of molecules, which fuel and generate all processes. This assumption largely neglected the activity of the cell as an integrated and integrating unit. However, the activity of the whole cell sets the framework for the molecular functions. Therefore, descriptions of molecular activities are not false, of course, but often one-sided because they disregard the processual context in which they take place and which is largely dependent on the agency of the coherent cells. Again, there is a concurrency of cellular activity and molecular reactivity.

As discussed before, diffusion, for example, has been taken as a major "mechanism" by which molecules interact within an aqueous solution of the cell's internal matter. The study of metabolism for example was predicated on the presumption that the internum of the cell was largely unstructured. This point was seldom made explicitly; cell structure was simply ignored (with a few exceptions) because the interest of researchers was focused on chemical reaction sequences (Agutter et al. 2000; Wheatley 1998). The concepts and techniques of organic chemistry, indispensable in studying metabolism, related exclusively to homogeneous solutions, not to multiphasic structured systems. If the cell were other than a "bag of solution," organic chemical approaches could not be applied to it and metabolism could not be elucidated.

Many details of intracellular structure had already been seen by light microscopists in the late nineteenth century, which should have established that the cell was not an unstructured bag of solution. However, the pioneers of biochemistry seem to have turned their backs on this evidence. The so-called internal structures of the cell were seen as artifacts caused by fixing, drying, and staining material for microscopy and were not considered as parts of the living material itself. The compromise notion, that the interior of the cell was a "cytosol" or a "colloidal solution," that is, a solution of large molecules and molecular complexes that could sometimes take on the characteristics of a gel, became established (Agutter et al. 2000; Wheatley 1998).

Under such an assumption, a "flow theory of enzyme kinetics" and comparable models have been developed to understand metabolic pathways within the cell. However, starting from this perspective, cell biology had to learn step-by-step that all processes within the cell are rather actively induced, regulated, controlled, and integrated by the cell itself. Nature gave a different answer than it had theoretically been expected. Some processes are induced by activities and sorting processes of the cell, using motor proteins along the cytoskeleton. Others are an assisted form of diffusion, in which the cell establishes well-directed conditions for the process. The inside of the cell is far more highly organized right down to the molecular level than was hitherto appreciated.

Cytoplasmic Transport

Today it is clear that eukaryotic cells create internal processual order by using protein motors to transport molecules and organelles along cytoskeletal tracks. A cell organizes its inventory of macromolecules and organelles precisely and actively. Setting reaction points and establishing the timing of transactions are of fundamental importance for cell behavior. The high degree of spatial as well as temporal organization of molecules and organelles within cells is made possible by proteins that transport components to various destinations within the cytoplasm.

Landmark discoveries of cytoplasmic transport have been, and continue to be made, through advances in microscopy. Intracellular motion was first observed in the alga *Chara* by Bonaventura Corti in the late eighteenth century, and chromosome movements were documented with remarkable accuracy by microscopists in the nineteenth century. The development of video-enhanced contrast microscopy in the early 1980s enabled the visualization of small membranous organelles and large protein complexes (Vale 2003). With this clearer view of the cell interior, the tremendous amount of directed cytoplasmic motion became apparent. The view of a "cytosol," which functions mainly along concentration gradients and diffusion, was gradually replaced by understanding the high degree of structural as well as temporal order.

The use of green fluorescent protein for tagging organelles, proteins, and RNA led to another wave of discovery of intracellular movement. The list of transported cargoes, which grows larger every year and touches almost every aspect of cell and developmental biology, now includes large membrane organelles (e.g., mitochondria, lysosomes), smaller vesicular or tubular intermediates in the secretory and endocytic pathways, lipid droplets, a subset of mRNA, protein building blocks for large macromolecular complexes such as cilia/flagella and centrosomes, and proteins involved in signaling and establishing cell polarity (Vale 2003). Signaling molecules are transported to their precise location of function (Dorn and Mochly-Rosen 2002).

Specific subsets of biochemical reactions in eukaryotic cells are restricted to individual membrane compartments, or organelles. Cells, therefore, move the products of those reactions between individual organelles. Because of the high density of the cytoplasm and the large size of membrane organelles, simple diffusion is grossly insufficient for this task (Barlan et al. 2013).

The most widely used function for intracellular transport involves molecular motor proteins that carry cargo directionally along a cytoskeletal track. Whereas prokaryotes contain cytoskeletal filaments, the cytoskeletal motors appear to be an early eukaryotic invention. Several types of cargo-transporting molecular motors emerged in unicellular eukaryotes, and this same ancient "toolbox" of motors expanded to meet the majority of transport needs of multicellular organisms. Molecular motors are also used for organizing cytoskeletal filaments, such as controlling their dynamics, collecting them into bundles, and causing filament–filament sliding (Vale 2003).

In general, the motor proteins belong to three classes—kinesins, dyneins, and myosins—which utilize two types of polar cytoskeletal filaments to transport cargoes (figure 4.14). Microtubules are long filaments that are typically arranged in a radial array with the plus ends near the cell periphery and the

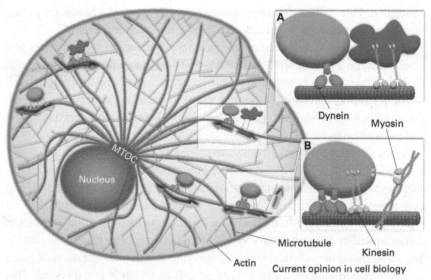

Figure 4.14
Membrane organelles require multiple motors and cytoskeletal filaments for their distribution. Long distance transport generally occurs along microtubules, via the molecular motors kinesin and dynein. Myosin motors move along actin filaments and contribute mostly to short-range transport of cargoes. Microtubules and actin also provide scaffolding where organelle interactions can take place (inset a), as attachment to a filament restricts three-dimensional diffusion of organelles to movement in one dimension. Organelles are often moved on filaments by multiple copies of motors, including motors of opposite polarity (inset b). The activity of these motors, and the way in which individual cargoes are transported, is likely regulated by specific factors on individual organelles to allow for rapid changes in distribution and motility. *Source*: Reprinted from Barlan et al. (2013, 484), with permission from Elsevier.

minus ends anchored near the cell center. Kinesins and dyneins move along microtubules and are responsible for most long-range movements of organelles and membranes. Actin filaments, in contrast, are shorter, and although the filaments themselves are polarized, they typically form a randomly oriented meshwork. Myosin motors that move along actin filaments mostly contribute to more localized short-range movements of cargoes. By utilizing two distinct types of transport networks, delivery of cargoes within a cell can be both efficient and precise. Because a given cargo is often attached to multiple motors of various classes, the precise mechanisms that control motor activity and directional bias can be complex (Barlan et al. 2013; Burute and Kapitein 2019).

Furthermore, while instrumental in the delivery of cargo by molecular motors, cytoskeletal filaments also act as compartments themselves by restricting three-dimensional diffusion of membrane organelles to movement in one dimension. This facilitates the interactions between organelles necessary for the transfer of molecules between different membrane compartments.

The regulation of molecular motors provides one way to specify when and where a cargo is transported or delivered. The arrangement of cytoskeletal filaments themselves is also critical for specific delivery of cargoes, as well as for providing dynamic scaffolding where organelle interactions can take place. Also, cytoskeletal filaments are continuously undergoing reorganization and redistribution by the cell.

A strong argument for the highly structured and active organization of this system within the cell is the evidence for bidirectional motion in several motor-cargo systems. The attached motors bind individually to the cytoskeleton and may have opposite preferential directions (Appert-Rollanda et al. 2015).

The importance of this extremely refined organization within the cell is demonstrated by a hypothetical scenario from Dorn and Mochly-Rosen (2002). They describe a hypothetical signaling enzyme X. Like all enzymes, X has a unique function in each different type of cell in which it is expressed. This function is determined by three characteristics: enzymatic activity (an intrinsic feature of the enzyme), the particular conditions of activation, and the identity of substrates. Normally, X is activated by the appropriate stimulus, finds its proper substrate, and acts. However, if inactive and X is not in its proper location, it may not be activated appropriately. On the other hand, if activated and X is not in its proper location, it cannot act on its normal substrates and may therefore act promiscuously on whatever substrate is available. Furthermore, activation of X may require that it is located in one cellular compartment, while its proper substrates are elsewhere. Under these conditions, X must be moved within the cell after activation. Specific targeting of enzymes to compartmentalized substrates is therefore not only necessary for proper enzyme function but also suggests a function for substrate specificity of closely related enzymes with similar catalytic activity. If enzyme X_1 is translocated to the nucleus upon activation, whereas related enzyme X_2 is translocated to the plasma membrane, they will act upon different substrates and have vastly different cellular effects.

One might have the impression that the activity is generated by the motor proteins, which physically is the case. The organization of the cytoplasm, however, its compartments and the arrangement of the cytoskeletal elements, in short, all the details of the spatial order and also the time order, are generated and arranged by the cell. In addition, the motor proteins are assembled by protein synthesis from DNA and RNA templates. Both the activity of the motor proteins as well as the regulation and organization by the cell are necessary. The cell has several proteins that are involved in regulation of the motor proteins (Barlan et al. 2013) and the microtubule organization can control motor-based transport (Bhaskar et al. 2007; Burute and Kapitein 2019). The overall activity, however, is only possible if these different levels are arranged in the appropriate way and thus are interdependent. If the overall integrity of the system is disturbed, problems result; also, if the motor proteins have a defect, the interdependent functions are disrupted. It is extremely improbable that the motor proteins themselves somehow "know" where to bring the cargo and when. They are clearly integrated and organized within the overall system of the cell and its organization.

This is exactly what Paul Weiss formulated in the 1960s, when he wrote "the cellular control of molecular activities" (Weiss 1968, 24; see historical overview in chapter 2). Molecules can be used to generate activity, but the cell controls it. Single molecules are not able to control anything in the cell.

One further example for intracellular activity is the organization of secretory functions in cells with high secretory activity, such as intestinal cells or hormone producing cells (Vliet et al. 2003). Generally, eukaryotic cells are highly compartmentalized into distinct membrane-bound organelles, a feature essential for fundamental as well as specialized cellular processes. To function, each membrane-bound organelle has a unique composition of proteins and lipids. Highly specific transport processes are required to direct molecules to defined locations and to ensure that the identity, and hence function, of individual compartments are maintained.

Proteins contain structural information that targets them to their correct destination. The organelles of secretory functions are involved in the sorting of proteins to a variety of intracellular membrane compartments and the cell surface. For example, proteins that are transported within the secretory pathway are either secreted from the cell, bound to the plasma membrane, sorted to lysosomes, or are retained as residents in any of the organelles.

Each organelle of the secretory pathway allows the transmission of a multitude of proteins while maintaining its own unique set of proteins, which define its structure and function. A striking feature of the Golgi apparatus, for example, is its ability to maintain a resident set of proteins against the flow of secretory cargo. How these organelles manage this differentiation has been a major conundrum of research (Vliet et al. 2003).

Over the past decades many of the cellular processes responsible for intracellular transporting and protein sorting have been described. This research has shown that all this intracellular trafficking is (1) an activity generated by the living cell; and (2) highly organized, directed, and controlled by the cell. In the past there has been much debate as to whether transport processes such as the conveyance of vesicles from the endoplasmic reticulum to the Golgi apparatus is selective or if cargo is transported as part of bulk flow processes. Today it is evident that all these processes are highly directed and that the cell leaves nearly nothing to uncontrolled diffusion processes.

Membranes

The plasma membrane of the cell is an especially remarkable example of how the cell organizes its molecular compounds. The modern image of a cellular membrane, which surrounds the whole cell, is that of a phospholipid bilayer with integrated proteins and some other substances such as cholesterol. The specific properties of the membrane depend on the molecular composition. First and foremost, this is the type of phospholipids that are integrated into the bilayer. The composition of phospholipids is essential for the physicochemical behavior of the membrane under certain conditions such as temperature. The membrane is constantly maintained in a special status of fluidity, avoiding hardening as well as dissolution. The intermediate state of fluidity is a result of the activity of the cell. If membranes tend to harden, the cell integrates more phospholipids with unsaturated fatty acids, and when they tend to dissipate the cell integrates more phospholipids with saturated fatty acids. Also, the cell is able to regulate membrane fluidity by regulating the amount of cholesterol being integrated.

Thus, the cell uses the chemical properties of the bipolar phospholipid molecules. Membranous functions depend on these properties, but to synthesize them and to organize them in a functional membrane underlies the constant activity of the cell. The molecular compounds are controlled, arranged, and organized by the agency of the cell itself. This is mainly the form 1 agency from section 4.4: nondirected agency. At the same time, however, the cell's agency depends on the integrity of the membrane and its components, again being a nonterminable interdependence. If this agency ceases, the membrane decays and the coherent organization of the phospholipids and proteins dissolves immediately.

It is just not the case that cholesterol, for example, regulates membranous fluidity, as it is regularly stated. It is the cell that integrates certain amounts of cholesterol to regulate this characteristic.

The organization of phospholipids in such a bilayer is not a result of mere physical properties of its compounds. During the synthesis of the biomembrane in the endoplasmatic reticulum of the cell, new lipid molecules are first integrated only into the cytosol side of the endoplasmatic reticulum's membrane (Alberts et al. 2017). Then it is necessary to translocate some phospholipids to the other side, to complete both sides of the new piece of membrane. This "flipping" does not occur spontaneously. Rather it is performed actively with the help of a membrane-bound phospholipid translocator (called a scramblase), which transfers lipid molecules from the cytosolic half to the lumenal half so that the membrane grows as a bilayer.

However, this translocator is not specific for particular types of phospholipids, so that the two sides of the piece of bilayer at first are identical. But in the cellular membrane, to which the piece is transported and where it is integrated, there is a difference between the inside layer and the outside layer. Here, another enzyme is active, one that flips specific phospholipids directionally from the extracellular to the cytosolic leaflet, creating the characteristically asymmetric lipid bilayer of the plasma membrane of animal cells. This enzyme is called flippase, and it is ATP dependent, which is the universal energy carrier in the cell. In addition, there too is a phospholipid translocator in the cellular membrane, which guarantees that both layers are inhabited by the same total number of lipid molecules.

Once again, there is an interdependence between the activity of the cell on the one hand, and the chemical properties of the molecular arrangements. It is not possible to reconstruct the properties of the membrane just from their molecular components without regarding the activity of the cell. Of course, this is obvious today and is recognized in modern membrane physiology. However again, there is no molecular control of membrane function, but rather a cellular control of membrane characteristics and functions.

Beyond this, the functions of the membrane, which are amazingly complex and varied, underlie the regulatory activity of the cell. In the transportation of substances through the membrane, diffusion has some importance. Range and amount of membrane passage is highly regulated. Today there is abundant knowledge about this regulation, and once more it demonstrates that the cell leaves nearly nothing to the physicochemical conditions.

Only some substances are allowed to pass freely by diffusion through the biological membrane. These are mainly small nonpolar molecules such as O_2, CO_2, N_2, and NH_3. For charged ions, charged organic molecules, and uncharged hydrophilic compounds, free permeation is already restricted or even impossible. Their translocation through the membrane mainly takes place through specific proteins.

Corresponding to an osmotic gradient, there is also for water a certain restricted possibility of diffusion directly through the membrane. However, the cell generates special proteins that enable and regulate water transfer through the membrane. After it had been clarified that the membrane is mainly composed of lipids, it was suspected that water movement across the membrane was in some way enhanced or facilitated by pores or channels, but the search to identify these channels was long and tedious (Brown 2017). The identification of water channels then began in 1992. Today these channels are known as aquaporins. Currently thirteen aquaporins have been identified in mammals, distributed in most tissues, but many more have been determined in other groups of animals and in the plant kingdom (Alberts et al. 2017; Brown 2017).

In addition to maintaining cell volume, rapid water exchange across cells enables tissues and organs to secrete and/or absorb water as part of their physiological function. Epithelial cells lining some kidney tubules are especially well equipped for this function. Kidney aquaporins form a highly organized network that facilitates the maintenance of water homeostasis. Other examples are aquaporins, which are involved in brain function, glandular secretion, skin hydration, sweat excretion, hearing, and vision.

Aquaporins do not transport water actively; rather the exchange takes place through osmotic water flow along a concentration gradient. But again, the generation and recruitment of the appropriate aquaporin within a specific membrane is an active function of the living cell. Some aquaporins are more water permeable than others, and some can transport other molecules in addition to water. Thus, the cell not only integrates aquaporins into the membranes where they are needed, but obviously can choose the appropriate type of aquaporin. In addition, there are assumptions that aquaporins are able to regulate water flux, thus exhibiting a further active component within the membrane (Verkman and Mitra 2000). Day et al. (2014) assume that the regulatory role of aquaporins, rather than their action as passive channels, is their critical function.

Thus, even within a process that basically draws on a physical principle such as osmosis, the cellular activity is the crucial organizer of the resulting function. The cell arranges the conditions for the facilitated diffusion. Other processes, which involve an ATP-dependent translocation of substances against its concentration gradient such as the sodium-potassium pump (Na^+/K^+-ATPase), are most clearly activities of the cell. The developed selective permeability and arrangement of molecular concentrations around the membrane, which are specific for each substance involved, is an essential element of self-determination and autonomy of the cell (section 4.3).

Further examples of how the cell actively organizes its components are the active transportation of molecules between the nucleus and the cytoplasm via gateways termed the nuclear pore complexes (Kabachinski and Schwartz 2015; Mor et al. 2014), the precise steps during mitosis and their regulation (Earnshaw and Pluta 1994; McIntosh 2016), or the active organization and reorganization of the genome by the cell (Shapiro 2013a, 2014).

The point to be made here is that the concurrency between the activity of the cell, the process (in the sense of Dupré and Nicholson 2018), and the biochemical molecular reactions is irresolvable. In addition, all this is organized in an extremely refined manner. The impression is that the more science learns about these processes, the more amazing and admirable the whole system appears.

4.6 Processing Information

Organisms accumulate, store, and process information and thereby use extremely refined information carriers. Genetic information is encoded in DNA, which is actively read by the cell and transferred to proteins and other information processes. In addition, however, all organic structures and processes have underlying information that is the basis of their organization. It can be understood as embodied information.

Information in Science

At present there is no uniform understanding of the term *information* in physical sciences, communication engineering, logic, and computer sciences, although it plays a crucial role in all these disciplines (Bawden and Robinson 2020; Marijuan 2004). Some physicists, however, hold that the principle of information is central for understanding the world and that it is a fundamental shortcoming that a precise definition is lacking in many scientific explanations (Baeyer 2003; Davies 2019; Stonier 1996, 1997). In some disciplines of physics even, the question exists as to whether the material world is wholly or in part constructed by information. They see a conceptual extension of the mechanical vision of matter and energy up to an information-related perspective. Baeyer describes that it is still a thorny problem of defining the concept of information. Engineers have at their disposal a variety of methods for measuring the amount of information in a message, but none to deal with its meaning.

Information concepts are playing an ever-increasing role in understanding the biological realm, conceptually as well as by way of the techniques of bioinformatics. This follows Schrödinger's initial insight that information may play a crucial role as a fundamental feature of living systems and may provide a link between the living and nonliving worlds (Bawden and Robinson 2020). Later insights include Gatlin's introduction of information theory concepts into biology (Gatlin 1972), and Marijuan's (2004) conclusion that living existence is "informational," that there is an "invisible hand" of information in biology. Fox Keller (2011) follows some traces of how information may be recognized as being effective in biological systems, especially on the molecular level.

Davies (2019) provides a recent review of progress toward a general theory of biological information. He makes a strong case for the central relevance of information in organisms and holds that "information does have a type of independent existence and it does have causal power" (47). He sees the reality of information as a fundamental entity in its own right and a key to bridge the principles of physics and biology.

Penzlin (2014, 269) summarizes:

The existence of living systems cannot be explained solely from their material and energetic processes and transformations, by physics and chemistry. Prerequisite for the functional order of living systems, for their organization, is an extensive exchange of information on all levels, within the cells as well as between the cells of a multicellular organism or between individuals. In addition, the ability of living organisms to register certain events in their environment in good time, in order to be able to react in a "useful" way, is one of the necessary characteristics of all organisms from bacteria to humans. Living systems are not only thermodynamic, they are also communicative systems. (translation by author)

Although a generally accepted overarching theory is currently missing, numerous studies can be observed that are expanding our understanding of information concepts in various areas of biology (Auletta 2010; Bawden and Robinson 2020; Farnsworth 2018, 2022; Farnsworth et al. 2013; Kauffman et al. 2008; Kofler 2014; Shapiro 2011; Walker 2014; Walker et al. 2017).

As Baeyer (2003) indicates for physics, there is also an essential distinction for the biological realm between the information content (its "meaning") and the information carrier. The structure of DNA, its chemical components, and the principle of how the information is encoded are well known today. But its meaning is different from that and is not understood by analyzing the carrier. However, studying transcription, translation, and the resulting proteins and their functions can generate more understanding of the meaning as well.

Walker et al. (2017, 2) pose the question: "Is information merely a useful explanatory concept or does it actually have causal power?" From the perspective of living organisms this seems to be the wrong question. Asking for a causal power is a question posed from the perspective of mechanical physics. It is once more the expectation that it might somehow be possible to dissect organic processes in cause–effect relations. Perhaps it is possible to reduce this issue to more fundamental or general questions: Does information bring anything about in organisms? Has it any relevance for function, structure, organization and for generating order? Looking at all the descriptions we have from genetics up to morphology (describing the structural organization of organisms) as well as individual development, the answer is clearly yes— information is of high relevance and influence, even if we cannot identify a *causal* effectiveness at some point. At least modern genetics, but also other areas of recent biological research, would not be possible without presuming information, even if it is still difficult to define it. But what definitely is describable is that *organisms process information*: It is stored and highly organized in DNA; it is used for the construction of proteins, which are involved in building structures, functions, and order in the organism; it is shared and exchanged between cells and between organisms; and it is transferred systematically to offspring. So, it can be said with certainty that organisms process information, similar to their processing of molecules and energy. "Living systems do not merely passively accumulate and store information, they also actively process it" (Walker 2014, 425).

DNA and RNA

Organisms deviate fundamentally from nonliving things by continuously gathering, concentrating, and organizing information. This is not only the case in DNA but also in higher levels of the cell and the whole organism.

The information content of DNA is well studied today, including many molecular details of its transcription and regulation. Open questions are whether this information provides something like a body plan or a blueprint for the whole organism and whether this information has a determining character for features of the organism (Lewontin et al. 1984; Moss 2003; Shapiro 2011; Strohman 1997, 2002; Weiss 2018). Today, the emerging picture of the DNA content makes clear that it provides a certain basic information, which the cell or the organism can use (and modify) for the production of specific proteins in specific locations and functions. The active part seems to be more on the side of the cell and the organism than primarily coming from the DNA itself. It has often been stated that DNA itself is an inert molecule that needs processing and usage by the system. Thus, increasingly formulations about genes (besides all the problems of defining them) has been shifting from gene activity to gene activation (Fox Keller 2000).

In recent years further levels of information have been identified in the field of epigenetics, describing principles such as chromosome condensation, DNA methylation, or histone modifications for example. In addition, some form of informational processing is involved in alternative splicing of mRNAs, which transfer the information from DNA to the cytoplasm. By way of RNA processing, different types of mRNAs and thus different resulting proteins can be generated. Additionally, other components (e.g., micro-RNAs) are generated, which have feedback functions for the regulation of gene expression. There are also extensive feedback functions to the DNA by proteins like transcription factors and so on, and within all these steps a transfer of information takes place. Within a very informative and up-to-date review of recent knowledge in genetics, Parrington (2015) calls this principle "multilayered information" (see also Farnsworth et al. 2013; Morris and Mattick 2014; Shapiro 2011; Weiss 2018).

Davies (2019, 64) writes:

The cell as a whole is a vast web of information management. . . . On its own, the information in the gene is static, but once it is read out . . . all manner of activity ensues. DNA output is combined with other streams of information, following various complex pathways within the cell and cooperating with a legion of additional information flows to produce a coherent collective order. The cell integrates all this information and progresses as a single unit through a cycle with various identifiable stages, culminating in cell division. And if we extend the analysis to multicelled organisms, involving the astounding organization of embryo development, then we are struck even more forcibly that simply invoking "information" as a bland, catch-all quantity, like energy, falls far short of an explanation for what is going on.

Longo et al. (2012) argued against the concept of information in biology because it encouraged genetic determinism. The problem, however, is the assumed determinism, not the notion of information itself. Recent data show that cells, as well as whole organisms, can operate quite flexibly with a given information content (Parrington 2015; Shapiro 2011). This may even be the case during evolution (Jablonka and Lamb 2005; Jablonka and Raz 2009; West-Eberhard 2003).

Versatility in Genetic Systems

Niklas et al. (2015) show that overall the function of genetic components during individual development might be much more malleable than was thought before. A fundamental assumption of developmental biology (the study of factors that are involved in individual development) during the twentieth century has been that gene-regulatory networks (circuits of interacting transcription factors and genes) are primary mechanisms controlling development. According to this assumption, at any time, the relative levels of transcription factors in an extended network determine the progress of development by regulating genes downstream. This conception of gene control in multicellular organisms proposes that gene-regulatory networks are deterministic systems exhibiting multiple stable states, specifying alternative but prefigured cell fates.

Niklas et al. (2015), however, hold that mounting evidence shows much more versatility in these systems. They focus especially on three processes in favor of this argument. One is that alternative splicing produces different protein isoforms from the same precursor mRNA to achieve differential translation. Two other elements of flexibility are that most transcription factors contain domains whose functionalities are context dependent, and that they are subject to post-translational modifications. Consequently, many transcription factors do not have fixed regulatory targets, they argue, so that gene-regulatory networks might not be as determined as thought before. Today, multiscale approaches increasingly explain how gene-regulatory networks operationally interact with the intra- and intercellular environments and thus are subject to change even in the absence of genetic or epigenetic alterations. Modifications such as alternative splicing, post-translational modifications, and context-dependent protein domains complicate gene expression and act synergistically to facilitate and promote time- and cell-specific protein modifications involved in cell signaling and cell fate specification; thereby they disrupt a strict deterministic phenotype mapping by gene-regulatory networks. The combined effects of these processes give proteins physiological plasticity, adaptive responsiveness, and developmental versatility without ineffectively expanding genome size. They also show how protein functionalities can undergo major evolutionary changes.

Overall, these are further hints to the assumption that the developmental result, the growing organism, is generated by the very process rather than predetermined in the genome (see section 4.13.). Consequently, Niklas et al. (2015) hold, that this system constitutes a *splicing code* that defines the protein isoforms and their versatility produced under different cellular conditions.

Embodied Information

Proteins contain information from the translated RNA in the form of the primary structure, and further information may be involved in building the final form of the protein (tertiary structure). Thus, the information of RNA does not vanish when it is translated into a protein. Rather, the information is now encoded—or embodied—in the protein itself. The basis of the specificity of the protein is this encoded embodied information. Structure and function of the protein, which are its organization, derive from its information content.

Expanding this principle, it can also be applied to the whole cell as well as the supracellular order. Basically, all structures of the organism must contain embodied information, whether it comes from the genotype or not. The organization of molecules in space is an embodiment of information. This "embodied pattern information" founds the structural organization of any organism, because the "functional information" is embodied in both the relations among the components (their placing relative to one another) and the possible interactions among them (Farnsworth 2018, 2022).

Farnsworth et al. (2013, 212) maintain: "Information is therefore not just stored in nucleotides: it is the whole biological system that embodies effective information." It can be assumed that such embodied information is present on all levels of the whole organism and is essential for its shape, structure, and composition, as well as for its functions.

Order is only possible by means of embodied information. In a remarkable paper Nurse (2008) stated that among the most pressing needs in biological analysis is the development of appropriate approaches to analyze the management of information flow within whole organisms. "[O]ur past successes have led us to underestimate the complexity of living organisms. We need to focus more on how information is managed in living systems and how this brings about higher level biological phenomena" (Nurse 2008, 224). Besides describing the logic circuits (which he calls "logic modules") that manage information in cells, Nurse states: "We need to know how information is gathered from various sources, from the environment, from other cells and from the short and long-term memories in the cell; how that information is integrated and processed; and how it is then either used, rejected or stored for later use. The aim is to describe how information flows through the modules and brings about higher-level cellular phenomena, investigations that may well require the development of new methods and languages to describe the processes involved" (424).

Nurse expects that the logic modules and circuits are combined into networks, and understanding how such networks operate in cells will be crucial. Complex networks in other areas, he states, have been well analyzed, and connections are often found to have diverse numbers of linkages between hubs in the network such that some hubs become crucial because they are highly connected to many other hubs (see section 4.1). Such networks are called "scale-free." It seems that biological networks derived from genetic, protein–protein and transcriptional interaction studies are also often scale-free. Nurse also expects that it is important to study logic modules on different organizational levels and that "studies at higher system levels are likely to inform those at the simpler level of the cell and vice versa. . . . This is the return to whole-organism and human physiology that many have argued is long overdue, but with a renewed emphasis on the logic of life and the management of information" (426).

Cavalier-Smith (2004) has described some inherited features of animal and plant cells as the "membranome," because lipids are not formed from DNA templates. An organism needs to inherit the membranome, and it comes complete with the fertilized egg cell. The precise composition of the cellular membrane, chemically describable by phospholipids, proteins, glycolipids, and glycopeptides, contains a robust order, which needs some organization. This is also a clear hint at a form of embodied information, however, still difficult to grasp in a precise way.

At present it is unusual to talk about such a form of embodied information. But already the necessity to search for information on further levels than just DNA, as it is done in epigenetics, shows that conceptual expansions are necessary. One can predict that, on all levels of the organism, information content is involved, forming its order and its functions. Knowledge of the principles that organisms use processing their information, and knowledge about embodied information on different levels will be essential in the future (Davies 2019; Kofler 2014; Nurse 2008; Walker 2014). In the same sense, Farnsworth et al. (2013, 205) formulate: "Every aspect of life may be regarded as a product and elaboration of the physical world, clearly made of the same matter and energy, ordered in space and time as is every physical system. What makes life special is not the material brought together to take part in living, it is the functional information that orders matter into physical structures and directs intricate processes into self-maintaining and reproducing complexes." Farnsworth et al. propose the description of a "functional information content" concerning patterns of molecules, cells, tissues, and other components.

Kofler (2014, 275) writes: "If it is correct to explain the movement of matter with an (unobservable) ability to deal with energy (power, fields) then it is also correct to explain the attribution/shift of information to matter with an ability to deal with information."

Further Information Systems

No cell is able to live permanently in isolation from its environment. It not only needs the exchange of substances and energy, it also must be able to receive and to process signals from its environment. This is already the case in single-celled prokaryotes and eukaryotes, which must react to changes within their surroundings. The bacterium *Escherichia coli*, for example, not only reacts to the pH-value and to the temperature of its environment but also to fifty different substances, having just four types of receptors (Ames et al. 2002). Many bacteria release signaling substances to their environment, which are perceived by bacteria of the same species and which can react to them. The variety of substances and pathways is bewildering.

In multicellular organisms the intercellular communication reaches a new dimension. The "holistic acting and reacting" (Penzlin 2014) of the cells involved needs a highly complex network of coordinated interactions between cells and organs. This transfer of information between cells can be carried out by chemical substances, thus becoming signals, which in most cases are transmitted via receptors to the different cells. Hormonal systems are systems of signals, which regulate and modify body functions and homeostasis.

Another major information system within organisms such as multicellular animals includes the nervous system, which is specialized in gathering, processing, and submitting information. Neurons and especially more complex nervous systems increasingly allow the organization of information for more and more complex realizations of biological persistence and functions. The immune system also works essentially with information about which substances are from the own body and which are foreign.

Reproduction is characterized by an inter-generational transfer, not only on the genetic but also the epigenetic level. And then there are all the information systems between different individuals. For example, information flow is essential within social insects such as ants or bees. The well-known bee dancing is just one out of endless examples that could be mentioned, and which are being increasingly described today in all their amazing complexity.

There is no need to describe these systems in more detail because the matter is well known from textbooks. However, the overall picture emerges that organisms exist within and by a tremendous interwoven net of information on different levels and with different implementations. Biosemiotics studies sign processes between the organism and its environment as well as within the organism. This may also be a perspective on information processing, with the attempt to understand the very content and meaning of the information in these processes (Hoffmeyer 2008, 2013; Sharov and Tønnessen 2021; see section 4.10).

Environmental Information and Choice

Grisogono (2017), writing from the perspective of a physicist, sees the capacity for autonomous agency as the hallmark of any living system. An autonomous agent, she describes, must be able to make a choice between at least two actions, in the sense of exhibiting different and appropriate behaviors under different circumstances. A living system has the ability to sense opportunities and threats and choose appropriate actions to preserve its existence and promote its reproduction.

But what exactly is happening when an agent makes a "choice"? Obviously, a choice implies that more than one action is possible. From a physics perspective this might suggest a system that was in metastable equilibrium, like a pencil balanced vertically on its point. It could fall in any direction, so many "actions" are possible. However, such a system would be at the mercy of noise, random fluctuations that would determine in which direction it "fell"—hardly a candidate for autonomous choice in its own interests! The key to autonomous choice is not in equilibrium physics but in that the agent must be a far-from-equilibrium system, poised in such a way that the internal states corresponding to different actions being initiated are reliably inhibited yet able to be triggered by essentially tiny signals that bear some correlation with a relevant aspect of the environment. Tiny, because it is not the energy or magnitude of the signal that produces the effect but the specificity of a pattern that can be interpreted as meaningful. Ambient light suddenly occluded provokes fleeing; sensing of a food molecule stimulates feeding; a particular sound pattern triggers freezing.

It is not hard, Grisogono elaborates, to imagine how such linkages between what can be sensed and what action is thereby triggered can evolve, and further to speculate how such linkages might in principle gradually evolve to become more complex and incorporate more conditionals—in other words, to evolve more information-processing capability—to provide more "intelligent" nuanced choices. Such an evolving complex system is characterized by a network of interactions through which inputs are processed and connected to the action mechanisms, and by displaying both high selectivity and sensitivity to certain inputs and robust insensitivity to others.

Davies (2019, 126) formulates: "Life may therefore be described as an informational learning curve that swoops upwards."

Until now, organisms have been described only rudimentarily from the perspective of information processing, although the term has been in everyday use since the 1950s. It is remarkable that molecular biology, which is expected to provide a firm chemical and physical basis for understanding life, introduced the notion of information early on and thus already relinquished a purely materialistic explanation. It became increasingly clear that information processes are essential in all life functions. With information, a new quality emerged that does not exist in the physicochemical terminology, which focuses on material interactions, atoms, molecules, and crystals, and about forms of energy and their transformation. Several authors noticed this situation, including Eigen (1987), Fox Keller (2011), Penzlin (2014), and Shapiro (2011).

Shapiro (2011, 4) even sees in recent developments a general change in biology toward an informational understanding of life: "The contemporary concept of life forms as self-modifying beings coincides with the shift in biology from a mechanistic to informatic view of living organisms."

4.7 Processing of Energy

A central property of organisms is that they process and transform energy to maintain the low-entropy state that characterizes their organization. As described in the section on autonomy (4.3), organisms generate and actively maintain a metabolic autonomy, which establishes a disequilibrium with respect to the surroundings. This disequilibrium concerns several factors such as molecular concentration (organisms concentrate and organize specific molecules within their boundaries), information (as described in the preceding section), and also energy. The energetic gradient toward the environment is highly regulated, as are most other factors as well.

Disequilibrium

Organisms have often been described from the perspective of thermodynamic disequilibrium, which especially focuses on the use and transformation of energy (Kauffman and Clayton 2006; Penzlin 2014; Prigogine and Stengers 1984). Thermodynamic equilibria occur spontaneously. In contrast, every living organism permanently prevents itself from relapsing into such a state of equilibrium.

This thermodynamic disequilibrium involves two aspects. First, energy-rich molecules are concentrated and integrated within the structures of the organism. Organic molecules are reduced compounds and therefore rich in energy. They are stabilized against the hydrolyzing and oxidizing influences from the environment. Second, energy for current usage is permanently provided in a flexible form by the cells to perform their work. This condition differs fundamentally from anything in the inorganic realm.

To prevent a relapse into a state of equilibrium, the organism does not just prevent degradation, but rather permanently balances degrading and synthesizing processes (figure 4.15). This balance and the resulting disequilibrium are not a fixed state, which exists for a certain time once it is established. Rather, this balance needs to be generated constantly. Therefore, its existence is rather a process than a thing (Nicholson and Dupré 2018).

Every organism exists in such a permanent process of degradation and synthesis, self-generation, and self-renewal. Even the degradation is an active process. Thus, most molecules and most organelles within the cell are permanently going through a process of renewal and replacement. Also, many cells within a multicellular organism (depending on the cell type) are being permanently replaced. Older or defective cells within many tissues are removed by the active process of apoptosis, whereas new cells are permanently being delivered by mitosis.

This is completely different from any human-made machine. With a machine, energy is just needed to operate it. Dupré and Nicholson (2018) describe that a car cannot function without fuel, but its existence and its structural integrity is not compromised when it is deprived of fuel. An organism, in contrast, is always acting, because it must remain permanently displaced from equilibrium if it is to stay alive. It is possible to leave a typewriter in an

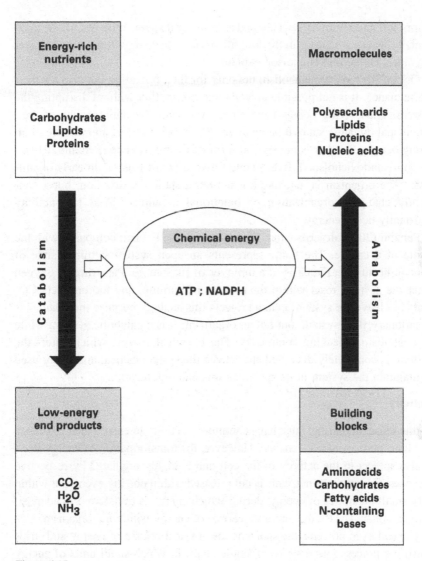

Figure 4.15
Balancing of anabolic and catabolic processes within the organism. *Source*: Republished with permission of Springer, from Penzlin (2014, 200); permission conveyed through Copyright Clearance Center, Inc., translation by author.

empty loft and return a year later and start using it again. Whereas, if one accidentally leaves a hamster in the loft, after one year there will not be a hamster anymore, Dupré and Nicholson explain.

Organisms have a metabolism not only for their operating but also for their maintenance. It is not possible to stop their energy flux without disrupting the system. Thus, living beings need energy even for the maintenance of their processual dynamic state of being alive. And this is not just an attribute of all living beings, it rather is the special form of their existence (Penzlin 2014).

Dupré and Nicholson (2018, 9) cite Edward Stuart Russel, an early organicist: "The organism is not, like a machine, a static construction, but a constantly changing organization of functional activities." And this activity constantly needs energy.

Penzlin (2014) describes that life has occasionally been compared with the flame of a candle. The flame represents an open system within a state of nonequilibrium on the basis of a turnover of substances. Paraffin and oxygen enter the system, react within the process of burning, and the products CO_2 und H_2O leave the system. Penzlin rejects this analogy because the "flame" is a stationary flow system, but not an organism. It is a catabolic system, while an anabolism is lacking completely. The chemical energy, which enters the system, is completely degraded and leaves the system as heat and is not used to maintain the system in its specifications and organization.

Activity

In this sense, organisms must have continuous activity to ensure their existence and their permanent self-renewal. However, the transformation of energy itself is also subject to the activity of the cell and is highly regulated by respective enzymes. Energy from nutrients is not released within one big event, but within very small quantities of energy, during which energy is transferred to adenosinetriphosphate (ATP), the universal carrier of energy within the organism.

The cell even prevents the spontaneous physical release of energy and strips down the process into a series of single steps, in which small units of energy become available. Contrary to a combustion, this does not happen within one big incident but is unfolded into many steps, which are controlled by enzymes. This guarantees that the energy that is released during the process is not just exploding, leaving nothing but heat. Rather, the major portion of the energy can be transferred in small amounts to carriers of energy such as ATP and some other compounds. This is what happens during glycolysis, in the citric cycle, and especially within the respiratory chain (oxydative phosophorylation). The transiently stored energy can be used in an extremely large variety of energy-consuming processes within the cell. Energy in such small quantities, which can be used well-directed and precisely within the organism, is beyond anything within the purely chemical and physical nonliving world. In addition, ATP is the connecting link between catabolism and anabolism.

Obtaining and dealing with energy is part of the agency of organisms (see section 4.4). Organisms are only to a lesser extent subject to passive energy fluxes. Plants spread out their surfaces to collect energy from sunlight. Animals search for nutrients, from which they extract certain molecules as well as energy. Major processes within the cell and within the whole organism serve the distribution and availability of energy. The agency concerning this dealing with energy is described here with the term *processing of energy.*

The cells of a multicellular organism are supplied with nutrients, but the transformation of energy into utilizable ATP is performed by the cell itself. There is no transfer of ATP between the cells. Yet the cell itself needs to perform the distribution of ATP within the cell. Within the inner membrane of the mitochondria, there is an ADP/ATP-translocator that provides the cytoplasm with ATP from the mitochondria and rapidly takes back ADP to the mitochondrial matrix so that it can again be used to generate ATP. This translocator works very fast and belongs to the active processes of metabolism. While only small amounts of ATP can be stored within the cell, its turnover rate is much larger. Within the human body, each molecule of ADP/ATP switches several thousand times per day between the mitochondrium and the cytoplasm. It has been estimated that a resting human has a turnover rate of ATP of about 1.7 kg per hour. During intensive exercise this amount can increase to 30 kg per hour. This again has a clear processual character.

This system of energy transfer by ATP/ADP within the cell is maintained on a level that differs from the thermodynamic equilibrium by the factor of 10^8 to 10^{10} (Penzlin 2014).

Organization and Neg-Entropy

The term *entropy* has been used in quite different scientific disciplines with different definitions. A simple understanding has been that the increase of entropy describes an increase in molecular disorder, which has been used especially in the thermodynamics of closed systems in physics (Davies 2019). The spontaneous changes within a closed system occur in the direction of increasing probability, and the thermodynamic equilibrium represents the macroscopic state of the highest probability.

Concerning living organisms, it has often been stated that organisms work against the general tendency toward entropy and thus somehow violate a fundamental physical law. Generations of authors have tried to solve this apparent conundrum. I do not intend to discuss these deliberations here;

I rather want to follow the description of "organization" by Heinz Penzlin (2014, 241 f.), which is quite the opposite of entropy and also has been called "negative entropy" (or "neg-entropy"). In chapter 3, I already discussed the principle of organization in a general manner. Here I want to continue this topic from the perspective of entropy.

Contrary to the tendency within the inorganic realm, Penzlin describes, there is an obvious tendency within the organic realm toward heterogeneity, diversification, differentiation, and lability. Life produces imbalances. Living beings are not only systems with a high potential energy compared to equilibrium systems of the same chemical composition, but represent systems of a high degree of order, which is maintained against all disturbing influences. The metabolism signifies an internal functional order, being oriented toward the preservation of the whole, which is called organization.

In physics, in which questions about function and purpose are irrelevant, the term *organization* in this sense is unknown. Penzlin claims that it is highly misleading when physicists use the term *organization* in the context of "self-organization," as they only consider the spontaneous generation of order from disorder under certain external pressures but not the autonomously generated state of a functional (teleonomic) organization.

Organisms not only represent organized systems but are self-organizing in a true sense, by means of which they generate a far-reaching autonomy. Environmental factors can influence the functions of life or even destroy them, but they cannot determine or cause their intrinsic processes. The term *self-organization* has been used in physics in the context of so-called dissipative structures, which do exhibit forms resembling organic forms. The phenomena seemed to promise an insight into how ancient and primitive organic forms might have been generated. It has been celebrated as a new paradigm, which resides within the universe. However, during the spontaneous creation of a certain order within dynamic systems away from thermodynamic equilibria, as for example in Bénard cells and in Belousov-Zhabotinsky oscillations, there is no action by the system itself. There are specific changes forced by physicochemical conditions, leading to some figures and structures, but no "self" and no "organization" is involved, as in living beings. Here, Penzlin sees a fundamental difference. Only living beings are really self-organizing, in the strict sense of the word. In living beings, order is not generated from chaos, but continuously order arises from order (Penzlin 2012).

Penzlin (2014) further asserts that living systems are not different from inorganic objects by having a different physics or chemistry, or having special substances, but by autonomously maintaining an intrinsic organization. Such organized systems generate new principles, which need to be studied in their own context. Organismic systems are organized in relation to a goal, to which everything within the system is subordinated. One goal is the self-maintenance of the condition of being alive.

Organization is necessarily a qualitative concept. It needs structure and information for its specification. Penzlin cites Jeffrey S. Wicken, who defined organization as "informed constraint for functional activity" (243).

Penzlin contends: "In the concept of organization are integrated all those properties, which characterize living systems as unique within our natural world, namely an autonomous, holistic behavior, i.e. purposeful behavior, purposeful in the sense of self-preservation. The organization provides the internal natural laws, which are necessary for the existence of living beings and determine their specific existence" (243, translation by author).

Bertalanffy (1952, 12) formulated it succinctly: "The problem of life is the problem of organization."

Earth as a System Far Away from Thermodynamic Equilibrium

An interesting and related effect is that life is the main source of generating a thermodynamic disequilibrium concerning Planet Earth as a whole. Kleidon (2010) describes that throughout earth's history, life has increased greatly in abundance, complexity, and diversity; at the same time, it has substantially altered the earth's environment, evolving some of its variables to states further and further away from thermodynamic equilibrium. For instance, concentrations in atmospheric oxygen have increased throughout Earth's history, resulting in an increased chemical disequilibrium in the atmosphere as well as an increased redox gradient between the atmosphere and the earth's reducing crust. He proposes to consider Planet Earth as a hierarchical and evolving nonequilibrium thermodynamic system, which has been substantially altered by the input of free energy generated by photosynthetic life.

With the generated free chemical energy due to photosynthesis, organisms can perform the work necessary to build persistent structures. Plants, for instance, develop leaves, stems, and roots. These persistent structures, in turn, enhance the ability to absorb solar radiation and—on land—enable trees to reach water deep in the soil. Therefore, organisms affect the extent to which incoming solar radiation is absorbed and the rate at which precipitated water is transpired back into the atmosphere, both affecting the prevailing environmental conditions surrounding the organisms. Furthermore, these two examples affect the environmental conditions of the photochemical processes of the organisms (e.g., the leaf temperature due to the latent heat of water) but also the environmental conditions at large, such as the water vapor concentration in the atmosphere.

Kleidon further explains that merely these two physical effects—the increased absorption of solar radiation at the surface and the increased ability to recycle water into the atmosphere in the presence of terrestrial vegetation—can already substantially modify the physical environmental conditions. At the local scale, these differences can be felt on a hot summer day when comparing the cool moist air in a forest to the hot and dry conditions of a parking lot. On larger scales, these effects modulate temperature and continental moisture recycling. Kleidon presents climate model simulations for a "green planet"—a world where rainforests were planted everywhere on land, contrasted to a "desert world," in which the effects of terrestrial vegetation were removed (Kleidon et al. 2000; Kleidon 2002). When evaluated in terms of what these contrasting physical conditions imply for biotic activity, it was found that the conditions of the green planet, as well as for the present-day, allow for substantially higher productivity. Such self-enhancing effects could be understood in evolutionary terms as a result of a "feedback on growth," as Kleidon proposes.

These examples are only a few of the many effects by which life alters the environment, which in turn alters the abiotic conditions in which life operates. These principles led Lovelock and Margulis (1974) to use the metaphor "superorganism" to describe our Planet Earth (GAIA hypotheses; see Ruse 2013).

The Spatiotemporal Organization of Matter, Information, and Energy: A Preliminary Synthesis

Having discussed now some central properties of living beings, the notion of process concerning organisms can be reconsidered. In chapter 3 I basically agreed with the approach of Dupré and Nicholson (2018), which moves the process view into the center of the notion of the organism. The organism, they say, is mainly process rather than substance. Substances and things are derived from processes, they are transient patterns of stability in the surrounding flux, temporary eddies in the continuous flow of process.

Here I want to expand this approach to a certain extent by placing the emphasis somewhat differently, as already indicated in chapter 3:

1. Organisms are characterized mainly by activity, as discussed in section 4.4 under the term *agency*. This autonomous activity is not in accordance with a general processual view of the world, which would include many nonliving but changing entities. Nonliving things lack the principle of autonomous *activity*, which is generated by organisms. A process can be passive; it can, in principle, be driven by some external influences or causes. Therefore, I would prefer to emphasize activity or agency, rather than process, for organisms. The alternative

would be to speak consequently about "active processes," which Dupré and Nicholson of course do in most formulations. However, activity, agency, or active processuality are only different words for the same principle, which is central for all living beings. The essential point is this: life is activity.

2. The organism is also substance. Of course there are molecules, and of course they build up physical entities, which we call their physical bodies. Thus, an organism is not only a process; at the same time, it is a thing that has a presence in the real world. It is an embodied entity, not only a process in a strict sense.

At the same time organisms contain energy as well as information. Thus, the existence of the organism includes substances, energy, and information, and all three are processed continuously by the active organism. *A central characteristic of an organism is its permanent processing and organizing of molecules (substances), energy, and information.* Neither molecules, nor the information content (as, e.g., from DNA) can take over the active part and direct, or even determine, an organism. The activity comes from the entire organism (minimally an intact cell), which processes molecules, energy, and information.

3. In chapter 3 I discussed that regarding process and change as the main principle that reigns in organisms, there might be a lack of a tool to understand and to describe the order of organisms. Every organism has a structure, a form, and an identity, even if these features are seen within constant dynamic changes. To realize that they are in flux does not abolish the existence of an order, in which they are dynamically arranged.

I argued that concurrently there are processes, in which everything changes permanently on the one hand, and continuities, at least for a given time, which make up the properties of each respective organism, even if these properties can change in the long run. If only the process aspect would be taken into consideration, there would be a theoretical neglect of the principle of *organization*. Organization, as discussed earlier, describes the structures, forms, and recurring features of organisms. It is the basis of morphological and physiological descriptions. According to this aspect, it is neither enough to describe the parts and molecules of an organism, nor just to describe the permanent changes of the parts and components as such. An essential question is how these parts and these processes are organized—how the components have been put together and how the changes are constrained in a particular way (figure 4.16).

Also, I argued that the key to understanding the relation between process and organization is to comprehend that both principles apply simultaneously at any given time: there is an organization, which is maintained by an active process. The process generates the organization, and the organization structures and directs the process. Neither is possible without the other. Processual

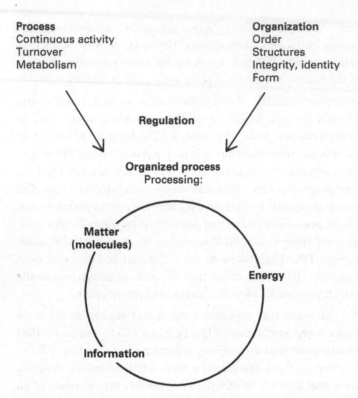

Process   **Organization**
Continuous activity   Order
Turnover   Structures
Metabolism   Integrity, identity
                                               Form

                        Regulation

                     Organized process
                       Processing:

              Matter
             (molecules)

                                          Energy

             Information

Figure 4.16
Process and organization.

turnover means that the very structure of every organism, its organization, unlike that of any machine, is wholly and continuously reconstituted as a result of its own active operations. Organization enables identity and order, and the constantly running process generates the organization.

In my view a coherent picture emerges when the following claim is made: A living organism is an organized, active process or a continuous activity, and what is being processed are substances, energy and information.

This is what biology is all about: biology studies substances such as molecules or tissues in their physiological dynamics; it describes how organisms make energy available and use it; and it describes how information is transferred from DNA to proteins, for example. Also, biology studies the bewildering varieties of forms, shapes, colors, types of adaptation, and so on, which are all expressions of the organization of the organisms. However, there is a fundamental difference to see all this up in front from the perspective of the process rather than (more or less tacitly) assuming that substances (today, mostly molecules) or genes are the hidden drivers of all these events.

In a slight extension of a citation from Dupré and Nicholson (2018, 17) it would be appropriate to say: Organisms, despite their apparent fixity and solidity, are not *only* material things but *concurrently* fluid processes. They are metabolic streams of matter, energy *and information* that exhibit dynamic stabilities relative to particular timescales. As *active* processes, and unlike *nonliving* things or substances, organisms have to undergo constant change to continue to be the entities that they are. (The words in italics are my supplements.)

4.8 Processes of Shape

Another specific property of organisms is their overall shape. Every organism and every part of an organism has a specific and characteristic appearance and manifestation, a form, a design. It is its spatial organization, being very specific for the respective species, but it also includes individual variations. Shapes are determined by their proportions, by the relationship of the parts to each other and within a superordinate overall composition.

The Importance of Shape

Different words are in use for this property, such as *organismal form, design*, or *design principles*. Often the German word *gestalt* is used. However, these terms indicate different meanings and different scopes of what they express. There is hardly any consensus about the respective content of these terms, so preliminarily they will be used in approximately the same sense to describe the overall appearance of the organism and its components, including their modifications.

In nature, organic forms, shapes, and gestalts show such a bewildering and impressive abundance that they need to be regarded and reflected in their own reality and relevance, as well as in their relation to other organismal properties. The seemingly endless variety of forms and shapes in nature belongs to the specific properties of life and possibly to one of its still unresolved enigmas (Goodwin 2000).

Primarily, shape or gestalt describes the exterior appearance. But it can also include the systematic and characteristic relationship of all distinguishing features of an organism, including all lower levels and substructures, which are always in a specific relation to the external appearance of the respective organism. Vertebrae exhibit a certain shape in close relation to their specific function within the spinal column, as well as a heart, a brain, or an epithelial cell. On the molecular level the form is crucial, such as in membranes or chromatin organization. The structure of a folded protein is critical for its function. Only if a protein forms the channel for sodium ions at exactly the appropriate location with exactly the appropriate charges will it perform its function as a sodium channel. Here also the form is crucial.

The appearance of shape also includes the functional differentiation of each organism, that is, its division into areas of specific activities and their respective manifestations. This is already the case in unicellular organisms, in which organelles perform specific tasks. In multicellular organisms, it is then organs that take over functions of respiration, digestion, or reproduction. The differentiation of the nervous system and blood vessels emanates the whole organism at the same time but reveals their specific activities. And externally, the whole organism is divided, for example, into head, neck, trunk, and tail. Their relationship to the whole is expressed in the fact that, although they are centers of activity, they radiate in different ways into the whole organism and are in turn influenced by it.

Shape is constantly changing during the lifetime of an organism. There are, for example, profound changes during the development from an egg to an adult animal, sometimes even including a profound metamorphosis such as in a caterpillar, which transforms into the imago of the butterfly, acquiring completely different shapes and appearances. Likewise, the shape of a tree is constantly changing when it grows year by year, starting from a shoot. The same is also true for an annual flower starting with its embryonic leaves, continuing with the leaves along the stalk, which themselves change in form and size, finally generating a blossom, a miracle which we can observe year by year in millions of specimens. When an adult animal or a human grows old, they again experience changes in their overall shape as well as in details. Therefore, shape basically is itself a continuing process of spatial changes, conversions, and degradations. What we perceive in a specific moment is merely a snapshot of these longer or shorter processes.

Morphology is the biological discipline that studies these phenomena. It studies the overall form of organisms as well as the setting, position, and order of organs and body parts. Comparative morphology has been very successful in analyzing phylogenetic relations, today being supplemented by cladistic methods and comparative genomics. In a wider sense the microscopic study of tissues in histology is morphological, because tissues have very specific forms of spatial organization.

Also, classical (comparative) embryology is a component of morphology. It studies different stages of the individual development, to describe the

sequence of morphogenesis and the principles of assembling and construction, as far as perceivable by structural changes. The knowledge of intermediate stages can be used for finding homologies between species and
groups. In addition, the study of malformations is a part of morphology
(teratology).

Goethe was among the first to use the term *morphology* for the study of
biological forms (Goethe 1981). He conceived form as a pattern of relationships
within an organized whole, a conception that is again at the forefront of
systems thinking today. He understood the term more in a dynamic, processual
sense, which could also be coined *morphodynamics*. This is in accordance with
the more recent processual view of life in general.

However, not everything that possibly appears is called a shape in this sense.
Pathology presents many structures that do not exhibit a shape, which is integrated into the overall building principles of the organism. Such structures
apparently lack the necessary information for subordination into the context
and functionality of the system. If the tumor of a sarcoma is compared with
the image of a healthy connective tissue, it has the appearance of a loss of
integration and regular shape. Such degeneration makes pathology possible
and shows how our understanding of the living shape depends on the presence
of not just any order but of a specific order.

Form and matter are not identical. If the form of an organism is destroyed,
only matter is left, as a formless mass, which demonstrates that form contains
a nonmaterial property. The term *form* is included within the word in-*form*-
ation. Obviously form expresses embodied information (see section 4.6).

Paul Weiss (1962) demonstrated this principle by a rather drastic example
(figure 4.17). He showed in a figure, side by side, a 6-day-old chick embryo
immersed in liquid, before (a) and after (b) having been homogenized by
crushing. As no substance has been lost or added during the procedure, Weiss
explains, the content of the vial before and after is the same in weight and
composition. An inventory of molecules, if it could be taken, would likewise
reveal no change. What has been lost is structural organization from the
highest level of the organism down to the order of whatever subsystems were
small or consistent enough to have escaped the disruptive force of the crushing
technique. He describes that organs and tissues have been broken up, so also
have the individual cells, their membranes, nuclei, and cytoplasmic systems.
"To get from a to b was easy. How to return from b to a—that is our uphill
problem" (Weiss 1962, 10). Subjecting the homogenate to centrifugation or
some other separatory measures by which the molecular scramble can be partly
unscrambled into distinct fractions (c) is no step back toward the lost order,

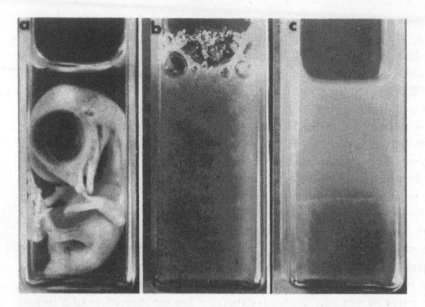

Figure 4.17
Chick embryo (a) intact, (b) homogenized, (c) fractionated. *Source*: Republished with permission of McGraw Hill LLC, from Weiss (1962, 9); permission conveyed through Copyright Clearance Center, Inc.

but rather a step toward a new and artificial order. The different components are all there, but the specific structural order, on which the functional capacity of the whole assembly depends, is lacking.

In biology, form and shape are used to acquire information about an organism. Taxonomy, for example, relies on this type of information. Thus, there is a huge informational content within the exterior appearance of organisms. We discern organisms according to their overall shape and can distinguish between a goose and a crane, between a horse and a donkey. A field guide summarizes this information, using pictures as well as descriptions. The informational content is transferred into words and drawings. However, the content itself is taken from the object itself. The biological object exhibits its form, its shape, its colors, in short, all its morphological characteristics.

Speciation during evolution is regularly linked to subtle changes in forms, the beaks of the Galapagos finches being a classical example. Thus, evolution deals with forms and transition of forms, again stressing the relevance of the spatial organization.

The Origination of Organismal Form

Müller and Newman (2003) make a strong case for the origination of organismal form. They state that it has been undervalued in evolutionary theory for a long time. In their introduction they describe how evolutionary biology arose from the age-old desire to understand the origin and the diversification of organismal forms. During the past 150 years, the question of how these two aspects of evolution are causally realized has become a field of scientific inquiry, and the standard answer is by "variation of traits" and "natural selection." The modern version of this tenet holds that the continued modification and inheritance of a basic genetic tool kit for the regulation of developmental processes, directed by mechanisms acting at the population level, has generated the panoply of organismal body plans. This notion is superimposed on a sophisticated mathematically based population genetics, which became the dominant mode of evolutionary biology in the second half of the twentieth century. Other major branches of evolutionary biology have concentrated on patterns of evolution, ecological factors, and, increasingly, on the associated molecular changes, whereby the preoccupation with the "gene" has overwhelmed all other aspects, and evolutionary biology today has become almost synonymous with evolutionary genetics.

These developments, Müller and Newman (2003) describe, have edged the field away from the original question, the origin of organismal form and structure, and what is being generated in a qualitative sense. They search for a plurality of causal factors responsible for the origination and innovation of organismal form. This is especially relevant because there still is a void of understanding of the relation between genomic structure and the phenotype (Parrington 2015; Shapiro 2011).

Müller and Newman pose a series of questions that arise from a focus on the evolution of organismal forms, that is, morphological evolution. Why, for instance, did the basic body plans of nearly all metazoans arise within a relatively short time span? Assuming that evolution is driven by incremental genetic change, should it not be moving at a slow, steady, and gradual pace? And why do similar morphological design solutions arise repeatedly in phylogenetically independent lineages that do not share the same molecular mechanisms and developmental systems? And why do building elements fixate into body plans that remain largely unchanged within a given phylogenetic lineage? And why and how are new elements occasionally introduced into an existing body plan? Some of these fundamental questions are presented in table 4.4.

Table 4.4
Open questions concerning morphological evolution

Burgess shale effect	Why did metazoan body plans arise in a burst?
Homoplasy	Why do similar morphologies arise independently and repeatedly?
Convergence	Why do distantly related lineages produce similar designs?
Homology	Why do building elements organize as fixed body plans and organ forms?
Novelty	How are new elements introduced into existing body plans?
Modularity	Why are design units reused repeatedly?
Constraint	Why are not all design options of a phenotype space realized?
Atavisms	Why do characters long absent in a lineage reappear?
Tempo	Why are the rates of morphological change unequal?

Source: From Müller and Newman (2003, 5).

Müller and Newman (2003) state:

Many of the phenomena on which these questions are based bear classical names . . . , giving the issues a seemingly old-fashioned aura. But hardly any of the problems specified by this traditional terminology are explained in the modern theory of evolution. Whereas the classical questions refer to phenomena at the organismal level, most can also be applied to the microscopic and even to the molecular level. All are linked by one common, underlying theme: the origin of organization. The nature of the determinants and rules for the organization of design elements constitutes one of the major unsolved problems in the scientific account of organismal form. (4)

Since the publication of their book (and similar ones), there is an increasing focus on these topics again. Thus, questions have been raised as to how much the phenotype itself may contribute to evolutionary changes, not driven by primary genetic variations (West-Eberhard 2003). Further questions inquire how much the activity of the whole organism contributes to evolution (Walsh 2015) and how much change in developmental trajectories during embryogenesis may lead to alternative phenotypes (Arthur 2004, 2011; Carroll 2005). These changes in research focus have been called "the return of the organism" (Nicholson 2014b). Some aspects of this new evolutionary approach have been discussed in chapter 2.

The Gestalt Concept

Breidbach and Jost (2006) discuss some aspects of the gestalt concept in morphology and emphasize that a gestalt is more than a collection of patterns. The gestalt concept expresses what is common in a collection of patterns and what characterizes them. They argue that even two individuals from the same species may exhibit rather drastic variations, describing the example of the different races of domestic cats. There do exist systematic morphological criteria allowing for a distinction between feline and canine carnivores, that is, between cats and dogs. These criteria, however, do not consist in single isolated properties. The definition of their basic types is more subtle and more fundamental. Certain relative criteria exist, drawing upon relations between sizes of different body parts, of their form, or upon their relative position within the whole body.

They describe that D'Arcy Thompson's attempt to capture the corresponding gestalt properties in a mathematical framework did not transcend the intuitive level. D'Arcy Thompson demonstrated how to transform the shape of organism X into the one of organism Y by a simple deformation. He used the exterior shape of the body of fish that, via stretching and deforming, he transformed into different forms (Thompson 1917). However, according to Breidbach and Jost (2006), he could argue only in terms of visual patterns and he did not achieve an analytical framework underlying the pictures he used for the illustration of his ideas. Although this work was important for the Anglo-Saxon tradition of the analysis of form, pattern, and gestalt, Breidbach and Jost argue further, the consequences of that situation presently show up, for example, for a biological morphology that conceives of gestalt as the result of a process of development and thus tries to define that gestalt by means of the mechanisms of development. Here, these form-building processes are not conceived as geometric structures, but as ontogenetic processes that can be analytically captured only by the methods of molecular genetics. Morphology is thus reduced to developmental genetics which, however,—and here we come full circle—presupposes a concept of structure. Thus, also under those assumptions, the gestalt concept cannot be replaced by a presentation of the gestalt forming process. So far, an alternative analytic gestalt concept has not been developed in morphology, although it seems to be crucial for the description of whole organisms and their organization.

Breidbach and Jost (2006) define "a gestalt as the invariants of a collection of patterns that can mutually be transformed into each other through a class of transformations encoded by, or conversely, determining that gestalt" (23).

Their definition contains two complementary aspects. One aspect (mainly in their words) is that different representatives of one and the same gestalt can be transformed into each other according to certain specific rules. Thus, a gestalt defines a similarity or equivalence class of patterns. The other aspect is that a specific gestalt also determines what is invariant under those transformations. This results in more than arbitrary collections of similarities. The transformations then identify what is typical for the collection of patterns representing the gestalt. Since each pattern is transformed as a whole, those invariants can only consist in internal relations.

Breidbach and Jost use geometric figures as easy examples. On the one hand, the gestalt of a unilateral triangle can be defined by exhibiting concretely one such triangle and defining the gestalt as the class of all figures that can be obtained from that given triangle through specific transformations. On the other hand, a unilateral triangle can be defined as a triplet of three points with mutually equal distances between them. These distances, however, are the invariants of the aforementioned class of transformations. The transformations and the invariants are dual aspects of the same gestalt. Thus, we can comprehend a gestalt either through the comparison of different patterns or through invariant relations within one pattern. Thus, to determine a gestalt, we can either describe the rules by which we can change a given pattern or describe which relations between the elements of that pattern must remain invariant.

The transformations capture how a pattern can be changed without affecting its relevant structure, and they thus allow for an exploration of the patterns representing the gestalt. The approach demonstrates what is not changed under those transformations, that is, the structure characteristic of the gestalt. The invariants are therefore relative relations, not absolute quantities. The individual elements of a pattern are undetermined, but not the relations between those elements. Thus, the gestalt is not a function of the sum of whatever patterns belong to it, but rather designates a structural property of the patterns in that class.

A more difficult endeavor is, however, to apply these considerations to organisms. They may contribute some general principles and represent a starting point, but organisms are much more complex than a triangle.

As shape, the gestalt in all living beings obviously is so important and decisive that the development of a science of form and shape is of central importance. Of course, this has been dealt with in many areas of biology, especially along comparative approaches. However, in recent years it has been pushed into the background. But in paleontology, for example, such comparisons are crucial to find patterns, trends, and evolutionary transitions (Carroll 1997; Clack 2012; Kümmell and Frey 2012; Kümmell 2020; Schad 1992, 1993; Shubin and Marshall 2000; Shubin 2008). A more general theory of gestalt, however, might be a further impetus to put the organism back into the center of inquiry (Schad 2020). Such a theory, however, probably needs much more elaboration. As it becomes clear today that the organism is not fully explainable by the analysis of the genome or the sum of molecular processes, the principles that rule the organism as a whole and how changes and transformations take place might regain interest in the near future.

Müller (2003) proposes that homology is a key concept for studying the organization of morphological order. The evolution of organismal form, he describes, consists of a continuing production and ordering of anatomical parts, and the resulting arrangement of parts is nonrandom and lineage specific. He advocates that homology is not merely a concept or a conceptual tool, as it is often understood, but rather the manifestation of morphological organization processes and thus a central feature of organismal evolution, whose explanation requires a theory of morphological organization. Such a theory will have to account for (1) the generation of initial parts, (2) the fixation of such parts in lineage-specific combinations, (3) the modification of parts, (4) the loss of parts, (5) the reappearance of lost parts, and (6) the addition of new parts. Müller describes how the continuous production and fixation of parts has resulted in the thirty-seven or so presently known extant body plans, and even more existed in the past. All minor clades are modifications of those major body plans, each characterized by distinct, hierarchical combinations of homologues, altogether represented in several million species. It is especially interesting that evidence increases that homologues remain constant over long, phylogenetic periods despite significant changes in the molecular, genetic, and developmental mechanisms that execute their realization. This indicates, Müller holds, that the position of homologues in the organizational hierarchy of the phenotype is more important than the pathways of their construction.

Form and Function

If the anatomical properties of a species are studied not only to describe them or to compare them with other species, but rather to understand and to explain them, it becomes obvious that the majority of structures and forms, and the whole of a gestalt, have a strong relation to their function. The shape of a bird is only understandable in relation to its way of living. On all levels, the overall organism as well as all elements of its organization, such as its feathers, the skeleton, its inner organs, and so on, are organized for flying. This is not the place to consider whether this might be an adaptation or an element of its autonomy of movement. However, there obviously is a strong connection between form and function.

The work of Scott Turner indicates this in a fairly instructive way (Turner 2007). Turner studied the physiology of termite chimneys, which are known for a very effective ventilation system, providing homeostasis concerning parameters like O_2 and CO_2, constantly being adapted to requirements of the growing colony (figure 4.18). The question was how do the termites manage to control this homeostasis? As it turned out, termites continuously transport material from points with a higher concentration of CO_2 to points with a lower concentration, which at once modifies the air current within the chimney. This amounts to a continuously adapted form fulfilling the regulatory functions in

Figure 4.18
Termite chimney in Kenia. *Source*: Kreuzschnabel/Wikimedia Commons, Lizenz: CC-BY-SA-3.0
(https://creativecommons.org/licenses/by-sa/3.0/de/legalcode).

an optimal way. The termites are equipped only with the necessary sensitivity and the behavioral repertoire needed at each point. For the chimney as a whole, this adds up to an "embodied physiology" or an "embodied homeostasis," simultaneously resulting in the form of the chimney.

Because of this embodied physiology, the mound also takes on a certain shape, a design, for which the intelligence is not in the individual termite. It is in the system as a whole: the termites together with their immediate environment. In this sense, self-organization can build up homeostasis and at the same time contribute to the phenotypic design of organisms.

Referring to the physiologist Claude Bernard, Turner calls this principle "Bernard machines." I would prefer "Bernard processes," because organisms are not machines. Turner describes how he revised his thinking about biological structures in a fundamental sense:

I had been thinking about biological structures in entirely the wrong way. I had been subscribing to the conventional notion that a living structure is an object in which function takes place. That's all wrong, I came to see. A living structure is not an object, but is itself a process, just as much so as the function that takes place in it. Even the convenient dodge that structure and function are inextricably linked is wrong, I decided. That implies that structure and function are somehow distinct. . . . But living structures are not distinct from the function they support; they are themselves the function, no different in principle from the physiology that goes on there. In this sense, the mound is not a physical structure for the function of ventilation, it is itself the function of ventilation: it is embodied physiology. (Turner 2007, 20)

Turner describes further examples of embodied physiology, such as the architecture of smaller blood vessels in complex organisms. The arrangement of blood vessels cannot be genetically determined in all details, but rather branches according to the necessities of tissue supply. Fibroblasts generate collagen fibers that stabilize a net within the basement membrane around blood vessels and provide mechanical attachment for the endothelial cells, while angioblasts form the hollow tube of the vessel. Since both, fibroblasts and angioblasts, function constantly (and actively!) within a certain status of antagonism, blood vessels (especially smaller ones) have a dynamic structure. This architecture is subject to change depending on the local needs for stabilization on the one hand and oxygen supply on the other hand. The vessel's architecture is stable when there is a rough balance of power between the two, but there are all sorts of things that can change the balance one way or another. For example, oxygen-deprived cells send out distress signals that tilt the competition toward angiogenesis. These induce fibroblasts to loosen their grip, allowing the vessel walls to part slightly, so that angioblasts can start to form new blood vessels. However, the effect of blood flow with respectively varying pressures and especially mechanical forces also contribute to the final structure of the system of blood vessels. Overall, the shape of blood vessels in a given area of tissues is to a large part a matter of function and process.

The blood vessels show considerable anatomical variability, which is not genetic. Although the major blood vessels of humans are generally conserved in their placement, the small vessels are not even bilaterally symmetric in the same person, and many vessels of moderate size, such as the coronary arteries, can show extreme variation in normal individuals. (Gerhart and Kirschner 1997).

The growth of new capillaries in a tissue is induced by specific protein factors such as the vascular endothelial growth factor and fibroblast growth factor, which are secreted by cells when deprived of oxygen. The outgrowth of capillaries out of a capillary bed into nonvasculated territory with such cells is a mixture of chemotaxis and random exploration. The final capillaries then stabilize according to the needs of the tissue. After the formation and growth of capillaries, the conversion of some of these vessels into arterioles and venules is largely dependent on the amount of blood flow, which in turn is a product of the size of the capillary bed in the tissue. Gerhart and Kirschner

(1997, 191) summarize that "hypoxia of the tissue produces angiogenic and chemotactic agents, which drive the random anastomosing assembly of the capillary bed, determining the size of the capillary bed, which in turn determines the level of blood flow, which in turn determines the caliber of the secondary vessels. Local conditions regulate the overall morphology of the vascular system and design is tied to need." Then they conclude that such exploratory systems generate a very large number of possibilities. There is no constraint to generating new forms of organization by gradual modifications of old forms during evolutionary transitions. Rather, these systems seem to be extremely flexible on the phenotype level.

Basically, the same principle is known from neuronal wiring within the brain. During childhood development neuronal structures are generated only in a rough manner, whereas the fine wiring is configured by usage and function. Thus, some systems develop their form according to their function. Very likely such interdependencies between form and function have been significant during evolutionary changes, Turner (2007) states.

It may be more appropriate to talk about "processes of shaping," because the shape at any single moment is just a window into the overall process. Life is always an active process, and shape is no more than a moment within the continuous flow of functions. As all this contains embodied information, the complex of matter, form, function, and information generates the process of shaping (figure 4.19).

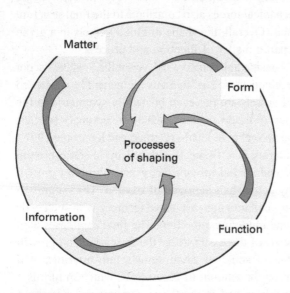

Figure 4.19
Schematic representation of the process of shaping.

Constructional Morphology

As it has been argued repeatedly now, features and characteristics of organisms can hardly ever be explained by single factors or causes. According to this frame of reference the research approach of constructional morphology may also be included in an organismic perspective to study shape and form of organisms. Constructional morphology recognizes that organismic morphology is not simply an adaptive response to selection pressure but incorporates phylogenetic and structural influences as well. Further factors such as chance and phenotypic response in the sense of individual adaptations have been added throughout this research field's history, so that it is a typical multifactorial approach that seems to be closer to organismic reality than any one-sided research program. Constructional morphology has been developed in zoology and especially in paleontology by Hermann Weber and Adolf Seilacher in Germany and was brought to the attention of English-speaking researchers by David Raup (Briggs 2014, 2017; Cubo 2004; Wake 1982).

According to Seilacher, form and shape of an organism are mainly influenced by three factors: a historic phylogenetic one, an ecological-adaptive one, and a constructional one. He illustrated his concept with a triangle describing a theoretical space. The position of any morphological character within the area of the triangle is given by the relative contribution of each one of these causal factors to explain the variation of this character.

The historic phylogenetic factor (historical constraint) depicts each structure's long evolutionary history, which constrains the possibilities for adaptation. For example, both brachiopods and bivalves have two hard shells. In articulate brachiopods both shells are generated by one single groove on the edge of the shell. All evolutionary changes need to depart from this configuration. Contrary to this, bivalves generate their shells from two grooves along the edges of the mantle. Therefore, they had, compared to brachiopods, the possibility to generate a siphon, coming from this double groove.

The ecological-adaptive factor (functional constraint, or adaptational constraint) generates a structure that is capable of performing its function within a certain environmental context in an optimal way.

The third factor describes constructional constraints, which are anatomical patterns or structures resulting from biomaterials, building blocks, growth, and development. Thus, the concentric rings on the exterior of a bivalve shell, the growth rings, are generated inevitably when calcium carbonate is added in batches from the edge of the mantle. This factor shows that some features of organisms are just constructional necessities.

One central assumption of Seilacher is that organisms cannot be optimal constructions. Because all three factors are simultaneously involved during morphogenesis, a construction can only be as optimal as the concurrent influences of these factors allow.

For paleontological questions, being the context in which the concept has been developed, this approach is of high value because it allows research to concentrate on the actual present fossil structures and to establish their morphodynamic principles of generation. In this sense it is distinctively phenomenon oriented.

Briggs (2017) states that the importance of Seilacher's insight was his departure from the thesis that all morphology reflects adaptation to a particular environment or mode of life as a result of selection acting on an evolving lineage. Seilacher considered constructional morphology to be a method of research rather than a theoretical framework. However, considering these different influential factors also says a lot about the properties each organism finally exhibits.

Seilacher was particularly interested in structures that have little or no adaptive significance and are simply the result of self-organizing processes or the nature of the materials available for construction. Today such principles are being discussed again from the perspective of molecular biology (Gerhart and Kirschner 1997). Many growth patterns, for example, are a product of fundamentally self-organizing mechanisms, which Seilacher considered autonomous to a degree "as reflected in variabilities that remain beyond genetic control" (Seilacher and Gishlick 2015, 2). The logarithmic spiral that underlies the morphology of mollusk shells is a classical example, which was the subject of early computer simulations by Raup. He showed how the range of the shell form in mollusks could be generated by varying just three parameters: expansion rate, translation along the coiling axis, and the distance of the curve from the coiling axis. This allowed a visualization of morphologies that can be simulated but are not encountered in nature and, more importantly, an exploration of the reasons why some areas of morphospace are occupied and not others (Briggs 2017).

Gould and Lewontin (1979) highlighted Seilacher's insight in identifying structural constraints as nonadaptive but necessary consequences of materials and designs. In their meanwhile classical "Spandrels of San Marco" paper (Gould and Lewontin 1979), they criticized the tendency of other evolutionary biologists and paleontologists to offer explanations of morphology, which assume that the individual attributes of an organism are optimized for particular functions, that is, that morphology is solely a product of natural selection—the so-called adaptationist program.

Briggs (2017) describes how Seilacher in 1991 expanded his triangle to a tetrahedron by introducing a fourth constraint, "effective environment," as the new corner and included an explicit dimension of time (figure 4.20). He named this broader concept "morphodynamics" to show that morphology is a product of evolution and that each fossil organism represents just one point on a

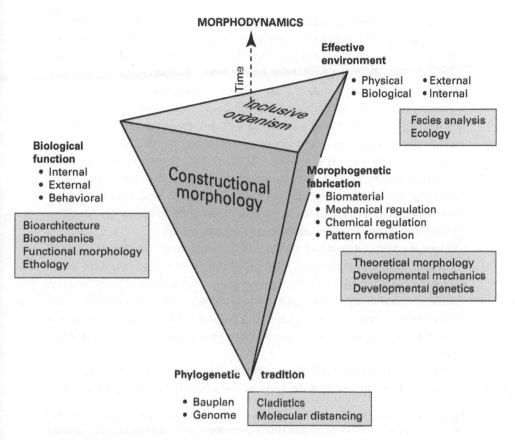

Figure 4.20
The conceptual framework of morphodynamics according to Seilacher and Gishlick (2015). *Source*: Republished with permission of John Wiley and Sons, from Briggs (2017, 204).

continuous time line. The addition of the fourth corner accounted for the recognition that organisms and their respective environment are interdependent and conjointly form an interactive system, which Seilacher called the "inclusive organism" (Seilacher and Gishlick 2015). Seilacher introduced the new term, *morphodynamics*, not only with the constraints or influences on morphology that define the corners but also with a variety of approaches that can be used to reveal morphodynamic relationships. He was familiar with the new concepts and research in genetics and evolutionary development. Thus, it is no surprise that his morphodynamics diagram included developmental genetics and developmental mechanics as approaches to interpreting the architectural and molecular aspects of understanding the role of phylogeny.

Today, the focus of evolutionary biology has shifted from constraints on morphology to the mechanisms that enable or explain morphological evolution—so-called evolvability (Love 2015). Major advances in research on evolutionary development have changed the emphasis from understanding controls on form to unraveling the processes that generate it. This corresponds to a definition of morphology David Wake gave in 1982: "Morphology is . . . the study of form, and in its purest state it deals with the materials of tissues, organs, and organisms as well as with the forces that mold them." (Wake 1982, 606). However, it is remarkable that Seilacher was able to include these approaches easily into a multidimensional concept. A broadened concept such as morphodynamics clearly complies with the various attempts of an organismic program and reaches far beyond such approaches that try to reduce the biology of form to any single factor, thus necessarily being one-sided. In recent years, especially the field of developmental prerequisites and processes, and their evolutionary changes, became important (developmental bias) (Jablonski 2020).

There are further significant approaches in morphology, which include interdependent factors to understand form and function of organisms, such as the approach by Dullemeijer and his "Leiden group of anatomists," who focus on functional morphology (Dullemeijer 1980; Dullemeijer and Zweers 1997; Wake 1982). Many of these approaches have been pushed into the background in recent years but might again be reconsidered more profoundly within an organismic program.

Wake (1982) summarized the reason why the study of form according to a multiperspective approach at the level of the whole organism is important:

Causes for the renaissance of interest in morphology are not hard to find. While the reductionist approaches of the middle part of our century have been enormously productive and progressive, there is a growing awareness of our need to study organisms more. While evolution occurs by means of changes in populations, it is the organism, after all, that evolves in an ultimate sense. It is the organism that lives in an environment. It is the organism that we most immediately perceive and want to understand. (611)

However, he also concluded:

For me the search for a science of form is still in its infancy, and despite the relative antiquity of the investigation we are far from an answer (612). What is to be explained is why there are so many kinds of animals, why they have so many forms, and why these forms function in such diverse ways. But perhaps an even more important corollary is—Why are there not more kinds, forms, and functions? (617)

4.9   Time Autonomy

Every process of an organism is organized within a specific, regulated, extended time structure. Most of this time regime is generated by the organism itself, which is called endogenous. Some functions can be synchronized with external periodicities and circumstances. Yet even then, the basic time organization of the process originates from the agency of the organism itself. Therefore, this endogenous time organization can be acknowledged as autonomous. Usually, a hierarchy of different periods at different levels is present in humans, animals, and in plants, so that time autonomy is a fundamental and decisive property of life processes.

The Duration of Processes

In physiology, we are used to seeing that many functions take a certain length of time. They usually have a specific duration, generated by the organism's own activity. Their duration can be relatively fixed or variable, but in any case, they have a certain passage of time, which is brought about by the function itself.

One example is the cell cycle. Depending on the organism and cell type, the cell cycle can be variable, but in a specific situation it takes a characteristic duration of a couple of hours, it "takes some time." The mitotic phase itself usually takes about one hour. Essentially, this duration also organizes the sequence of events.

Another example is the heartbeat. The human cardiac cycle has a specific duration and sequence of events, in which the heart contracts and relaxes with every heartbeat. The period of time during which the ventricles contract is known as systole, while the period during which the ventricles relax is known as diastole. The atria and ventricles work in concert, and the coordination in time ensures that blood is moved efficiently through the heart as part of circulation. The normal rhythmical heartbeat is established by the heart's own pacemaker, the sinoatrial node, and thus has a specific autonomy. Here, an electrical signal is created that travels through the heart, including the atrioventricular node, causing the heart muscle to contract. The transduction time of this electrical signal has a specific duration. The resting heart rate can be influenced by a variety of factors. Exercise and fitness levels, age, body temperature, basal metabolic rate, and a person's emotional state can all affect the heart rate. However, there is always a self-generated chronological order of events with a specific duration. This is the time organization of the heart.

Further examples from physiology are the passage of nutrients through the intestines, signal transductions in nerves, and the female menstrual cycle. Throughout each 24-hour day, plants' leaves can change their orientation, human body temperature rises and falls, fungi determine their sporulation, and activity levels fluctuate. Every physiological function takes its specific time. Looking at the variety of animals and plants, the specific lifetime is also basically a self-generated length of time and varies extremely.

Also, the embryonal development of living *organisms* requires the precise coordination, duration, and chronology of all basic cellular and molecular processes in space and *time*.

Biological Rhythms

Because most of these functions repeat continuously and are subject to characteristic variation in their course, they are described as biological rhythms. The discipline that studies these rhythms is chronobiology (Dunlap et al. 2004; Hildebrandt et al. 1987; Koukkari and Sothern 2006; Moser et al. 2008; Rensing 1973).

By definition, *rhythm* is a change that is repeated with a similar pattern (Koukkari and Sothern 2006) or a periodic occurrence of specific physiological changes. The word *similar* is important here because rhythmicity in organisms usually involves some variability and modulation (Schad 2006). In most cases, these rhythms involve repeated changes of the respective function or value. Chronobiology describes them as oscillations, including specific frequencies, amplitudes, and lengths of duration (figure 4.21).

A *period* represents the time required to complete a cycle. *Frequency* is the number of cycles in a unit of time and is the reciprocal of the period. For example, a leaflet that moves up and down four times each hour displays a frequency of four cycles per hour, which is the same as a period of 15 minutes. Biological rhythms are usually distinguished or grouped in reference to the length of their period.

Humans, like all other organisms, have a rhythmic order underlying their life (Furst 2020; Hildebrandt et al. 1987, 1998; Moore-Ede 1982 et al.; Moser et al. 2008). Change, not constancy, is the norm for life, and the rhythmic

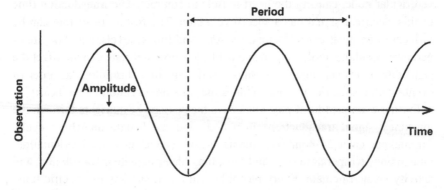

Figure 4.21
Basic form of biological oscillations.

timing of change is a central aspect of living beings. Life is not only a process in itself, but it is a *hierarchy of rhythmically ordered processes*, which are coordinated and orchestrated in an often extremely complicated fashion. An organism can be recognized as an organized system of rhythms.

Because these processes have certain time structures and a multilevel organization, this principle has also been called *time-Gestalt* (time shape) (Hildebrandt et al. 1998; Schad 1997, 2006). This term complements the morphologically oriented term *Gestalt* described in section 4.8., which indicates predominantly the organismal form or shape. Also, shape is to be found at different levels of organization. Yet, as described in that chapter, spatial form needs also to be understood in terms of processes (processes of shape), so there is a clear overlap here.

Furst (2020, 295) explains:

Thus, in addition to having a form-in-space, an organism also possesses a time-Gestalt or a form-in-time of higher order. When the momentary changes in form are known, e.g. the growth process of an embryo or a plant, one can reconstruct the entire formative process in-time, say by time-lapse photography. Of note, the common characteristics of an organism such as growth, nourishment and excretion, reproduction, breathing, regeneration, wound healing, apoptosis, and rest-activity cycles, to name a few, are all organized in time. Their temporal compartmentalization makes it possible to run unhindered in the limited space of an organism. The movement of blood, lymph, and other bodily fluids is therefore regulated by a highly differentiated time matrix or "time body" subject to cycles, repetitions and reversals that go beyond the conventional, linear time. Periodic time (rhythm) is the medium in which evolves the stream of life.

All known variables of life, be they levels of potassium ions in a cell, stages of sleep, or the opening and closing of flowers, have either directly or indirectly been found to display rhythms. These rhythmic changes of life represent an enormous network of biological rhythms, passed on from one generation to the next. The range of periods for biological rhythms is broad, extending from cycles that are measured in milliseconds to cycles that are more than 100 years in length. Furthermore, adaptations of organisms relative to geophysical cycles, such as the solar day, seasons, and tides, attest to the evolution of certain types of rhythmic timing (Dunlap et al. 2004; Koukkari and Sothern 2006). "The rhythmic nature of life influences the very existence of organisms, commencing before conception and extending beyond death. Rhythms may be the most ubiquitous, yet overlooked, phenomena of life" (Koukkari and Sothern 2006, 4).

When the discipline was young during the 1970s, the phenomenon of biological rhythms did not at all match with ordinary mechanistic thinking. Chronobiologists of that time report how they were met with suspicion; at conferences they were asked, "Do you really believe in rhythms?" But today chronobiology belongs to the established disciplines of biology and human physiology and has considerable relevance in medicine. Nevertheless, the importance of this property is still underestimated today, so that Furst (2020, 295) remarks: "The notion that virtually all biologic phenomena are subject to a hierarchy of rhythms is yet to find its way into the standard texts of physiology."

Autonomy of Biological Rhythms

As described, most of these rhythms are endogenously generated by the organism itself and maintain certain frequencies and amplitudes. A critical feature of this principle is that they are relatively unaffected by temperature; that is, they are temperature compensated. This presents a significant explanatory challenge because biochemical reactions usually proceed more rapidly at higher temperatures. Thus, *rhythmicity is an autonomous, actively generated function of the organism*.

Autonomous also means that organisms are not always fully synchronized with external periodicities. The endogenous daily rhythm in humans typically tends toward a length of 25 hours, with some individual differences. The actual frequency then needs to be synchronized with the external day. Therefore, chronobiology describes "circadian rhythms." Today it is clear that most, if not all, physiological functions in humans exhibit a circadian rhythm (figure 4.22; Dunlap et al. 2004; Hildebrandt et al. 1987; Koukkari and Sothern 2006; Moore-Ede et al. 1982; Moser et al. 2008). All rhythms, which are called "circa-," are endogenous and continue when external zeitgebers disappear.

Although the circadian rhythm is the most well-known oscillation in organisms, there are also oscillations of different frequencies, below and above the daily rhythms. They superimpose each other, resulting in a complicated time pattern within the organism (figure 4.23; Hildebrandt 1979; Hildebrandt et al. 1987). If the circadian rhythm is regarded as a standard, biological rhythms can further be divided into ultradian rhythms with a period of less than 24 hours, and infradian rhythms with a period longer than a day. Ultradian rhythms are usually related to cell functions. Indeed, presently about 400 different ultradian rhythms have been identified (Eckert 2000). Most of these ultradian rhythms are clearly endogenous.

Infradian rhythms are widespread in animals and plants. The moon influences various physiological functions of many animals living in the intertidal zone via the moonlight or through the generation of high and low tide. Circatidal rhythms, which are related to tides, usually have a length of 12.4 hours. Circalunar rhythms correlate with the lunar cycle of 29.5 days and effect reproduction in many mammals (Endres and Schad 2002). Circannual rhythms are related to the year of the earth with 365 days and are most obvious in strict

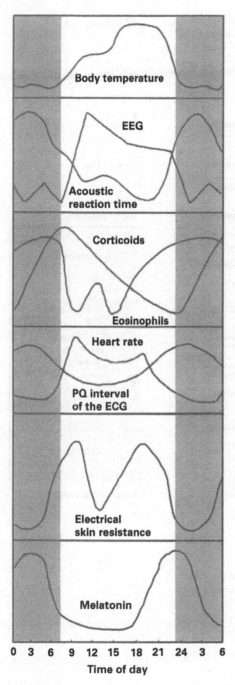

Figure 4.22
Some circadian rhythms of humans.

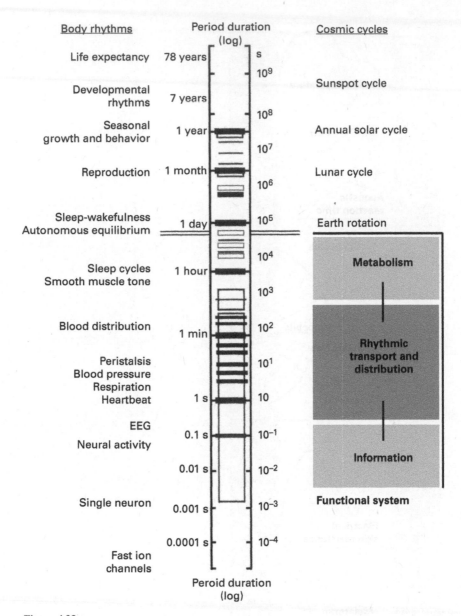

Figure 4.23
The spectrum of human biological rhythms. *Source*: © 2008 IEEE. Reprinted, with permission, from Moser et al. (2008).

seasonal cycles, which are effective on nearly everything from coat color to hibernation, migration, and reproduction of animals (Eckert 2000).

Regulation of Circadian Rhythms

As the knowledge of circadian rhythms in animals and humans increased, a search began for a central organ, which would generate these processes. A structure in the anterior hypothalamus has been identified as the locus of the circadian clock in mammals including humans, the suprachiasmatic nucleus (SCN). Genes and proteins have also been identified in the cells of the SCN, which were related to this clock, and there was hope that it might be possible to find the molecular components that are responsible for the phenomenon. Bechtel (2010b) describes the history of this research and the associated concepts and explanations. In these early explorations, a linear causal model was initially assumed (figure 4.24, top).

Further research showed how intertwined the different components of the system are within the organism. First, there are peripheral oscillators, which are able to generate a circadian rhythm themselves and that provide feedback to the SCN. Also, it was discovered that genes and proteins, described in the SCN cells, were present in peripheral oscillators as well. The SCN also

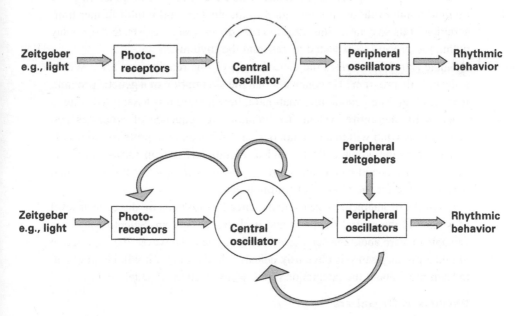

Figure 4.24
(top) Classical linear causal model of the circadian system; (bottom) integrated circadian system. *Source*: Redrawn from Bechtel (2010b, 323).

receives stimuli from sensory organs, and in turn it influences them as well. Thus, within the oscillatory system there are clear interdependencies and feedback processes working concurrently, and the SCN has a coordinating function, rather than being the causal origin of the oscillation.

Bechtel (2010b) presents an alternative representation of this system (figure 4.24, bottom), which shows some of the existing interdependencies. This example again shows how such a research area grows over and above its own original paradigmatic prerequisites, guided by the phenomena themselves. It also depicts how necessary the analysis of the components is, being the prerequisite for developing an integrative concept.

The SCN itself comprises approximately 10,000 cells on each side of the brain. Each cell has the genes and proteins to generate the rhythm. However, when single cells from the SCN were experimentally cultivated, a substantial variability between the rhythms of individual cells was detected. Only groups of a few hundred cells generated the exact rhythm (Mohawk et al. 2012). Thus, only their synchronization delivered the exact rhythm to coordinate the circadian system of the whole body.

The same principle can be traced observing the genes involved. When genes are transcribed, the transcript leaves the nucleus and enters the cytoplasm, where it is the template for proteins, which are involved in generating the rhythm. Some of these proteins return to the nucleus and inhibit further transcription of these genes. Thus, here too a feedback system exists, and the delay within the process is expected to generate the rhythm.

Noble (2008a) describes the consequence of this relation and asks: Where is the genetic program? He concludes that the assumption of a genetic program is misleading. The genetic information on how a protein is assembled is not a program for the entire system. To maintain the sequence of processes that generate a rhythm within a cell, not only the genes and the proteins are necessary but also the whole cell, the nuclear membrane, certain forms of mRNA and microRNAs and so on. It is more a gene–protein–lipid–cell network that is responsible for the effect, not simply a gene.

Networks of interactions generate the circadian rhythms on the cellular level as well as in the central oscillator and in the entire system. Scientifically, it is valuable to learn about the components of the system. However, the expectation to find a primary cause is obviously delusive. In the future, it will be important to learn more about the integration of the whole system (Bechtel 2010b).

Rhythms as Organizers

Buzsáki (2006) describes a bewildering spectrum of self-generated oscillations in the mammalian brain and identifies them as the brain's fundamental organizer of neuronal functions. They vary from extremely slow oscillations with periods of minutes to very fast oscillations with frequencies reaching 600 Hz (called ultrafast oscillations) (figure 4.25). They superimpose each other and many are performed simultaneously.

To recognize that organic functions—in the brain and throughout the body— oscillate in a self-generated way provides another processual view of organisms. However, within such oscillations the processes are highly structured and coordinated. Rhythmicity is the order and manifestation of organic processuality and thus essentially supplements the process view. This form of multilayered organized processes cannot be resolved into individual causal processes. Processes at the molecular and genetic level, as well as at the level of cells and organs, are integrated and organized within and by rhythms. In this sense, the study of oscillations and rhythmic functions is an eminently organismic approach, which has been acknowledged by many scholars of chronobiology (Buzsáki 2006; Duboule 2003; Hildebrandt 1979; Hildebrandt et al. 1998; Schad 2006).

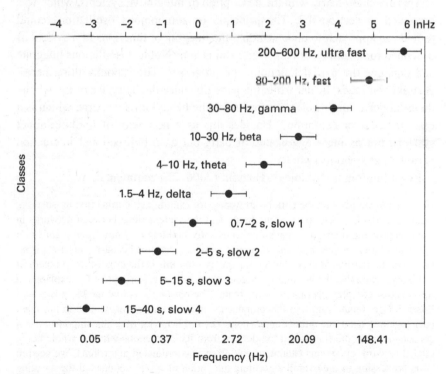

Figure 4.25
Multiple oscillators of several classes form a hierarchical system in the cerebral cortex. *Source*: From Penttonen and Buzzati (2003, 149), reproduced with permission. Copy taken from Buzsáki (2006, 114).

After a brief discussion of complex systems far from equilibrium, Buzsáki (2006, 13) concludes:

Complexity can be formally defined as nonlinearity, and from nonlinear equations, unexpected solutions emerge. This is because the complex behavior of a dynamic system cannot easily be predicted or deduced from the behavior of individual lower level entities. The outcome is not simply caused by the summation of some agents. The emergent order and structure arise from the manifold interactions of the numerous constitutes. At the same time, the emergent self-organized dynamic, for example, a rhythm, imposes contextual constraints on its constituents, thereby restricting their degrees of freedom. Because the constituents are interdependent at many levels, the evolution of complex systems is not predictable by the sum of local interactions. The whole is based upon cooperation and competition among its parts, and in the process certain constituents gain dominance over the others. This dominance, or attractor property, as it is called in chaos theory, can affect other constituents such that the degrees of freedom in the system decrease. Such compression of the degrees of freedom of a complex system, that is, the decrease of its entropy, can be expressed as a collective variable.

This matches clearly with the description of integrative systems, which was presented in section 4.2. The principle of autonomous oscillations could provide an approach to understanding the integrative functions of a system as described for example by Paul Weiss and Dennis Noble. Oscillations integrate and organize the partial functions and processes. The "spooky thing here," Buzsáki continues, is that while the parts do cause the behavior of the whole, the behavior of the whole also constrains the behavior of its parts, which is a case of "circular causation." He sees this as a new way of thinking about systems, not as mere aggregates of parts but as a bidirectional interaction between parts and the whole.

Regarding brain physiology, Buzsáki (2006, 25) comments:

How far can we get with the bottom-up strategy of examining neurons first in isolation, local networks in small slices of the brain, and then interactions between networks in conveniently anesthetized preparations, constantly building on knowledge gained at a lower level and moving up? This approach provides comfort because causal explanations may be reached at every level—separately. And this is the crux of the problem. It is almost certain that the bottom-up strategy alone will never provide a full explanation for the most complex operations of the brain. The reason . . . is that the brain is a nonlinear device: break it up into its components and you will never be able to put them back together again into a functional whole. The full behavior of each component is not contained within the component but derives from its interactions with the whole brain. Global network operations cannot emerge from uncoordinated algorithms. We need to be in possession of the overall algorithm, the "brain plan", to understand the meaning of local processing.

This autonomous creation of time processes is of central importance not only within the adult organism but especially during embryogenesis. This has been highlighted in a special issue of *Science*, beginning with an editorial titled "Time for Chronomics?" (Duboule 2003, 277): "Animal development is, in fact, nothing but time. From the cell cycle to the beating of the heart, our own lives are composed of a multitude of microscopic and molecular oscillations. . . . The goal of the developmental clock is not simply to mark off time, but to integrate and unify the myriad temporal signals received from throughout the organism. . . . The spatial construction can be understood only in the light of time."

Time Integration

The different rhythms, such as circadian, ultradian, and infradian, are highly ordered and coordinated among each other within a single organism, generating its time-Gestalt. Schad (1993, 1997, 2016) considers this ability of an organism to integrate the self-actively produced processes of different frequencies and durations as a central property of living organisms. For him, life is even constituted in its time integration, and periodic phenomena are actual carriers of vital functions.

Schad describes this principle of integration not only for the chronobiological physiology of the individual organism. Rather, he also discovers a form of integration in the evolutionary dimension. According to this, all organisms can usually be found to consist of a combination of different features that are joined together like a mosaic, and these features can often belong to different evolutionary stages.

In human evolution, for example, it is well described that the evolution of limbs and uprightness occurred earlier than the evolution of brain size and related skull shape (Klein 2009). Some of the early human fossils already have limbs and a vertebral column, which, simply said, differ little from that of modern humans. At the same time, however, they possessed a skull with an extremely low and primitive curvature and proportionally small size. Limbs and cranial vault belong, so to speak, to different evolutionary stages.

According to Schad (1993), something similar can be found in many other organisms. These are also a mosaic of different evolutionary stages and can carry features of their predecessors, as well as evolutionarily distinct progressive features, and those, too, which are genuine transitional features of their own evolutionary stage. The respective organism integrates this mosaic to a respectively viable whole. Thus, it integrates characteristics that belong to different evolutionary periods.

In general, it has been shown that characteristics (or elements of them) in individual lineages do not always change at the same time but usually at different times. This often results in the mosaic character of organisms. This is a phenomenon known in paleontology as *heterochrony* (McKinney and

Namara 1991; McNamara 2012). Schad therefore proposes that the ability to integrate these temporal dimensions—both in individual physiology and evolutionarily—be considered as a specific feature of organisms.

In modern evolutionary developmental biology, heterochrony is any difference in the timing or duration of a developmental process in an organism compared to its ancestors or other organisms. This leads to changes in the size, shape, characteristics, and even presence of certain organs and features. This can create morphological innovation. Heterochrony can be divided into intraspecific heterochrony, the variation within a species, and interspecific heterochrony, phylogenetic variation, that is, the variation of a descendant species with respect to an ancestral species (McNamara 2012; Reilly et al. 1997). The presence of these phenomena shows that the time order can be involved in largely different dimensions of organismic existence, which is probably still seriously underestimated. Also, variations in the speed, at which different body parts develop in an individual organism, can also be called heterochrony (Marshall 2021).

In the embryonal development of vertebrate animals, a "segmentation clock" has been discovered (Müller and Hassel 2006). Regular pulses from a group of genes (with a key gene named *Hes7*) and their protein products induce the generation of body segments, and different speeds of these oscillations can be involved in different species. They run, for example, about twice as fast in mouse embryos as they do in human embryos. It is assumed that this difference in timing contributes to different morphological outcomes, but it is still unclear, what exactly induces these differences. There is no master gene nor other singular factor, which could determine the speed of the oscillations, so that researchers suggest that "it's in the cell's environment" (Marshall 2021).

Because time, or at least living within a specific time, is so important for organisms, Baquero (2005) proposes that time in organisms might have a reality in itself. He views time not merely as a regulatory component but instead as a real and creative constituent of nature, and for this reason, an object worthy of research in the natural sciences. He promotes the hypothesis that time might even be a replicating entity, which would be important for understanding evolution.

Now in summer, in front of our terrace at home blooms evening primrose (*Oenothera biennis*), a large stately specimen with many branches. During the day, only a few faded flowers from the previous night hang on it. But just before dusk begins, several flower buds thicken. Then, one after another, single bright yellow flowers burst open within a few seconds. Before night comes and the night butterflies become active, the plant has a whole bouquet of bright flowers on its inflorescence. The process is precisely organized into the times of day and demonstrates to us every evening the temporal organization of this plant—just one example out of millions of time organizations among organisms.

4.10 Sensitivity and Affectability

Every living being has abilities to perceive its internal states as well as certain conditions of its environment: they have a certain sensitivity. Already unicellular living beings can respond to such sensations with self-determined actions and reactions. This is described as affectability. Such an ability cannot be attributed to a crystal. A living organism is irritable by the fact that external stimuli are not only transmitted by physical force but can be received by sensory organs, so that a self-active interaction with the environment takes place in comparison with internal states. Therefore, sensitivity and affectability are fundamental properties of every living being.

Sensitivity Evolves

Early in evolution, primeval unicellular organisms undoubtedly possessed rudimentary detection systems to sense, for example, nutrient and temperature gradients, to react to variations in light or to avoid noxious stimuli in their environments. With the advent of multicellularity, eukaryotic organisms required progressively more sophisticated organs to respond to an ever-larger array of external cues and to coordinate activities between different cells. Also plants such as sunflowers can sense heat and light, so they can turn toward the sun's rays. Increasingly complex signal transduction networks evolved in higher organisms to control intricate physiological processes such as cellular differentiation, embryonic development, memory, and learning. Complex and extremely refined organs such as the vertebrate eye or the vestibular organ developed.

Individual cells as well as multicellular organisms often receive many signals simultaneously, and they then integrate the information they receive into a unified action plan. But they are not just targets. They also send out messages to other organisms both near and far.

Sensitivity to stimuli arises simultaneously with the development of the first form of autonomy, which leads to a distinction between inside and outside (section 4.3). This basic principle of sensitivity is constantly augmented with further evolutionary developments of organisms. As autonomy increases (Rosslenbroich 2014), the ability to experience the environment expands simultaneously. Highly specialized sense organs give more and more extensive and precise information about processes and conditions of the environment. Here again, the principle of *concurrency of polar characteristics* can be identified. While organisms emancipate to a certain degree from vagaries of the environment and gain in self-determination and flexibility during major transitions in evolution, they concurrently connect with the environment by enhancing the intensity, quality, and amount of sensations from it.

In amniotes, the clade including modern mammals (Synapsida), reptiles (Reptilia), and birds (Aves), the evolution of sensory perception took place in a stepwise manner after amniotes appeared in the Carboniferous. Fossil evidence suggests that Paleozoic taxa had only a limited amount of sensory capacities relative to later forms, with the majority of more sophisticated types of sensing evolving during the Triassic and Jurassic. Alongside the evolution of improved sensory capacities, various types of social communication evolved across different groups (Müller et al. 2018). The highly developed physiology of recent birds and mammals enables high performance of sensory perception. This also refers equally to the perception of the external environment as well as to the internal world.

Therefore, along with the emergence of elementary forms of inwardness, subjectivity already arises too, in that now the organism interprets the experienced stimuli in relation to its respective situation and reacts in a self-determined way (Kather 2012). Separation is a basis for self-determination of reaction. *Signal perception, irritability, affability*, or *stimulus sensitivity* are common terms in biology, whose application to inanimate things does not yet make sense. With the distinction between the inner and outer world, the mere physicochemical influence is replaced by the ability to receive, distinguish, and emit different signals. It is an exchange of information with the environment, which already contains a rudimentary criterion for the distinction between different stimuli.

Experiencing the Environment

The form in which organisms gain information from their environment and from their internal status, and in which they communicate, is usually called signals. The study of signaling pathways, chemical as well as neuronal, is a major field of modern biological sciences.

Jakob von Uexküll was among the first to focus on the aspect of signal perception in animals (Penzlin 2009; Uexküll 1909, 1973). He stated that every organism perceives a specific selection of signals out of many influences from its environment. He was particularly interested in how living beings perceive their environment and argued that organisms experience life in terms of species-specific spatiotemporal subjective reference frames which he called *Umwelt*, according to the signals and factors being relevant for the organism. This specific Umwelt (environment) is distinctive from what he termed the *Umgebung*, which would be the living being's surroundings. Umwelt may thus be defined as the perceptual world, in which an organism exists and acts as a subject. The Umwelt of an animal is a selection of its surroundings—namely, that which is grasped by the senses of the respective animal and can thus become relevant for its behavior. It is the section that interests the animal. The organism is the precondition for the surroundings to become an environment after all (Lindholm 2015).

By studying how the senses of various organisms like ticks, sea urchins, amoebas, jellyfish, and sea worms work, Uexküll generated theories on how they may experience their world. His Umwelt theory is a biological theory about the subjective experiential worlds of animals.

Because all organisms perceive and react to sensory data as signs, Uexküll argued that they were to be considered living subjects. This argument was the basis for his biological theory, in which the characteristics of biological existence could not simply be described as a sum of its nonorganic parts but had to be described as a subject and a part of a sign system. In this sense he developed an aspect of an organismic understanding.

The organism possesses sensors that report the state of the Umwelt and effectors that can change parts of the Umwelt. He distinguished the effector as the logical opposite of the sensor, or sense organ. Sensors and effectors are linked in a feedback loop, a "functional circle" (*Funktionskreis*). The modern term *sensorimotor* used in enactive theories of cognition (Fuchs 2018; Thompson 2007) encompasses these concepts.

Uexküll's understanding of evolution departed from Darwin's ideas of environmentally driven natural selection. He thought that organisms were active players in evolution, and that embryogenesis and animal behavior were mediated by the meanings that the "subject" associates with various objects (Sharov et al. 2015a; Sharov and Tønnessen 2021).

Biosemiotics

Biosemiotics is the approach within today's biology that most systematically advances Uexküll's point of view. The main idea of biosemiotics is that life and semiosis, or sign exchange, are coextensive (Sharov et al. 2015b; Sharov and Tønnessen 2021). It is assumed that life has a semiotic nature, because it is based on an endless interpretation of environmental cues and transfer of life-related functional meanings vertically across generations and horizontally to neighboring organisms.

Signal transduction in cells is described in many details. Receptor molecules stretching all the way through the cell membrane recognize and bind to specific molecular surfaces outside of the cell, and this binding subsequently causes changes at the inside end of the receptor molecule, which in turn serves to initiate cascades of internal processes within the cell. There are also very detailed descriptions of such signal transductions for whole organs, organ-systems, and whole organisms (Cantley et al. 2014; Kramer 2015).

Semiotic processes help organisms to perform their functions, preserve their habits and pursue their agendas throughout generations. Signs carry meaning, and life carries properties that enable meaningful interpretation of perceived signs. One implication of this view is that life has certain mindlike properties, enabling such meaningful interpretation of signs. This point of view contrasts with the traditional dualistic approach in science, which distinguishes sharply between mind and matter (Hoffmeyer 1996, 2008; Sharov 1992; Sharov et al. 2015b; Sharov and Tønnessen 2021).

Biosemiotics is understood as an evolutionary science because it explores the gradual emergence of the capacity to deal with signs. Semiosis in bacteria is qualitatively different from the semiosis in eukaryotic cells and even more radically different from the brain-based semiosis of animals. In general, evolution is seen as a process of continuous interpretation and reinterpretation of hereditary signs alongside other signs, which originate in the environment or the body.

Sharov et al. (2015b) state that semiosis evolved with life and gradually increased in complexity and organization. They see this process of evolution not merely as a quantitative affair but as a process paved with important qualitative inventions, so that at some points in time both the nature of life and the character of signs were restructured.

In this very basic sense, biosemiotics attempts to study the relation of phenomena such as life and mind. The mindlike properties found elsewhere in nature, however, can be very different from the human mind. In this sense there are broad overlaps between the ability of organisms to process information (section 4.6) and the property of subjective experience and consciousness, which will be discussed in section 4.11.

In addition, it must be emphasized that organisms have an active role in reshaping sign relations, and that organisms are active in shaping their own lives. By altering sign relations, organisms influence what they recognize and how they respond to it. Organisms also change sign relations in other species. By and large, we all affect each other. Therefore, there is another overlap here in regard to the property of agency, which has been discussed in section 4.4.

Biosemiotics and Agency

Hoffmeyer (2013, 152) states that biosemiotics "is based on an understanding of agency as a real property of organismic life, a property that is ultimately rooted in the capacity of cells and organisms to interpret (whether consciously or unconsciously) events or states as referring to something other than themselves or, in other words, the capacity to interpret signs. These signs need not be emitted with a purpose of communication, in fact by far most signs are not part of a sender-receiver interaction but are simply important cues (internal or external) that organisms use to guide their activities."

Hoffmeyer (2013) elaborates that science had a profound difficulty to accept organisms as entities with an autonomous agency because of its historical foundation in Newtonian explanations. Within this framework, science has never ascribed a genuine constructive capacity to unitary agents. Such agents, whether atoms, genes, embryos, or individuals, were always understood as passive bodies to be steered by forces beyond their own control. Darwin wished to be the "Newton of the grass blade" and although natural selection, as Darwin himself knew, is nothing like the austere gravitational forces, there nevertheless are parallels. Most important, by posing natural selection as the essential mechanism of evolution, Darwin created a perfectly externalist theory, a theory that seeks to explain organisms and their adaptations exclusively in terms of challenges and influences from their external environments. The necessary variations were explained as a result of chance fluctuations, later translocated into the genes. To this day, Hoffmeyer states, mainstream biology ascribes variations solely to mutational events, and routinely plays down the generative role of the whole autonomous organism itself. This is what Walsh (2015) called the disappearance of the organism as a fundamental unit of life (section 4.4). Only very recently has this standpoint begun to change in some parts of evolutionary biology, as summarized in section 4.12.

Then Hoffmeyer asks: Why does common sense feel that the creatures of this world, and humans in particular, possess agency?

When pressed, biologists may respond to such questions by alluding to our human propensity to project our own humanness upon the poor creatures of this world. We may think of them as agents, but they are not true agents. Rather, they are programmed by natural selection to act as if they were agents in their own right. When further pressed, the same explanation may be given even for human agency. We think we act out of a free will, but this again is only an illusion; instead, natural selection has formed us to experience a freedom of will, and in reality we are programmed by our genetic inheritance to feel that way. So, ultimately, the old Cartesian machine conception prevails—only, contrary to Descartes, the machine-thinkers of today do not think it necessary to equip humans with special res cogitans-like properties. (Hoffmeyer 2013, 149)

He then further explains that recent developments in molecular genetics, cell biology, and developmental biology imply that an essential link is missing in the modern synthesis. The production of phenotypic variation depends on internal dynamics that cannot be determined by selection pressures from the environment alone. Instead, an interactive process must be considered, in which the activity of embryonic systems and whole organisms enter the equation. Or, in other words, agency cannot be accorded solely to the selective forces: individual entities exhibit agency. There is a good reason, Hoffmeyer holds, why evolutionary biologists have shrunk away from taking this step. It implies saying goodbye to externalism, and it thus involves a break with deep-rooted preferences of science. By according agency to individual organisms, a creative element is introduced into the world, which has been forbidden ever since the Newtonian revolution.

This antinomy may also be the deeper origin of the sharp controversies around the new approaches to evolutionary biology called "Extended Synthesis of Evolution" (Bateson et al. 2017; Corning 2020; Laland, Uller, et al. 2014; Pigliucci and Müller 2010), and it is also the origin of the disregard of organismic approaches since the beginning of the twentieth century (see chapter 2).

Hoffmeyer describes that semiosis often assumes a weblike character, where the interpretant may release further sign processes by inducing the formation of new interpretants in the system itself or in other systems. To illustrate this, he depicts the process of male sex determination in humans. Around the seventh week, certain embryonic cells called epithelial sex cord cells, for unknown reasons, begin to express a gene (termed *SRY*) located at the Y-chromosome. The unknown signal, which induces the activity of the *SRY* gene, needs to be related to the appropriate timing within embryonal development as well as to the morphological context. Thus, it is context dependent. This results in the production of the *SRY*-factor (formerly called testis-determining factor [TDF]) and perhaps in the activation of the expression of a few other genes. From there on, apparently, the male sex determination process is performed by these transformed cells. What we see is that the organism acquires its male determination through a series of steps, whereby semiotically competent cells "read the messages" made available to them in part from their internal genetic makeup, and in part from the external biochemical context set by a multitude of cues (molecular signs) derived from neighboring cells or from other embryonic tissues. These contextual cues are received at specific receptors located in the plasma membrane. Even though all these processes may sooner or later be fully characterized at the biochemical level, Hoffmeyer remarks, this will not by itself exhaust our need for explanation. For obviously, we are not dealing here with a haphazard mixture of biochemical processes, but with a delicately organized system of processes and signals. What we really want to know and understand is the logic of the organization of these biochemical processes.

In my view, the semiotic interpretation introduces two important points here. First, it indicates that the signaling pathways contain a meaning, a content. They need the molecular vehicle described by biochemistry, but the essential thing is their informative content. "Semiotic causation" is thus "bringing about things under guidance of interpretation in a local context" (Hoffmeyer 2013, 158). The second point is that agency here resides within the cell, the tissue, or the organism, not within the genes, which are necessary nonetheless. The activity of the cell interprets the genome at the appropriate moment and location. Signification, then, is the capacity of agents to use signs as guides for activity and is a fundamental property of all living organisms.

Hoffmeyer (2013, 162) is convinced that a

semiotic understanding of animate nature will potentially influence science and culture in important ways. Above all, it will strengthen our human feeling of relatedness to the other creatures of this world and our belonging in the biosphere. The image of animals and plants as stupidly obedient slaves of simple survival schemes will dwindle and be replaced by an understanding of, and an admiration for, the marvelous semiotic interaction loops through which organisms pursue their interests. Living beings are not the senseless and ignorant machines that science has taught us they are, and in the long run this can well have profound implications for how we treat natural systems.

4.11 Subjective Experience

Some very basic and rudimentary forms of subjective experience and decision-making may already be present within single cells (prokaryotic and eukaryotic) as well as in primordial multicellular animals. This may have been a starting point for the evolution of more sophisticated forms of experience such as consciousness, cognition, and mind. If this is the case, then subjective experience belongs to the elementary properties of life, at least within the animal kingdom.

A Challenging Property

The topic of subjective experience and consciousness in animals or life in general is difficult. For a long time, it was hard to discuss the phenomenon at all in a scientific context. The experience of consciousness is of course familiar in humans, and common sense attributes something comparable to higher animals around us, such as dogs, horses, and many birds. However, scientifically this seemed to be inappropriate because it hardly seems to be possible to study and substantiate the phenomenon in objective terms. However, in recent years this attitude has changed fundamentally, although there are very different views and theories on this question (Birch et al. 2020; Feinberg and Mallat 2016; Godfrey-Smith 2016, 2020; Ginsburg and Jablonka 2019, 2020; Noble et al. 2013; Reber 2019). In the last years, many scientists have dismissed the view that only humans have consciousness.

Subjective experience can be regarded as a basic nonelaborated form of consciousness (Godfrey-Smith 2016, 2020). It may form a baseline, a minimal form of consciousness, containing several comparable qualities. However, boundaries and transitions between subjective experience and consciousness can hardly be defined and identified, so that the usage of both terms is inconsistent.

In general, "a conscious system—an experiencing subject—has a subjective point of view on the world and on its own body" (Birch et al. 2020, 1).

A related although distinct term is *cognition*. It describes primarily the mental operational activity in learning and problem solving in humans, but it is being increasingly attributed to some higher animals. One pioneer in the scientific description of cognition in animals was Donald Griffin (Griffin 1976, 1984, 2001; Griffin and Speck 2004). Today the field of cognitive ethology is a fruitful and rapidly developing field of science, increasingly unmasking all the amazing mental and behavioral abilities in animals. It is much more difficult, however, to trace cognitive abilities down to the evolutionary past and to primordial animals.

A recent issue of the *Philosophical Transactions of the Royal Society* with the thematic focus on "basal cognition" makes a strong case for deep origins of cognitive abilities (introduced by Lyon et al. 2021). Lyon et al. write that much about the evolution of cognition is not known yet. But latest considerations and knowledge clearly point to an origin in the earliest cells and most primitive organisms. At the same time, interesting behavior is being discovered in a wide variety of organisms without a nervous system as well as organisms with simple nervous systems. Therefore, they argue, there is no reason to assume that cognitive abilities are restricted to animals with a nervous system or even more complex brains:

We believe that the study of cognition is on the cusp of a seismic shift similar to the Copernican and Wegnerian revolutions. If we truly recognize, in a biologically realistic fashion, the deep evolutionary inheritance of cognitive behaviour—individually and collectively, in both unicellular and multicellular organisms—a great deal of data that currently resist understanding will be more comprehensible and their implications less obscure. Or so we aim to demonstrate. Evidence that cognitive concepts such as "sensing", "memory", "learning", "communication" and "decision making" can be applied nonmetaphorically to the behaviour of bacteria (for example) is extensive and growing. (Lyon et al. 2021, 2)

Cognition in the theme-centered issue is focused on processes by which animals acquire, process, store, and act on information from the environment to become familiar with, value, and interact with features of their environment to meet existential needs. Indeed, such characteristics are found in any autonomous organism. They accompany organisms from their evolutionary beginning.

Obviously, the question of cognition still needs to be distinguished from self-experience or sentience, but at present there is no answer to the question of how they are related. If self-experience does occur somewhere in the evolutionary series, it could have its origin in forms of basal cognition. At the very least, one can postulate that these qualities are closely intertwined, or are one and the same in the first place and are viewed under slightly different perspectives in research. This, however, opens up an essential area of investigation that has been neglected up to now. It is once again the appeal to start with the actual phenomena and to take them seriously instead of parking difficult phenomena on a siding. The complex of phenomena of consciousness, cognition, and self-awareness is real and calls for an explanation (Godfrey-Smith 2020; Kather 2003; Lyon et al. 2021; Nagel 2012; Reber 2019).

No attempt will be made here to differentiate between terms such as experience, sentience, cognition, mind, and so on. In general, it would be interesting and helpful to develop a differentiation of these phenomena, but this is not the right place and must be left open here. Only an attempt shall be made to suggest different degrees of consciousness and cognitive abilities without being able to lay claim to precise definitions or delimitations. Further discussion is a matter for future research, which is just gaining momentum on this topic.

Neither will an attempt be made to provide an overview of the related literature, which is quite diverse (a good overview is provided by Reber 2019). Rather, it must suffice to describe a few basic phenomena, which suggest that essential elements of subjective experience and basal cognition are already qualitatively connected with life itself. It will be proposed that Godfrey-Smith (2020, 26) is correct when he formulates that "The background to the evolution of the mind is life itself. . . . The start is the cell," and that Reber (2019) is right with his postulation of a "Cellular Basis of Consciousness."

Two directions of argument are possible for this: one is top down and the other is bottom up.

The Top-Down Argument

The quality, which we humans experience as consciousness, can—with some caution—also be attributed to animals. This seems to be describable for higher animals such as other primates, dogs, horses, crows, parrots, and so on (Allen and Bekoff 1997; Bekoff et al. 2002; Bekoff 2007; Byrne 1995; Godfrey-Smith 2016, 2020; Gould and Gould 1994; Griffin 2001; Griffin and Speck 2004; Stamp Dawkins 1993). Especially the field of cognitive ethology has developed an impressive body of research approaching related questions in recent decades.

Famously, in 2012, a group of prominent neuroscientists formally recognized the importance of the issue and signed the "Cambridge Declaration on Consciousness" (Low 2012):

The absence of a neocortex does not appear to preclude an organism from experiencing affective states. Convergent evidence indicates that non-human animals have the neuro-anatomical, neurochemical, and neurophysiological substrates of conscious states along with the capacity to exhibit intentional behaviors. Consequently, the weight of evidence indicates that humans are not unique in possessing the neurological substrates that generate consciousness. Non-human animals, including all mammals and birds, and many other creatures, including octopuses, also possess these neurological substrates.

Nonetheless, when it comes to other vertebrates, such as fishes and amphibians, this is difficult to judge, because the scope of their internal experiences is quite far removed from ours. They have, however, emotions such as fear; they also experience pain. They are sentient and have some form of subjective experience, all of which can be studied by systematic research today.

It is even more difficult to decide about consciousness in invertebrates, but here too some specific phenomena are describable. In the literature especially cephalopods (as mentioned in the declaration) and arthropods are discussed as having some form of conscious experience (e.g., Birch et al. 2020; Ginsburg and Jablonka 2019).

It seems to be complicated or even impossible to define a dividing line between those living beings with and without such abilities. Different efforts have been made to find criteria for such a dividing line. The overall impression is that, although these attempts are well able to describe differences in the extent of such qualities, they are rarely able to exclude that more rudimentary forms of conscious experiences are present in other (and especially more basal) animals.

This was Hans Jonas's argument (1992), when he stated that if humans are related to animals, then animals are also related to humans and are in various degrees bearers of that inwardness, of which the human being is conscious. At which point then in the row can a line be drawn with good reason, with a "zero" of inwardness on the side turned away from us and with the beginning "one" on the side turned toward us? Where else, he asks, than at the beginning of life can the beginning of inwardness be set?

Charles Darwin also used this argument when he asked whether animals have consciousness. His ideas about evolutionary continuity—that differences between species are differences in degree rather than kind—lead to the firm conclusion that if we have something, "they" (animals) have it too (Godfrey-Smith 2020; Smith 2010). Traits, skills, functions, processes, and forms evolve over geologic time, and their foundational traits and functions are seen in earlier species in the phylogenetic panoply. An important part of evolutionary biology is to search for the roots of components of evolved animals, and so there should also be a quest for the roots of consciousness.

This is also one of the arguments Godfrey-Smith (2016) uses to propose a deep origin of mind and consciousness. He quotes William James, who argued in 1890 to take "continuity" seriously in understanding the origin of mind, and to realize that the elaborate forms of experience found in us derived from simpler forms in other organisms. Consciousness surely did not, James claimed, suddenly irrupt into the universe fully formed. "The history of life is a history of intermediates, shadings-off, and gray areas," Godfrey-Smith extends the argument. "Much about the mind lends itself to a treatment in those terms. Perception, action, memory—all those things creep into existence from precursors and partial cases. Suppose someone asks: Do bacteria really perceive their environment? Do bees really remember what has happened? These are not questions that have good yes-or-no answers. There's a smooth transition from minimal kinds of sensitivity to the world to more elaborate kinds, and no reason to think in terms of sharp divides" (Godfrey-Smith 2016, 77).

He says, however, that this gradualist attitude makes a lot of sense for qualities such as memory, perception, and so on, but that it is more difficult to evaluate this precisely for something like subjective experience.

We humans experience consciousness in the first-person perspective, that is, the personal inner perspective, and we can be quite sure that this is a real phenomenon (even if some scholars try to persuade us that it is an epiphenomenon). However, we cannot experience it directly in other persons. There is no direct way of perceiving consciousness in somebody else. Nonetheless, there are plenty of possibilities to experience indirectly that the other person must have consciousness just as we do. On the one hand, one can easily realize this in the way I myself experience consciousness: this must probably also apply to other people, because it would be very unlikely that I am the only person in the world who is equipped with the experience of such consciousness. Furthermore, I experience gestures, facial expressions, and body postures of the other person, finally an infinite number of signals, which I know from myself and which are connected with certain conscious processes within me. Thus, I presume that similar experiences of a consciousness should also take place within another person. We can intensify this approach, however, if we develop our capabilities for empathy into a kind of sensory organ. This requires practice and works only relatively well, but it can lead to a very intensive way of perceiving others. Beyond that, by means of language we can exchange our feelings and experiences and communicate them extensively.

In the same manner, we can master the difficult task of understanding animals. With everything we know about their biology and their way of life, and with all the possibilities of observing and empathizing, we must strive to comprehend as much as possible of their world, which is initially alien to us (de Waal 2009).

Many ethologists have done this in a quite impressive way regarding mammals, birds, and octopuses. Even in more remote animals we can at least surmise what might be going on within them or describe differences in behavior and reactions to conclude indirectly on their internal status. Often, many different observations need to be put together to achieve a secure scientific observation. After all, ants become frantic when disturbed, and snails retreat into their shells. This is related to some kind of subjective experience—of whatever quality.

The Bottom-Up Argument

Even single cells already have multiple means of sensing their external environment, and they use signals for interaction with other living beings in their vicinity. Furthermore, they are also capable of perceiving their internal status, such as the supply of nutrients or gases, osmotic status, or temperature. In addition, cells have the ability to reconcile the external situation with their internal status. They are able to interpret their perceptions. This is a prerequisite for reacting appropriately to given circumstances and to act according to their individual needs and interests.

For example, if there are harmful substances in the environment, many cells are able to move away from them and to move toward areas with lower concentrations of those substances. The single-celled amoeba responds to its environment, approaching chemical gradients in its immediate vicinity that come from a food source, and withdrawing from potentially damaging noxious chemicals which contact its cell membrane. The amoeba can even respond to a source of light, and it reacts differentially to hot and cold temperatures. Other organisms experience the source of light, which they need for photosynthesis. The green alga *Volvox* (having a simple multicellular organization), for example, executes considerable vertical migrations to make use of the sunlight during the day and to gain more nutrients at night.

Organisms react like that not just because the substance or the energy is available, but because they discern that these have a relevance for their internal status and situation. Single cells, as well as multicellular organisms, are able to discern and experience themselves in relation to internal and external conditions and factors. Of course, it may be very simple and direct in primordial organisms.

This type of sensing and subsequent processing of signals according to an internal perspective and status is a very basic and rudimentary ability of *subjective experience*. It is the ability to sense oneself in relation to internal and external conditions. The experiencing subject has a subjective point of view on the world and on its own body. According to Birch et al. (2020), such a subject has selective attention and exclusion, intentionality, integration of information over time, an evaluative system, agency and embodiment, a registration of a self/other distinction, and many more faculties.

Unlike dead objects, a living being senses, however rudimentarily, a difference between itself and the world. Without the opening to the world, a living being would have no relation to itself; and without the sensation of itself, it would not perceive the world. One can indeed construct inanimate systems that control their movements through causal feedback mechanisms. However, in the case of living beings, the relation to themselves and to the world is determined by an experience: there is a sense of "self."

The processes that are involved in these phenomena can be described as "basal cognition," as mentioned earlier. Lyon et al. (2021) describe that the evidence that cognitive concepts such as "sensing," "memory," "learning," "communication," and "decision making" can be applied nonmetaphorically to the behavior of bacteria and unicellular eukaryotes is constantly growing. In addition, the infrastructure for capacities typically associated with brains long predates the evolution of neurons and central nervous systems. Lyon et al. argue that taking modern evolutionary and cell biology seriously now requires the recognition that the information-processing dynamics of "simpler" forms of life are part of a continuum toward human cognition. These authors define basal cognition as the fundamental processes, which enable organisms to track some environmental states and act appropriately to ensure survival (finding food, avoiding danger) and reproduction long before nervous systems, much less central nervous systems, evolved. "Life at every level of development demands tough choices. This is what basal cognition is for. When nervous systems first evolved, this is also what they evolved for. As it is used here, basal cognition also describes a toolkit of biological capacities involved in becoming familiar with, valuing and exploring, exploiting or evading environmental features in the furtherance of existential goals" (Lyon et al. 2021, 5).

Reber (2019) makes a strong case for an early origin of basic forms of consciousness. In his very thoughtful and convincing book, he argues that single-celled species, procaryotic as well as eucaryotic, already have a rudimentary form of consciousness, which then advances throughout the evolution of the animal world. In his view, life creates consciousness, and it does so without a nervous system. At the beginning there is a very tiny but determinable mental existence. Because he assumes that the phenomenon has its first origin in the living cell, he calls his approach the cellular basis of consciousness (CBC).

CBC is mainly predicated on two assumptions: "(I) The origins of mind and consciousness will be found in the simplest, single-celled organisms, and (II) Consciousness, subjectivity, phenomenal experience or, if you prefer, sentience is an inherent feature of living organic form(s)" (Reber 2019, x).

Reber bases his arguments on recent research, which increasingly describes the surprisingly sophisticated systems of unicellular organisms for sensing, communicating, conditioning, and in many cases also learning and remembering, including immunological memory. In addition, there are studies that show that a unicellular organism perceives its own physiological form and thus has some representation of its "self," which is distinct from the sense of others, so that there is a basic form of decision-making. Therefore, he argues, single cells are already more than stimulus-response reaction automata, whose behaviors can be described in simple mechanical terms.

This process becomes more and more elaborated throughout the evolution of the animal world, which generates nervous systems, brains and elaborated sensory organs. He explains: "All experience is mental. All organisms that sense, perceive, and behave have minds; all have consciousness. There is only one category of sentience, although there is a vast plethora of types within it, each representing the particular kinds of phenomenal experiences characteristic of that species and expressive of its evolutionary roots and current ecological niche" (Reber 2019, 12).

He observes a multiple appearance of particular forms of consciousness, that made it difficult for many observers to realize a common principle in all this, for which he argues: "Because consciousness, sentience, phenomenal experience, subjectivity—whatever one wishes to call it—is so pervasive, so richly displayed across the vast panoply of species and environments and ecologies from the crushing depths of the mid-ocean rifts to the highest, rarified plateaus of the Andes and Himalayas, from species to species to species it's far from obvious that all are mere tokens of a singular type" (Reber 2019, 115).

Regine Kather (2003) explains how the inner vitality of an organism is gained at the price that it can also be destroyed. The environment can become dangerous for the organism that has a certain autonomy from it. A program or a machine is indifferent to its existence. The fact that an organism fights, flees, or resists for its survival implies that it somehow senses satisfaction or thwarting of needs, being or not being as a difference. Striving for self-preservation is not only a function of the organism, but at the same time its vital interest. It has a value. Because of the inseparable connection between self-perception and world-perception, the reaching out into the world is not a blind reaction to random events. It has a direction; it is guided by the need to live. The environment is evaluated, it acquires a meaning for a living being that wants to survive. To be or not to be are no longer of equal importance. Kather calls this "qualified perceptions." Perceptions are interpreted and acquire a meaning for the organism.

As described in section 4.3, unicellular organisms already separate themselves in relation to their environment and form an inner space. This is a first form of emerging autonomy. The formation of a closed membrane space, defining an inside-outside asymmetry, must have been a decisive step on the path leading to the appearance of living systems. At the same time, a rudimentary internal representation of what is going on outside of the system is established. This last quality may first be a sort of measuring process and may soon have generated the beginning of interiority and self-determination. However, if this is the primal form of autonomy and inwardness, there is no reason not to look for the origin of a very initial, simple form of subjective experience here.

With the formation of nerve cells in multicellular animals, the extent of qualified perception, self-determined processing, and self-determined reaction increases, but the basic principle is already present even in early forms, which do not yet possess nerve cells.

The more complex the experience and the higher the level of subjective experience, the more differentiated the references become to oneself, to other living beings and to the world. The increase in self-awareness is accompanied by ever more comprehensive representations of the environment. As individuality becomes more and more pronounced, social roles also become more differentiated.

From Subjective Experience to Mind

It has now been mentioned several times that there are different degrees of subjective experience and cognition. This spectrum can be extended up to abilities such as learning, memory, and acting from insight. However, it is not yet possible to precisely mark boundaries and transitions.

Ginsburg and Jablonka developed an approach for studying the evolutionary origins of consciousness, which they call the unlimited associative learning (UAL) framework. The central idea is that there is a distinctive type of learning that can serve as a marker for the evolutionary transition from nonconscious to conscious life (Birch et al. 2020; Ginsburg and Jablonka 2019). They define the term *unlimited* as follows: the possibilities for learning are sufficiently open-ended, such that there is no serious prospect of all the possible associative links that could be produced actually being formed by a real organism. A system with a capacity for UAL can, within its own lifetime, learn about the world and about itself in an open-ended way.

Based on a review of the literature on animal learning, they conjecture that UAL is present in most vertebrates, some cephalopods, and some arthropods (including honeybees and fruit flies). In these taxa it is also possible to identify the brain regions underlying processes of integration of several inputs, dedicated memory systems for the storage of compound precepts, value systems dedicated to motor programs, and sensory-motor associative areas. If UAL is used as a transition marker for consciousness in this sense, it hints at three separate events as the origin of consciousness (Birch et al. 2020): one in the vertebrate lineage, one in the arthropods, and one in the cephalopod mollusks. If UAL is present in all or most extant vertebrates, this points to an origin event in the Cambrian period. A similar hypothesis has been reached via a different route by Feinberg and Mallat (2016).

Importantly, Birch et al. state that UAL is only a positive marker: it can tell (in their demarcation) which animals are conscious, but it does not aspire to tell us which are not. Here is not the place to discuss whether UAL can in some form discriminate between nonconscious and conscious beings, as is expected by the authors, although formulated cautiously. However, they obviously describe forms of experience that have already gained some complexity and sophistication. Therefore, in a tentative way here, a spectrum of degrees of the ability in question may be postulated (figure 4.26). First, subjective experience generates the common basis of internal sentience, then in more elaborated animals—especially those with more complex nervous systems—consciousness might be identified, and then in higher animals, further abilities such as flexible behavior, memory, self-awareness, and intelligence become possible (in the figure provisionally called mind).

If basal cognition is included as a somehow related basic quality of subjective experience, this would propose a hierarchically nested model for the phenomena in question. Again, here is not the place to discuss details of these relationships. It shall only be stated that mental abilities most likely have a deep origin.

Feinberg and Mallat (2016) study the phenomenon of consciousness from a neurophysiological background. Basically, they associate it with the origin and evolution of nervous systems, beginning during the Cambrian period and the subsequent generation of brains. In their introduction, they discuss the difficulty of studying consciousness from a scientific perspective. They describe the classical "core problem" of neurological studies: How is the material brain able to produce a nonmaterial phenomenon? Essentially, they state that many aspects of this relation will necessarily remain unexplained, because consciousness can only be studied from the first-person perspective and everything else, concerning animals for example, needs extrapolation from observed phenomena.

The problem is that when it comes to explaining sensory consciousness in terms of physics, chemistry, or even neurobiology, certain aspects of subjectivity always appear to remain unexplained. As philosopher Joseph Levine put it, a mysterious "explanatory

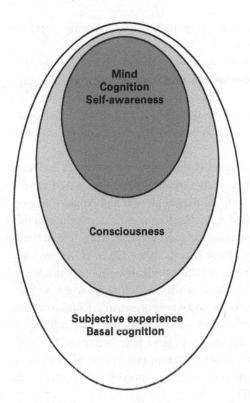

Figure 4.26
Hierarchical representation of experiential and cognitive abilities.

gap" remains between the physical properties of the brain as we know them and the subjective experience that the brain thereby creates. He argues that no matter how detailed the objective explanation of the neural pathways that create subjective experience is, something is always left out—namely, the personal "experience" itself. (Feinberg and Mallat 2016, 2)

The question that arises here is whether an organismic approach could overcome some of the obstacles inherent in studies of this kind, or rather—as in other topics that have been touched on throughout previous chapters— enable some new meaning and comprehension within an organismic context. If the brain is not only seen as a material brain, but as a living brain, including its systems aspects, its self-generated activity, and the fact that it not only transports information along its neurons and axons, but is itself organized by embodied information, it should become clear that much more is needed to study the brain than just physical, chemical, and neurobiological properties.

And there are impressive approaches today that are overcoming previous limitations of neurobiological research.

These include, for example, the investigations of Buzsáki (2006), who found a hierarchy of rhythmic oscillations of the most diverse wavelengths, thus drawing attention to the fact that brain functions are apparently also a function of time (time autonomy). Other disciplines take the systems principle seriously with respect to the brain and point far beyond the widespread reductionist approach (Auletta et al. 2013).

The impression remains that a bridge between material processes and subjective experiences could be built if such investigations were based on the actual properties of life. This can only be hinted at here and would make up a whole research program. Nonetheless, research programs developing in this direction already exist. This is especially the case with the embodiment theories, which have been increasingly elaborated in recent years. However, they do not consider only the brain. Crucial for the approach is that they acknowledge the brain as embodied in a whole organism and consciousness as produced by the interdependent system "brain and whole organism." Consciousness would then not only be a neuronal function, even though the brain is likely to have a central position in this, but a property of the whole organism. Embodiment theories also emphasize that a more pertinent concept of life would be necessary, and that such a concept has not been developed so far (Fuchs 2018; Noble et al. 2013; Thompson 2007).

In a very fundamental way Nagel (2012) stated that a scientific worldview can only be appropriate when its basic assumptions include the possibility to explain the emergence of consciousness and mind: "The existence of consciousness is both one of the most familiar and one of the most astounding things about the world. No conception of the natural order that does not reveal it as something to be expected can aspire even to the outline of completeness" (53).

Nagel does not see this as a given in the conventional scientific worldview. Again, here it would go too far to discuss this extensive and difficult topic. Certainly, no attempt shall be made to design an alternative worldview. Yet, with the components discussed here—that life is to be described on the basis of its unique properties, that these properties include from the outset those that are not primarily traceable to material interactions, and that from the simplest living beings onward a subjective experience is involved—perhaps some steps have been taken toward a solution of the problem raised by Nagel. "What interests me is the . . . hypothesis that biological evolution is responsible for the existence of conscious mental phenomena, but that since those phenomena are not physically explainable, the usual view of evolution must be revised. It is not just a physical process. . . . For a satisfactory explanation of consciousness as such, a general psychophysical theory of consciousness would have to be woven into the evolutionary story" (Nagel 2012, 50).

I do not claim that I am providing such a theory, but perhaps some building blocks.

4.12   The Ability to Evolve

With the last considerations of section 4.11, the next property of living beings has already been introduced: the ability to evolve. The energetic-metabolic-informational state of disequilibrium, which organisms generate repeatedly during ontogeny and adulthood, can also be modified over time. Organisms went through consecutive changes and transitions, which not only enriched the different life-forms in the sense of diversification but also lead to the generation of new systems states. Beyond various forms of adaptation, there also have been qualitative modifications and innovations with fundamental changes in structure and functionality. All these changes have been brought about by real living forms and groups of organisms, so that the ability to change, to generate new systems states, is in the long run an essential and fundamental property of living beings.

The Extended Evolutionary Synthesis

Throughout the twentieth century the standard model of the evolutionary process has been the synthetic theory of evolution, which concentrates on gene frequency dynamics in populations of organisms, driven by natural selection. The core of the theory is population genetics, and it postulates that the dynamics on this level also account for all other phenomena in evolution, including the major transitions that have occurred throughout evolution. The empirical basis and key concern of the population genetic approach is the measurement of trait variability in populations. It attempts to explain adaptive variation and speciation by calculations of fitness.

Müller (2017b, 3) summarizes the key tenets of this approach in a condensed form as follows:

(i) all evolutionary explanation requires the study of populations of organisms; (ii) populations contain genetic variation that arises randomly from mutation and recombination; (iii) populations evolve by changes in gene frequency brought about by natural selection, gene flow and drift; (iv) genetic variants generate slight phenotypic effects and the resulting phenotypic variation is gradual and continuous; (v) genetic inheritance alone accounts for the transmission of selectable variation; (vi) new species arise by a prevention of gene flow between populations that evolve differently; (vii) the phenotypic differences that distinguish higher taxa result from the incremental accumulation of genetic variation; (viii) natural selection represents the only directional factor in evolution.

Applying this model of evolution, current evolutionary theory is predominantly oriented toward a genetic explanation of variation. It performs well regarding the issues it concentrates on, providing predictions and explanations on the dynamics of genetic variation in populations, on the gradual variation and adaptation of phenotypic traits, and on certain genetic features of speciation.

However, beginning around the year 2000, biosciences found more and more phenomena that were not compatible with the view that all evolutionary changes and transitions are reducible to these population dynamics with accumulated genetic variations. Doubts that the assumed random process would be able to create order within the evolutionary process had often been formulated, but now several new enigmas arose with the growing knowledge of comparative molecular biology. It became increasingly difficult to explain the immense diversity of life despite its deep and pervasively similar molecular architecture. Modern genetics, and especially the augmented availability of whole-genome analyses verified that there is a deep homology of core genetic elements. Such elements have hardly changed during evolutionary transitions but are integrated into new situations and functions of the developing organisms.

Increasingly tangible insights into the origin of evolutionary innovations have emerged. Although the picture is still fragmentary, the findings contain several surprises. Symbiosis, for example, delivers a new state of a system within a single macroevolutionary step and probably has a function in a number of transitions in addition to the generation of the eukaryotic cell (Margulis and Sagan 2002). Thus, there seem to be systemic shifts in evolution and not just gradual processes, which Gould (2002) also emphasized from a paleontological perspective. Other examples come from cell biology, comparative genetics, and developmental biology, showing that novelties can be brought forth by new combinations of conserved structures and functions. The genome, at least in some parts, is obviously not so much a result of random mutations but of conservation of core functions together with new arrangements and duplications of building blocks (Carroll et al. 2005; Gerhart and Kirschner 1997; Kirschner and Gerhart 2005), and these combine with epigenetic functions (Jablonka and Lamb 2005). These results, together with the paleontological description of evolutionary patterns such as heterochrony (McKinney and McNamara 1991; McNamara 1990; Schad 1993) or convergence (Conway Morris 2003), are beginning to trigger a new stage in the evolution of evolutionary biology itself (Erwin 2000; Jablonka and Lamb 2005; Pigliucci and Müller 2010; West-Eberhard 2003).

Therefore, many evolutionary biologists from different disciplines are calling for an extension of the synthetic theory (Laland, Uller, et al. 2014; Laland et al. 2015; Laubichler and Renn 2015; Müller and Newman 2003; Müller 2017b; Noble 2013b, 2017b; Pigliucci and Müller 2010; West-Eberhard 2003). They criticize that population genetics is assumed to be the privileged type of explanation for all evolutionary phenomena, thereby negating the fact that, on the one hand, not all of its predictions can be confirmed under all circumstances, and on the other hand, a wealth of evolutionary phenomena remains out of consideration. For instance, the theory largely avoids the question of how the complex organization of organismal structure, physiology, development, or behavior actually arise in evolution, and it also provides no adequate means for including factors that are not part of the population genetic framework, such as developmental, systems theoretical, ecological, or cultural influences. Corning (2020) provides a short and precise overview of topics in this field.

Therefore, a broad discussion on an extended evolutionary synthesis, which seems to be necessary regarding these new developments, is under way. For the time being, one culmination in this development was a large international conference at the Royal Society in London in the year 2016 with the title "New Trends in Evolutionary Biology." The contributions have been published in a special issue introduced by Bateson et al. (2017).

Müller (2017b) explains that a renewed and extended theoretical synthesis aims to unite pertinent concepts, which have emerged from the novel fields, with elements of the standard theory. The resulting theoretical framework differs from the latter in its core logic and predictive capacities. Whereas the synthetic theory and its various amendments concentrate on genetic and adaptive variation in populations, the extended framework emphasizes the role of constructive processes, ecological interactions, and systems dynamics in the evolution of organismal complexity as well as its social and cultural conditions. Single-level and unilinear causation is replaced by multilevel and reciprocal causation. Among other consequences, Müller reasons, the extended framework overcomes many of the limitations of traditional genecentric explanation and entails a revised understanding of the role of natural selection in the evolutionary process. All these features stimulate research into new areas of evolutionary biology.

The following paragraphs will summarize some of the ideas and concepts, which are being discussed presently. It is a young field that only recently started to develop in a larger scale, and it has some diversities too. However, the most important issue is that a more diverse theory building, examining different perspectives, is possible now.

New Concepts in Genetics

Modern genetics demonstrates that there is something wrong with the assumption of the genome as a puzzle of accumulated random changes (Niklas et al. 2015; Shapiro 2011, 2013a, 2013b, 2017). First, there is a great deal of conservation in the genome. Genetic components involved in basic cell functions are old building blocks, which have only changed minimally during evolution. Second, it is becoming less clear what a gene really is. From one transcript, often several or even many different messenger RNAs can be assembled by way of the splicing process, and these are often the basis for many various proteins.

Another point is that there is seldom a one-to-one translation from genotype to phenotype. There are complicated self-sustaining feedback loop systems with multiple interacting genes and gene products, such as daily and seasonally cyclical/rhythmic changes in physiological states. To realize a phenotype, usually a cascade of interactions between DNA information and regulative functions is at work. Today, geneticists are beginning to think in terms of genetic networks composed of tens or hundreds of genes and gene products, which interact with each other and together affect the development of a particular trait. Geneticists today recognize that in the majority of cases, the development of a trait does not depend on a difference in a single gene.

Development involves interactions among many genes, many proteins and other types of molecules, the whole organism, and the environment in which an individual develops. Thus, the "genes," whatever they will turn out to be, do not determine organisms, although they contain indispensable information within the network.

In their influential book, Jablonka and Lamb (2005, 7) make a strong case about the new concept of the genome, as recent discoveries challenge the old ideas about what genes are:

No longer can the gene be thought of as an inherently stable, discrete stretch of DNA that encodes information for producing a protein, and is copied faithfully before being passed on. We now know that a whole battery of sophisticated mechanisms is needed to maintain the structure of DNA and the fidelity of its replication. Stability lies in the system as a whole, not in the gene. . . . And because the effect of a gene depends on its context, very often a change in a single gene does not have a consistent effect on the trait that it influences.

These new ideas about genes and genomes are having an increasing impact on evolutionary theory building. If a gene has meaning only in the context of the complex system of which it is a part, the standard way of thinking about evolution, in terms of changes in the frequency of one or more isolated genes, needs to be questioned. It is more appropriate to focus on changes in the frequency of alternative networks of interactions rather than on the frequencies of individual genes.

Also, these ideas and data are a strong argument in favor of the integrative systems model presented in section 4.2, not only as a model for single organisms but also as a framework for understanding evolutionary changes within these systems. The overall picture suggests that modifications can be generated on different system levels of the organism. Besides primary changes on the gene level, organismal systems seem to be able to integrate genetic building blocks into new situations and functions. Often the respective building blocks then undergo no (or only minor) changes, and they are used in a new context or at a different time in development (Müller 2017b; Noble 2011a, 2017b).

Jablonka and Lamb (2005) call for a renovation of evolutionary theory at the start of the twenty-first century. They argue for recognition of an "evolution in four dimensions," rather than focusing on just one. In addition, to take into account the regulative character of genetic components, they plead for adding a perspective, wherein three other inheritance systems also play essential roles in evolutionary change. These other systems are the epigenetic (organic systems outside, or in addition to, the DNA that can affect genetic expression, development, and biological function), behavioral (by means of social attention and learning), and symbolic (via language and other forms of symbolic communication) inheritance systems. Epigenetic inheritance is found in all organisms, behavioral inheritance in most, and symbolic inheritance only in human beings. Jablonka and Lamb state that some hereditary variations are nonrandom in origin, that some acquired information is inherited, and that evolutionary change can result from instructions as well as selection.

Epigenetic Inheritance

Most revolutionary are the findings in recent years about the further levels of inheritance beyond the level of DNA itself, which is called epigenetic inheritance (Jablonka and Raz 2009; Shapiro 2011). A person's liver cells, skin cells, and kidney cells look different and function differently, yet they all contain the same genetic information. With few exceptions, the differences between specialized cells are epigenetic, not genetic. They are the consequences of events that occurred during the developmental history of each type of cell and determined which genes are turned on and how their products act and interact. The remarkable thing about many specialized cells is that they not only can maintain their own specific phenotype for long periods of time but also can transmit it to daughter cells. When liver cells divide, the resulting cells are liver cells, not muscle cells or neurons. Thus, although their DNA sequences remain unchanged during development, cells acquire information that they can pass on to their progeny. This information is transmitted by way of what are known as epigenetic inheritance systems.

At present, the most well-known principles of epigenetic inheritance are chromatin marking systems. These involve the degree of condensation of chromatin (chromatin that is more condensed is less accessible to the factors needed for transcription), DNA methylation (methyl groups are attached to some bases, also regulating the access of transcription factors), and the modifications of the nucleosomal histones. Another process is that small RNA pieces are able to regulate post-transcriptional processes in the cell and thus contribute essentially to what is finally assembled.

Jablonka and Lamb (2005) argue that these principles of epigenetic inheritance are important prerequisites in the evolution of complex organisms, when developmental decisions have to be transmitted to daughter cells and the long-term maintenance of tissue functions depends on stable and transmissible cell phenotypes. They also argue that epigenetic variations can be transmitted not only in cell lineages but also between generations of organisms. Although genetically identical, organisms could have evolved because they passed on some of their epigenetic characteristics and thus provided an additional source of variation. Evolution can occur through the epigenetic dimension of heredity even if nothing is happening within the genetic dimension.

In addition, epigenetic variations might be generated at a higher rate than genetic ones, especially in changed environmental conditions, and several epigenetic variations might occur at the same time. Furthermore, they might not be blind to function because changes in epigenetic marks probably occur preferentially on genes, which are induced to be active by new conditions. This could increase the chances that a variation will be beneficial. The combination of these two properties—a high rate of generation and a good chance of being appropriate—means that adaptation by way of the selection of epigenetic variants might be rapid compared with changes via genetic change.

The discovery that some induced epigenetic states are inherited so phenotypic variants can persist even if the inducing conditions do not, is adding a whole new dimension to the question of evolutionary innovations. It seems that at least short-term evolution does not depend so much on new mutations but on epigenetic changes. They might introduce new variants or just unveil genetic variants already present in a population. In addition, it is perhaps possible that heritable epigenetic variants can do a retaining job until genes catch up.

According to Jablonka and Lamb (2005), this has general consequences on how we think about evolution. Because they give weight to the epigenetic, behavioral, and symbolic dimensions of heredity, evolutionary change does not have to wait for genetic changes. These can follow. Phenotypic modifications might come first.

It also implies that evolution can be rapid because often an induced change will occur repeatedly, and in many individuals simultaneously; therefore, it becomes likely that some levels can produce variations targeted toward special needs of the organism in its environment.

Phenotypic Plasticity

West-Eberhard's (2003) broad overview of developmental plasticity and evolution led her to suggest phenotypic plasticity (the capacity of an organism to generate different phenotypes using the same genotype) as one of the key factors of evolution. Like other authors, she argues that reducing the processes of development as well as evolutionary change to the genomic level is not appropriate. She rather sees evolution as phenotypic change involving gene frequency change and not just as gene frequency change alone.

For her, the secret to understanding evolution is first to comprehend phenotypes, including their development and their responsiveness to the environment. The phenotype is characterized by plasticity, which allows for the evolution of variations. Only later might this be fixed within the genome. In this context, genes are followers rather than leaders in evolutionary change.

West-Eberhard considers environmental induction a major initiator of evolutionary change. Evolutionary novelties can result from the reorganization of preexistent phenotypes, and phenotypic plasticity can facilitate evolution by accommodation. The term *accommodation* is central to her approach. She differentiates between phenotypic accommodation and genetic accommodation. Phenotypic accommodation is the adjustment, without genetic change, among variable aspects of the phenotype, following a novel or unusual— external or internal— input during development. Genetic accommodation is a genetic change that affects the regulation or form of a new trait.

Thus, an evolutionary change might follow several steps: First, a novel input (a mutation or an environmental change) affects some individuals or a population. Because of inherent developmental plasticity, a phenotypic accommodation of the novel input occurs; consequently, a novel phenotype emerges. If the novel phenotype is advantageous, natural selection fixes it by stabilizing its appearance through an alteration of the genetic processes, so that genetic accommodation has occurred.

Facilitated Variation

As already described, increasing knowledge in comparative molecular biology produced a huge surprise: Where the synthetic theory expected the most variation, on the level of genes and their products, there is far-reaching conservation. This is the starting point for the discussion by Gerhart and Kirschner, who developed the theory of facilitated variation (Gerhart and Kirschner 1997; Kirschner and Gerhart 2005). Conservation means that even distantly related organisms use similar processes for cellular function, development, and metabolism. Each process, with many protein components working together, contributes to the phenotype. When a process is conserved, most of its protein components are conserved. Details of metabolism are the same in certain bacteria and humans, basic cell organization and functions are similar in yeast and humans, and some developmental strategies of fruit flies are strikingly similar to those of human beings.

There are also new features that had no forerunners in more ancestral organisms, so that organisms are a mixture of conserved and nonconserved processes. However, novelty in the organisms' physiology, anatomy, or behavior arises mostly by the use of conserved processes in new combinations, at different times and in different places and amounts, rather than by the invention of new procedures. The surprisingly small number of genes for humans and complex animal forms reflects the anatomical and physiological complexity that can be achieved by the reuse of genetic products.

The conserved processes are fundamental cellular processes: They operate on many levels within the development and functioning of the organism. Gerhart and Kirschner call them "the core cellular processes."

Central to the argument of Gerhart and Kirschner is that these conserved processes facilitate rather than constrain evolutionary change. These processes have been conserved, they suggest, not only because changes would be lethal but also because they have repeatedly facilitated changes of certain kinds around them. Many of the conserved core processes have the capacity to be easily linked in new combinations. New linkages can occur with a minimum requirement on genetic change and hence can happen readily. They can arise with little or no change of the units themselves.

Thus, some genetic—and possibly much epigenetic—variation is needed to integrate conserved components into new heritable functions. Genetic changes might especially have occurred in regulatory regions of genomes. Gerhart and Kirschner call the increasing possibilities of integrating conserved functions into new combinations and the involved regulatory changes to give new outputs of the conserved processes "facilitated variation." This increased during evolution so that there was an "evolution of evolvability."

One most obvious example of this principle on a systems level next higher than the genome and the organelles is the eukaryotic cell. It is essentially always the same cell type that delivers the building block for multicellular life. These cells might have many specializations, such as being neurons, liver cells, or epithelial cells, but they all have the same basic equipment. Novelty usually comes about by the deployment of existing cell behaviors in new combinations and to new extents, rather than in their drastic modification or the invention of completely new ones.

What changes on the cellular and molecular level, the argument of Gerhart and Kirschner continues, are regulatory components: small features of proteins, RNA or DNA, that allow a regulation of time, circumstance, and degree of activity in the processes. These are often involved in controlling the linkage and activity of the processes as well.

The eukaryotic cells within a multicellular organism especially gain functions for their regulation within the whole system. Multicellularity is characterized by the emergence of cell specializations within the same organism: complex physiology, complex spatial organization, embryonic development, and complex life cycles. All of these depend on an elaborate interaction of cells with each other via systems of signals and responses. Thus, the mostly conserved processes within the cell must have a regulatory linkage to extracellular and intracellular events. Regulatory linkage is defined as the complex association of conditions and responses. For example, cells in multicellular organisms acquire many signaling molecules at their membrane to make them receptive within their tissue and within the whole organism.

Within such multicellular systems, the organism's anatomy, physiology, and behavior are only remotely connected to the DNA sequence, Gerhart and Kirschner argue. In between, there are all these complex processes of growth, development, and metabolism. A change in the DNA sequence can therefore be only indirectly correlated with a change in anatomical and physiological characteristics of the organism. Gerhart and Kirschner maintain that currently our understanding of this connection is not sufficient enough for us to predict the phenotypic consequences of most genetic changes. We can identify genes that predispose a person to cancer, but we cannot draw a perfect correlation between the gene and the disease. Given the remote connection between the DNA and the phenotype, we have no way of knowing how often random DNA modification can produce useful results for the organism. Without an adequate understanding of how DNA changes are interpreted by the organism, we cannot recognize how important they are during evolution.

Further principles the authors identify as components of evolvability are "exploratory processes" (many complex processes are not determined by the genome but have self-organizing features according to their functionality, such as the fine wiring in the brain); "weak linkage" (features of information systems, where signals of low information content evoke complex, preprogrammed responses from the core process); and "compartmentation" (different compartments can develop different functions from the same basic organization).

In their view, the capacity for facilitating variation has itself evolved as the core processes of organisms have accumulated "more adaptive and robust behaviors." Evolution does not proceed on a random generation of dysfunctional phenotypes, which usually results in lethality and only by accident gives rise to an advantageous trait. Lethality is mostly an issue when genes are mutated that encode components of the conserved processes. These mutations are eliminated by selection in each generation. Exempting those, the population accumulates genetic variation because of the robustness of physiologically adaptable processes, and the individual brings forth phenotypic variation in response to genetic change or environmental change, which is predisposed to be less lethal (see also Newman 2002).

Gerhart and Kirschner (1997, 140) accentuate the importance of the phenotype for evolutionary changes:

Evolutionary biologists . . . need to explain why organisms have changed. Evidence of conservation, to a first approximation, suggests that they are looking in the wrong place. The difference between birds and mammals is not going to be found in the structures of their muscles or nerves, their types of collagen or their microtubules. It will be found in their wings, their feathers, their sweat glands, their hair, and the organization of their brain cortex. Change has occurred principally in the organization of tissues and in the evolution of novel physiological and embryological mechanisms.

Similar to the approach of Walsh (2015), which was discussed in section 4.4, Gerhart and Kirschner (1997, 252) consider organisms as being more actively involved in their evolution: "On the side of generating phenotypic variation, we believe the organism indeed participates in its own evolution, and does so with a bias related to its long history of variation and selection."

In the same sense, Shapiro (2013a, 2013b, 2017) even proposes the view that cells have agency in generating organisms with new genome configurations and that this is important during evolution. The genome has traditionally been treated as a read-only memory, subject to change by copying errors and accidents. Instead, he writes, we need to change that perspective and understand the genome as an intricately formatted read–write storage system, constantly subject to cellular modifications and inscriptions. Cells operate under changing conditions and are continually modifying themselves by genome inscriptions. These inscriptions occur over distinct timescales such as cell reproduction, multicellular development, and evolutionary change, and involve a variety of different processes at each timescale. According to Shapiro, research dating back to the 1930s has shown that genetic change is the result of cell-mediated processes, not simply accidents or damage to the DNA. This cell-active view of genome change applies to all scales of DNA sequence variation, from point mutations to large-scale genome rearrangements and whole-genome duplications.

The geneticist Barbara McClintock, who during the 1930s and 1940s first observed that whole domains of genetic material move around the genome, called the genome "a highly sensitive organ of the cell," which is constantly monitored, corrected, and restructured by the cell to maintain its functionality (quote from Noble 2017b). Today, it has clearly been described that organisms can reorganize their genome. McClintock could not have anticipated the extent to which her idea would be confirmed by the sequencing of whole genomes, Noble (2017b) explains. Comparisons between sequences in completely different species of eukaryotes demonstrate that the evolution—at least of certain proteins—must have involved the movement of whole functional domains. This is far from the idea of slow progressive accumulation of point mutations, but rather is a function within the system.

In addition, the physiological processes, by which events in domains near the cell surface signal to the nucleus to control specific gene expression levels, have now been studied in detail. There is no longer any mystery in understanding the highly specific transmission of information to the nucleus that can control gene expression and its organization. Noble and Noble (2017) assume that behavior is one factor that can rearrange existing functionality and is thus able to direct evolution, so that evolution is not just random.

Niklas et al. (2015) present strong evidence for the plasticity of regulatory networks during development. Alternative splicing of mRNA transcripts and other processes above the level of DNA allow for a high versatility in the usage of DNA elements. The occurrence of alternative splicing increased steadily during the last 1.4 billion years of eukaryotic evolution, and it has been assumed that organismic complexity increased especially due to an increasing information content on this level. The authors assume a sort of "splicing code," which defines the protein isoforms produced under different cellular conditions. This would be consistent with the view of evolution at different systems levels.

However, there is still a challenge for life sciences today to understand how all the regulatory processes and control circuits operate to make complex reproductive, repair, and morphogenetic processes come out right in the face of changing circumstances (Shapiro 2013a). Thus, the gap of knowledge between the genotype and the phenotype has not been bridged to a full extent. Yet, one thing can be said: The genome is a matter of high regulation, integration within the cell and the organism, and its organization is subjected to the agency of the cell. Thus, the genome is embedded within a multilevel hierarchy of systems. Triggered by empirical results, today genetics and evolutionary biology are becoming much more organismic than they had been during the twentieth century.

Gawne et al. (2018) emphasize the fact that biology is a science of levels, which by its very nature precludes the possibility of explaining evolution as well as development of phenotypes in terms of single factors. Understanding such processes almost always requires the integration of findings from numerous hierarchical levels. Biological causation, they state, is multivariate and multi-level, and any attempts to reduce this complex to one single factor will fail. The available data suggest that most phenotypes result from nonlinear interactions. Nonlinearity is present in almost all biological processes at all levels of organization.

Bernard Processes

Turner's (2007) concept, which we heard about in section 4.8, contributes another piece to the jigsaw puzzle of evolution by pointing to the importance of physiology on the level of the organism, including its immediate environment, as it builds up homeostasis to generate stability and regulation.

One of his examples is the termite mound. The mound not only ventilates the air but also regulates the gaseous composition within the colony. The mound captures wind energy at a particular rate matched to the colony's metabolism, which makes it an organ of homeostasis. During the growth of the colony, the termites adjust the rate of the mound's wind-powered ventilation by continually adjusting its structure to keep pace with the colony's growing respiratory demand.

As described, a simple principle underlies the complex process of building the chimney within the mound: The termites transport soil from areas of high carbon dioxide concentration to areas of low carbon dioxide concentration. In so doing, they maintain the structure of the mound as a whole in a way that allows it to perform its respiratory and regulatory functions.

Thus, a plan for the building does not exist within the termite so that it would know what to do at each and every point. The termite, rather, works according to the necessities of its direct environment, for which it is equipped with the necessary sensibility and the behavioral repertoire to operate appropriately at the respective point. For the mound as a whole, this sums up to a system that Turner (2007, 27) calls a "form of embodied physiology" and "embodied homeostasis," because it is simultaneously structure and function (= *Bernard processes*, see section 4.8). Further examples Turner refers to are tendons and muscle systems, arterial trees, and bones, which are typical samples from within single organisms that demonstrate similar general principles.

Evolution brought forth such self-organizing systems, so this principle appears to be an essential instrument for evolutionary change. Persistent environments are created by systems of Bernard processes, having a process-based form of heritable memory. Therefore, self-organizing systems might be of enormous significance in evolution and might in a certain sense also be intentional, as Turner assumes.

Turner suggests that something analogous to intentional planning goes on in many other domains of the biological world when homeostatic processes, themselves endowed with a form of reactive plasticity, help to direct the proper functioning of organisms.

Turner (2007, 1) summarizes: "My thesis is quite simple: organisms are designed not so much because natural selection of particular genes has made them that way, but because agents of homeostasis build them that way. These agents' modus operandi is to construct environments upon which the precarious and dynamic stability that is homeostasis can be imposed, and design is the result."

Like some of the other theories I mentioned previously in this chapter, Turner focuses on the phenotype, the largest level in the model of Weiss. Turner's point is that Bernard processes have a kind of flexibility, which means that the action of a system cannot possibly be understood in terms of genetically specified rules. Some genetic prerequisites are necessary, of course, but then the system works on its own and builds physiological stability, a principle Gerhart and Kirschner (1997) call "exploratory systems."

With this concept, Turner comes close to what I describe as autonomy. This is already expressed by the term *Bernard processes*. With his theory of the milieu interieur, Claude Bernard belonged to the first scientists pointing to the principle of organisms becoming independent from vagaries of the environment. Basically, what Turner describes are the means by which organisms gain autonomy.

Evo-Devo

The field of evolutionary developmental biology (evo-devo) compares the processes of development of different organisms to study the evolutionary relationships between them and to discover how developmental processes evolved. It addresses the origin and evolution of embryonic development, how evolutionary modifications of developmental processes lead to the production of novel features, the role of developmental plasticity in evolution, and the developmental basis of homoplasy and homology (Arthur 2004, 2011; Carroll 2005; Gilbert et al. 1996; Müller and Newman 2003; Müller 2007; Pigliucci and Müller 2010; Riedl 1978; Wagner and Laubichler 2004).

Since the 1990s, the contemporary field of evo-devo has gained impetus from the discovery that cascades of gene regulation are essential during the development of the embryo, and that small changes in the regulation networks might lead to extensive changes in phenotypes. Evo-devo demonstrates that during evolution developmental processes altered, creating novel structures from old gene networks. In other cases, conserved processes are involved in new structures and functions, so that unexpected "deep" homologies exist, as in *Hox* genes, for example.

The *Hox* gene cluster, or complex, as a classical example of evo-devo, belongs to the so-called tool kit genes. They are transcription factors containing the homeobox protein-binding DNA motif, which functions in patterning the body axis. By specifying the identity of particular body regions, *Hox* genes determine where limbs and other body segments will grow in a developing embryo or larva. Another example of such a toolbox gene is *Pax6*/*eyeless*, which induces eye formation in many animals, also when they are only distantly related to each other. Evo-devo also studies how development itself evolved.

Development of the embryo might generate variations by self-regulation, either spontaneously or induced by internal or external stimuli, including environmental ones. Thus, morphological form and complex structures such as body plans may be generated through describable internal changes within the system itself.

Findings suggest that the crucial distinction between different species, orders, or phyla might be caused less by differences in their genes than differences in spatial and temporal expression of conserved genes. The implication that large evolutionary changes in body morphology are associated with changes in gene regulation, rather than the evolution of new genes, suggested that changes in "switch" genes might play a major role in evolution and might be induced on different levels within the systemic hierarchy of the organism.

The proposal of West-Eberhard (2003) on developmental plasticity is also a focus of evo-devo and is derived from the recognition that phenotypes are not uniquely determined by their genotypes. Evolutionary changes in development might proceed by a "phenotype-first" route, with genetic change following, rather than initiating, the formation of phenotypic novelties.

The study of developmental processes is further extended by the more general idea of heterochrony. This term describes changes in the timing of a developmental process as relevant as a process of evolutionary change. Therefore, there can be dissociations of the development of different organs or body parts within an organism or also of the whole organism compared with the normal course of development and closely related species. Hypermorphosis, for example, involves a delay in the offset of a developmental process. Paedomorphosis (or juvenification) is a change in which the adults of a species retain traits previously seen in juveniles (Gould 1977; McKinney and McNamara 1991; McNamara 1990; Schad 1993). Heterochrony basically describes features of phenotypic plasticity, but there can also be heterochronic changes on the genetic level when the timing of gene expression is altered.

Niche Construction

Building on the work of Lewontin, Mayr, and Waddington, Odling-Smee formalized and proposed niche construction as another significant evolutionary principle (Jones 2005; Laland and Sterelny 2006; Odling-Smee 2010; Odling-Smee et al. 2003; Sterelny 2005). Niche construction, the building of niches by organisms and the mutual dynamic interaction between organisms and environments, was a long-accepted concept but has not been taken further into account as a major factor in evolution because adaptation was the predominant notion. Adaptation, in one version of its different meanings, considers the environment as a given factor with which the organism has to cope. The environment is only seen as the agent of selection, determining which variants survive and reproduce. However, Lewontin (2000) demonstrates that there is more in regard to the environment that complicates—or better enriches—the way in which we have to think about the role of the environment in evolution.

Often, the organism itself is responsible for selecting its environment and for constructing some aspects of it. This is relevant for evolution as well. Lewontin has been stressing the importance of niche construction for years, but only recently have more biologists adopted his ideas and expanded them.

Laland et al. (2017, 1) formulate:

Organisms modify and choose components of their local environments, a phenomenon known as "niche construction". Animals construct nests, burrows, webs, dams, pupil cases; select habitats, microhabitats, mates, foods, oviposition and nesting sites; and build and provision nursery environments for their offspring. Plants modify the temperature, moisture level, cycling of nutrients and chemicals in the soil, alter atmospheric gasses, create shade, induce condensation from fog, alter wind speed and manufacture allelochemicals. Fungi, protists and bacteria play diverse roles in the decomposition of vegetative and animal matter, weathering, soil production and/or photosynthesis, while bacteria and protists also show microhabitat choice. Niche construction is a universal feature of living organisms.

The standard view on natural selection sees direct environmental influences on genotype fitness. Laland et al. (2017) describe that this view is a linear perspective that is exemplified by the hammer and nail example. Appropriate use of a hammer drives the nail into some substrate such as wood, but nails do not reciprocate by propelling hammers. Chickens and eggs, on the other hand, represent cyclical causation. Chickens produce eggs, and eggs hatch into chickens. Neither could exist without the other.

Niche construction creates feedback situations between organisms and their environment, and it is one source of the fit between organism and environment (Sterelny 2005). This, in turn, is supposed to influence evolutionary dynamics. It is expected that this includes a prominent role for phenotypes. Niche construction behavior might produce alterations that persist across generations and space, such that they, themselves, are a factor creating the pattern and strength of selection. In some cases, these changes are extreme: Living in mounds influences every aspect of termite morphology, physiology, and behavior (Turner 2007).

Organisms can influence the environment in ways that affect not only their own lives but also the development and lives of their descendants, forming a sort of ecological inheritance. Organisms transmit to their offspring altered physical and selective environments and a niche choice as they affect their offsprings' lives by choosing where they will live and breed (Jablonka and Lamb 2005; Odling-Smee et al. 2003; Sterelny 2005).

Even bacteria are ecological engineers because the products of their metabolism diffuse into the environment and transform it, changing the situation of their neighbors and descendants. These organisms contribute centrally to the conditions that make life possible. Other famous examples are the beaver, whose "inherited" dams provide the environment for new generations of beavers, or earthworms changing the properties of the soil in which they and their descendants will grow, develop, and be selected.

Differentiating all these relations with a much higher resolution will be a crucial advance beyond the externalist picture, in which a lineage is seen to accommodate to the environment. It will contribute to the problem of ecological theory, Sterelny (2005) points out, concerning the extent to which ecological communities are integrated systems rather than mere aggregates of individual agents that happen to live and die adjacent to one another. The concept does make it clear that the individual aggregate conception of communities understates the range of potentially important and stabilizing interactions between organisms.

One of the most prominent examples for the significance of niche construction are earthworms (Laland 2004; Laland, Odling-Smee, and Turner 2014). Through their burrowing activities, their dragging organic material into the soil, their mixing it up with inorganic material, and their casting, which serves as the basis for microbial activity, earthworms dramatically change the structure and chemistry of the soils in which they and the myriad of other organisms live, often on a huge scale.

Symbiosis and Horizontal Gene Transfer

The hypothesis that certain organelles of eukaryotic cells, in particular plant chloroplasts, evolved from bacteria had already been proposed by several researchers in the late nineteenth century. However, this relationship did not generate much interest for evolutionary theorists. In the 1960s, Lynn Margulis (Sagan) summarized the then-available data on the similarity between certain organelles and bacteria (in particular, the striking discovery of organellar genomes) and came to the conclusion that not only chloroplasts but also the mitochondria evolved from endosymbiotic bacteria (Margulis 1993). The concept was heavily rejected in face of the predominance of ideas of gradual change in evolution. However, subsequent work, in particular, phylogenetic analysis of both genes (i.e., those contained in the mitochondrial genome and those genes encoding proteins, which function in the mitochondria and apparently were transferred from the mitochondrial to the nuclear genome), turned the hypothesis on endosymbiosis into a well-supported theory.

The major evolutionary role assigned to effectively unique events like endosymbiosis is, of course, incompatible with gradualism. Nonetheless, it was finally incorporated into evolutionary thinking without further rumblings within the general theory. People returned to their former agenda without realizing that a fundamentally new nongradualistic principle for the origin of evolutionary innovations had been introduced. It looks like a cuckoo's egg within the synthetic worldview.

Later, Margulis extended her theory of symbiogenesis and postulated similar relationships between organisms of rather different phyla or kingdoms as an essential driving force of evolution. Genetic variation, she proposes, occurs mainly as the result of transfer of nuclear information between bacterial cells or viruses and eukaryotic cells. Thus, symbiosis is considered as a central factor of the evolutionary process in general, stressing the importance of cooperative relationships between species (Margulis 1999; Margulis and Sagan 2002). Again, her theory is receiving heavy criticism, but the future will tell.

Similar to the hypothesis on endosymbiosis, there were other considerations under way during the second half of the twentieth century, which slowly undermined the view of gradual infinitesimal changes as the only material of evolution and mainly came from genetic research. Among these can be counted the neutral theory (Motoo Kimura), the concept of evolution by gene duplication (Susumu Ohno), the discovery of mobile elements ("jumping genes," i.e., genetic elements that were prone to frequently change their position in the genome, first described by Barbara McClintock), as well as unexpected insights into the genomic organization of viruses and bacteria (Koonin 2009).

Today, because of increasing knowledge from whole-genome sequencing, the extent of larger genomic exchanges becomes evident. Horizontal gene transfer between prokaryotes seems to be the rule rather than the exception (Boto 2010; Doolittle 1999; Koonin 2009; Shapiro 2010, 2011). The rate of horizontal gene transfer seems to differ for different genes depending on their functions. Eukaryotes are different from prokaryotes with respect to the role played by horizontal gene transfer in genome evolution. The genomes of eukaryotes exhibit more stability and thus robustness or autonomy. Especially, endosymbiosis made substantial contributions to the genomes of the host cells, as many genes of the symbiont were integrated into its nuclear DNA. Thus, there is no reasonable doubt that the gene complement of eukaryotes is a chimera comprising functionally distinct genes of archaeal and bacterial descent. Also, later in the evolution of plants and animals, there were several occurrences of horizontal gene transfer. This, however, is still under debate, at least concerning its extent and its significance for evolution.

Other substantial reorganizations such as gene and whole-genome duplications; large deletions, including loss of genes or groups of genes; and various types of genome rearrangements, which are becoming increasingly well documented today, contribute to a dynamic view of the genome.

The observation of extensively occurring horizontal gene transfer leads to a fundamental generalization: The genomes of all forms of life might be collections of genes with diverse evolutionary histories. Thus, it is possible that on the genetic level the tree of life is a sort of network rather than a branching tree (Koonin 2009).

However, this possibly does not affect so much the usual treelike representation of the history of life in regard to whole organisms because the role of the genome within organisms obviously has to be corrected in a profound manner.

Kitano and Oda (2006) connect their considerations about increasing robustness through evolution with these symbiotic events. In section 4.3, I showed that the term *robustness* parallels my definition of autonomy. Kitano and Oda argue that biological robustness fosters evolvability, and that selection tends to favor individuals with robust traits; thus, robust systems progressively adapt to become more robust against the environment in which they are embedded. They suggest that, over evolutionary time, robustness against external perturbations was enhanced by adding diverse new functions to the input and output components of the organism, and that many of these new functions were gained through symbiosis. They argue for a "self-extending symbiosis" as a process to further enhance robustness. Self-extending symbiosis refers to phenomenon by which evolvable robust systems continue to extend their system boundary by incorporating foreign biological forms (genes, microorganisms, etc.) to enhance their adaptive capability against environmental perturbations. Thus, robust evolvable systems have consistently extended themselves by incorporating "nonself" features into tightly coupled symbiotic states.

Looking at the history of evolutionary innovations, they describe some of the major innovations as the result of the acquisition of nonself into self at various levels, and that horizontal gene transfer facilitates evolution by exchanging genes between different species that have evolved within different contexts.

In early cases of symbiosis, such as during the generation of the eukaryotic cell, the relation between the host and the symbiont became close. In later cases, such as the acquisition of the gut flora in vertebrates, the mutual dependency is comparable, although the physiological relation is not that strong.

Kitano and Oda also argue in the sense of modern systems theory that different degrees of symbiosis add additional systemic layers within an organism. They see a general tendency in the continuous addition of external layers to the host system by means of symbiotic incorporation of foreign entities. They cite genomic studies that revealed that the bacterial flora even manipulates host gene expression to establish mutually advantageous partnerships, and that host gene expression changes according to the composition of microbes in the flora. Not only does the bacterial flora affect the host, but also the host affects the activity and composition of the flora through its immunological responses.

In this sense, Kitano and Oda see self-extending symbiosis as a fundamental process of significant evolutionary innovation that adds greater levels of robustness and functionalities to the species.

Theory of Increasing Autonomy in Evolution

Research in our institute added a further aspect to these considerations, already mentioned in section 4.3. It revealed that there are not only adaptations to given environments but also changes in the capacity for autonomy in individual organisms in the sense of a relative emancipation from the environment. Especially during the major evolutionary transitions, organisms gained in stability, robustness, self-regulation, homeostasis, and flexibility. The direct influences of the environment were gradually reduced and a stabilization of self-referential, intrinsic functions within the respective systems was generated. In higher animals and in humans, this includes the potential for more flexible and self-determined behavior. These processes are described as changes in relative autonomy because numerous interconnections with the environment and dependencies upon it were retained.

A set of resources can be involved to change autonomous capacities: changes in spatial separations from the environment, changes in homeostatic capacities and robustness, internalization of structures or functions, increase in body size and changes in the flexibility within the environment, including behavioral flexibility.

We followed this pattern on different levels of evolution, from the evolution of the eukaryotic cells, through the origin of multicellularity, the generation of animal phyla, to the evolution of vertebrates including birds, mammals, and humans. An overview is provided in Rosslenbroich (2014).

Evolution Becomes Organismic

What is of importance in our context here is that with many of the new concepts and considerations, the approach to understand evolution is becoming much more organismic (Müller 2017b; Noble and Noble 2017; Walsh 2015). It develops far beyond the singular focus on the genetic level to an integrated systems view of evolution. It regards several typical properties of organisms as described in the present book. Especially the following properties are involved in these new considerations.

Integrated systems evolve
Regardless of where the initiation of an evolutionary change comes from, what changes is the whole organism. Therefore, biologists need to regard evolution from the viewpoint of organisms as entire systems. The focus of attention is on the dynamics of an entire complex, self-regulating systems that constantly exchange matter and energy with their surroundings (Laubichler and Renn 2015). These are systems that have the capacity to implement features of their own repertoires in response to the external and internal conditions they encounter. The systems perspective incorporates cell circuitry and molecular networks into a more integrated view of cell and organismic activities. It accords explanatory primacy to the dynamics of whole systems over the capacities of their parts (Noble 2011a, 2013b, 2017b). If the organism's anatomy, physiology, and behavior are only remotely connected to the DNA, as Gerhart and Kirschner argue, and all the complex processes of growth, development, and metabolism are in between, evolutionary changes can come about on several of these units. Evolution is most likely driven by multilevel feedback processes (Noble and Noble 2017).

Interdependence
Several of the newer considerations are based on the biological principle of interdependence and circular causation, which has been characterized in section 4.1. One of these considerations concerns the construction of phenotypic characters, in which causation not only flows from the lower levels of biological organization, such as DNA, "upward" to cells, tissues, and organisms, but also from higher levels "downward," such as through environmental- or tissue-induced gene regulation. If both directions are involved concurrently, there is an interdependence between these levels rather than any linear causation from only one of these levels.

Another aspect of interdependency lies in the fact that populations of organisms are not relegated to being passive recipients of environmental selection pressures but exert influence on their surroundings through various forms of active niche construction. The respective relation of an organism to its environment is multidimensional and reciprocal. The type of relation an organism has with its environment is more like that of a network. Networks are described extensively in ecology, and many ecologists—though not all—see ecological systems as integrated and regulated systems, of which organisms are integrated parts. In many cases, evolutionary changes come about within such networks.

The principle of adaptation to environmental conditions, which is stressed so much in conventional theories of evolution, is of course still part of this interdependence. However, it needs to be regarded in its relation to all the other factors involved.

Thus, a major feature of the extended synthesis is that causation not only runs one way but assumes reciprocal relations between its participating components, both in the relationship of populations with the environment and in the generation of heritable phenotypic architectures (Müller 2017b).

Autonomy
The ability of a certain relative autonomy and robustness toward external and internal influences is a central property of individual organisms. This ability undergoes major changes throughout evolution and is part of the formation of complex and highly flexible organisms. Thus, regarding changes and transitions in the capacity of autonomy makes understanding evolution much more organismic.

This does not contradict the principle of interdependence between ecosystems and adaptations of organisms, because changes in autonomy can only be relative, and different forms of relations will be established by different levels of autonomy.

Agency
Several newer considerations on evolution include insights into the activity of organisms within evolutionary transitions. This is quite obvious for the relation to the environment within the niche construction theory. However, if changes on the level of the phenotype are relevant for evolutionary transitions, then every activity of the organism in its way of life and its circumstances can be involved in this process. All three forms of agency, as proposed in section 4.4, may be relevant, as recent discussions suggest (Noble and Noble 2017; Walsh 2015; West-Eberhard 2003). Especially the role of behavior as a source of innovations has been discussed for a long time (Bateson 2004; Mayr 1988).

Variation in time processes
Especially the findings in evo-devo demonstrate that differences in species, orders, or phyla might be less based on variances of their genes but rather on

differences in the spatial and temporal expression of conserved genes. Thus, within development the point in time at which a certain gene is expressed appears to be essential for the phenotypic product. It is thus a function of time autonomy. Time processes and their relative autonomy are another essential property of organisms, as described in section 4.9.

Processes of shape

Perhaps less understood at present is the role of the shape of an organism (*gestalt*; see section 4.8) in evolution. However, if today it proves to be that the phenotype is more important than has been realized before, the characteristics of its shape, as well as its changes, constraints, and potentials, may play a major role in understanding evolution. Form, size, and the arrangement of body parts are in interdependent relations to functions and processes throughout life.

The importance of shape is well understood in paleontology (or rather a matter of course), as the field must rely on the study of whole bodies or their fragments, because tissues, cells, or molecular residues are usually not available. Thus, paleontological reconstructions of evolutionary changes are based on anatomical changes and their shape, including the reconstruction of functions. Therefore, traditionally there have been more paleontologists, who tended toward an organismic concept of evolution than people from other fields, either explicitly or not.

The Return of the Whole Organism in Evolution

If the principle of individual biological integrity, as it has been described in previous sections, is taken into consideration, it must be noted that organisms are subjects that evolve, not objects that are forced to change by external factors (Kather 2003; Walsh 2015; Weingarten 1993). This is not only the case when organisms are actively involved in the evolutionary changes, but it is also the case when evolution takes its starting point from random variants. When such variants are able to deal better with the given environmental conditions by making some changes, it is still the living subject that produces them. In this case, too, the organism is not just an object that is induced to evolve. Basically, it is the subject in its continuous endeavor to maintain its life and to preserve the species.

In summary, the process of biological evolution differs fundamentally from mere physical and chemical processes. Extensive conversions also take place within the nonliving world, as they are described in geology and cosmology. However, the generation of new systems as in organismic evolution, which leads to new self-determined abilities and possibilities of subjects, including the long-term continuation of this very process, does not take place.

The emerging view of evolution dislodges the gene from its privileged position it was given by the synthetic theory and restores the organism back to the center of the evolutionary stage (Bateson 2005), however, without underestimating the importance of the genome. In this organism-centered view, organisms rather than genes are the primary agents of evolution. In this sense the extended evolutionary synthesis is rooted in the organicist tradition (Baedke 2019; Müller 2017b; Nicholson 2014b; Noble 2013b; Noble et al. 2014; Walsh 2015).

4.13   Growth and Development

The process of growth and differentiation is an essential constituent of all multicellular organisms. Plants, fungi, and animals not just grow larger, but they undergo a complex developmental process of rearrangements to generate their characteristic form. How a single fertilized egg cell transforms itself into a complex organism is still a miracle and a mystery (Purnell 2012; Travis 2013; Wolpert 1991).

Development as System Process

Morphological and histological studies in embryology had already been collecting a wealth of knowledge when molecular biologists promised to uncover the causes of the observed processes. The predominant view, especially during the second half of the twentieth century, was that development essentially is a linear reading out of an information that is encoded in the DNA of the cell's nucleus. The phenotype of an organism (i.e., its traits and behavior) has been studied as the outcome of a developmental "program" coded in its genotype. This is essentially a deterministic view: an individual organism's set of genes (its genotype) determines that individual's physical traits and behaviors (its phenotype), so it is possible to know what the organism's features will be just by knowing its DNA sequence.

Although this research delivered a major step forward in understanding details, now doubts are increasing that the cascades of genes and proteins, read out from the genome, alone are able to bring about the generation of macroscopic forms, or that they can be identified as the primary cause and origin of the whole process (Fuente and Helms 2005; Gilbert 2014; Newman 2002; Nijhout 1990; Purnell 2012; Radlanski and Renz 2006). In a broad approach out of different perspectives Minelli and Pradeu (2014, 4) question this "neo-preformist" view of a "genetic blueprint" or a "genetic program" and discuss non-genecentric perspectives on development.

The genetic program was initially formulated in terms of a multiplicity of developmental genes controlling the process, through which an egg eventually gives rise to an adult animal. The rapidly expanding body of experimental evidence soon directed attention beyond the patterns of expression for single genes to the cross talk of macromolecules, through which the expression of developmental genes is controlled. Finally, developmental genetics moved a further step forward, describing whole gene networks deemed to be responsible for specific aspects of the organization of complex organisms (Minelli and Pradeu 2014).

Increasingly, this research showed that not only the genes themselves are important but also the characteristics and behavior of their products, which establish a sort of spatial order by their distribution within the egg, the whole embryo, or parts of it. Transcription factors, gradients, and signals between cells turned out to be equally important and are able to control the expression of the genes. Today, it can be claimed more generally that the tissue context, the position, and the time period are equally important. Genes act as resources to be activated by signals out of the context. Thus, not only genes or genetic networks control certain steps of development but also many factors control the use and recruitment of genes.

Many regulatory genes, which are recruited in this way, operate in diverse situations, so that the same gene can have very different functions at quite different times and places. A characteristic of many developmental modules, such as the *hedgehog* or *engrailed* pathways, is that they are repeatedly deployed in different functional contexts. For example, the *engrailed* pathway is involved in establishing the anterior–posterior compartment boundary in *Drosophila* and butterfly wings but is also used in establishing the eyespot organizer in butterfly wings (Laubichler and Wagner 2001).

Therefore, genes do not simply contain information for the respective organ. Some contemporary authors propose that genes are predominantly *mediators* in the generation of tissues and organs rather than the main cause for the generation of structures (Fuente and Helms 2005; Radlanski and Renz 2006). Laubichler and Wagner (2001, 55) formulated: "Similar molecular interactions, such as the binding of specific transcription factors to similar stretches of DNA thus lead to quite different phenotypic results. Any explanatory generalization that can account why particular modules and regulatory pathways are employed in the development of certain kinds of phenotypic patterns will have to include the systemic properties (spatial, regulatory, dynamical) of higher level (cellular, organismal) biological entities as well as their evolutionary history."

Thus, genes are not determinants of development; they are part of the system, and they depend on being integrated into the system (Dupré 2012; Noble 2006). Development is not a linear reading out of a code or program

but a systemic process of feedback interactions between genetic and nongenetic templates, cells, and tissues that mobilizes physical and informational
properties at different scales. Hence, development is a systems relation in
which no component is informationally privileged (Müller 2017b).

One tool the organism uses for this integration is generating a protein gradient within a certain region, so that cells differentiate along such a field according
to the concentration at each single point. The system generates a signal, which
induces the differential expression of genes. Of course, for the generation of
the protein gradient a gene was necessary as well, but when it is available it
is used as a signal by the system. The continuous interactions between genome
and gradients within the system, or between genome and nearby tissues, enable
the respective steps. Thus, genes work only within a certain context and are
only one factor within a cascade of interactions and interdependencies (Dupré
2012; Gilbert 2014; Willmer 2003).

Today, additional factors are known to contribute to these networks. These
include, for example, physical organizational effects. Data indicate that different
forces such as tension, pressure, and differential adhesion influence the expression of developmental genes and the differentiation into cell types. More
generally, cells in embryos have the ability, via contractile and protrusive
activities, to exert forces on one another and on the ECM they produce. Often,
tissue structure is involved and is mediated by signals between cells and the
ECM. These elements contribute to the regulation of cellular proliferation,
migration, differentiation, and apoptosis. Apparently, geometrical and mechanical signals contain information about tissue and organ properties on the whole
(Forgacs and Newman 2005; Fuente and Helms 2005; Newman 2012; Piccolo
2013; Purnell 2012; Radlanski and Renz 2006; Wozniak and Chen 2009).

Newman (2014) presents a whole list of examples for such nongenetic physicochemical factors and proposes a multiscale nature of developmental systems,
which can be productively analyzed in relation to one another as well as to the
genetic factors. He sees a connection here to the "law-of-form" tradition extending back at least to Johann Wolfgang von Goethe, Richard Owen, William
Bateson, Brian Goodwin, and others. He attempts to update this law-of-form
approach by linking it to modern knowledge of nongenetic and genetic factors.
However, because developmental processes are also historically determined, the
lawlike behaviors they exhibit are inherently hybrid and complex, he explains.

Here we encounter an overlap with some aspects, which have been discussed in section 4.8, that shape has an autonomous relevance in the organism
that needs to be studied within the entire context.

Once again, time processes are involved. It is not only important for regular
development *where* a certain gene is expressed and *where* certain tissues are

generated but also *when* this takes place (Duboule 2003). Some genes are activated early, others later, which has been called differential gene activation. Probably there is a certain time in which a gradient works as a gradient so that cells react to it (competence period) (Kicheva et al. 2012). In some cases, rhythmical pulses also play a role in the activation of genes, as during the generation of somites in the chick embryo (*hairy*-oscillator; Gilbert 2014).

This all leads some authors to assume that during morphogenesis the cells do not follow an intrinsic genetic program; rather they react to different situations into which they maneuver themselves by growth (Radlanski and Renz 2006). The context of space and time is just as important as the genetic information itself. The first differentiation, which occurs in the early egg cell, is one of the most evident examples for the significance of the context. Although the DNA within the nucleus contains no information about the specific position and function of a developing cell, there is much information within the system on the whole, for example, in form of a local distribution of certain cytoplasmatic factors (Müller and Hassel 2006; see figure 4.12 in section 4.2).

At the same time, this is a good example for the integrative systems principle as discussed in section 4.2: Cytoplasmatic factors within the entire system determine which genes are activated within the nucleus, and the resulting gene products are used to establish new positional information for further developmental steps. Thus, a processual interdependence is established between these levels, and this interdependence initiates development. It is neither the genome nor the context alone but rather their mutual interaction, a fact that has been well established in recent developmental biology and again points to the principle of concurrency. Jaeger et al. (2008) offer an important extension of the theory of positional information, which they call the "relativistic theory of positional information." They understand this theory in contrast to the classical theory of positional information. According to the classical theory, the establishment and interpretation of positional values are independent of each other: the cells are "passive," in that they simply measure the morphogen gradient but do not themselves influence the developmental field. However, according to these authors, the classical theory fails to account satisfactorily for important phenomena, including size regulation and developmental robustness. Using diverse experimental data, established in particular in *Drosophila* and in neural tube patterning in vertebrates, Jaeger et al. show that positional specification actually depends on regulative feedback from responding cells. Most fundamentally, the relativistic theory of positional information suggests the existence of a dynamic metric, which allows cells to measure their relative position within a developing field that itself changes in response to the activity of those cells. The authors then describe this metric in more detail and argue that it is possible to explain and predict how cells during early development, far from being simply passive "receivers" of positional information coming from a morphogen gradient, participate actively in the process through feedback mechanisms that affect the developmental field.

Developmental Systems Theory

Already in the 1980s, Susan Oyama developed the theory that DNA does not contain the entire information necessary for building an organism in the sense of a body plan. In her developmental systems theory, DNA is only one of several factors, even if certainly a most significant one. Neither DNA sequences nor any other factors can be privileged as the bearer of ultimate causal control in the developing organic system. Rather, the whole complex of factors—whether related to cellular morphology, the dynamics of biochemical transitions, external circumstances, previous developmental history of the system, or the DNA sequences available—is considered as equally important for explaining the occurrence and the regularity of developmental steps. DNA only transmits information about some possibilities of how to build a protein. The organismic form is then constructed during the developmental process and is a result of interactive processes taking place between the parts and processes of the organism and between elements and processes in the environmental context. Thus, as the organismic form is constructed during ontogeny, as opposed to being preformed in the genome, Oyama calls her approach "developmental constructivism," and she also expands it beyond the time of embryonic development, so that every organism can be seen as self-constructing during its entire lifetime (Downes 2001; Dupré 2012; Oyama 2000; Oyama et al. 2001; Rehmann-Sutter 2002).

Constructive development refers to the idea that the development of organisms does not result from the execution of a genetic program but rather from multiple interactions between many factors within the developing system, as well as between the system and the environment. Such a view has eminent evolutionary implications, insofar as development is conceived as a process that facilitates the emergence of phenotypic variation. The classical conception of developmental bias, which reduces the range of possible variations, is replaced by a vision according to which development produces phenotypic novelty and affects evolutionary trajectories.

In this regard, Lewontin (1991, 63) states: "[T]he organism does not compute itself from its DNA. A living organism at any moment in its life is the unique consequence of a developmental history that results from the interaction of and determination by internal and external forces."

Recent research in modern genetics increasingly vindicates at least the basic assumptions of these considerations and also shows that there are again processes involved as they have been described in previous sections of the present book.

Development as Process

A general consideration reveals that the complete life cycle is a process. All organisms undergo a characteristic series of structural and functional changes over the course of their lifetime. They do not stay the same from the moment they come into existence, but rather they develop progressively over time, acquiring certain properties and capacities and losing others along the way. This is not only true for development from the fertilized egg to an adult organism but rather for the whole time of existence of the organism.

Within a process ontology (Dupré and Nicholson 2018; Holdrege 2017), the life cycle itself is the primary reality of any organism. The traditional understanding of organisms as things or substances tends to privilege the adult stage of the life cycle, because this is the period during which the organism most closely resembles a thing, by virtue of its relative stability. It is, however, the life cycle itself that constitutes the organism (e.g., a frog; figure 4.27). In this sense, it is incorrect to speak of an egg developing into a frog, because the egg is a temporal part of the developmental trajectory that is the frog. What we perceive as an adult organism is only a snapshot of any given moment in this continuity, without any sharp boundary demarcating a certain stage. Although an adult frog is very different from a tadpole, the developmental progression from tadpole to frog is smooth and gradual. This continuity also includes the time when an organism is growing old and once more is making several more or less characteristic changes.

Griffiths and Stotz (2018) discuss the developmental systems theory (DST) as a process theory. What makes DST a process theory, they claim, is that it seeks to explain developmental outcomes as the result of a dynamic process, in which some of the interacting factors are products of earlier stages of the process rather than as the result of the arrangement of preexisting factors in a static mechanism. Even when factors exist independently of the developmental process, they are drawn into it and made part of a developmental system by the unfolding process. It is the process that defines the system. The developmental *process* replicates itself across generations, making use of persistent resources as well as of resources created by earlier cycles of that process. This process can be described as a life cycle because it encompasses the entire period between conception and death.

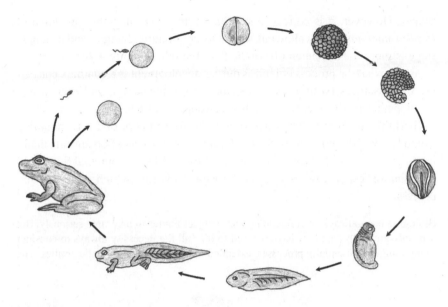

Figure 4.27
Developmental process of a frog.

This point of view also provides a processual definition of the individual: "A biological individual is a process that may intersect with other organismic processes, but it has a principle of identity that marks just this series of events out as one biological individual." Thus, DST describes identity as a continuity of organization (Griffiths and Stotz 2018, 229).

However, the description of a process as such is not enough either, because it tends—at least formally—to neglect the order within the process, which leads to a highly specific organization. The exemplary life cycle of the frog goes through several characteristic types of organization, which is especially interesting in such organisms that perform a metamorphosis during their development. In frogs, this involves the organization of the tadpole. Through metamorphosis then, the organization of the frog is generated. At each stage, a certain shape is realized, which has a decisive reality in the life-fulfillment at the respective stage of development.

Even more profound is the metamorphosis of the butterfly, which goes through the stages of the caterpillar, the pupa, and the imago. Each of these stages has a special organization, with a special shape, special functionality, and particular behavior. This means that it is not a process as such. Rather, the process generates different forms of organization with extremely specific shapes. However, it is correct to say that at any stage it is this specimen of *Papilio machaon*, the swallowtail, within its continuous changes, and no single stage is more the specimen of *Papilio* than the other (figure 4.28).

Again, what the process organizes during development is a complex consisting of substances (with a specific shape too), information, and energy, and development involves changes in this interdependent complex.

Griffiths and Stotz (2018) describe some forerunners of DST, including Conrad H. Waddington, one of the most prominent pioneers of organismic thinking in general (see chapter 2). In "The Evolution of Developmental Systems," Waddington especially focuses on the aspect of time which underlies any process:

Biologists have always been forced by their subject matter to take time seriously. But it is only gradually that they have realised to the full the necessity always to consider living things as essentially processes, extended in time, rather than static entities. The

Figure 4.28
Metamorphosis of a butterfly (*Papilio machaon*). *Source*: Courtesy of Jens Stolt/Alamy Stock Foto.

day-to-day activities of living things are carried on by processes which occur anything from a fraction of a second to an hour or two . . . [the organism] is gradually carried along through another series of processes, those of development, each phase of which occupies a time which is fairly long compared with the life-cycle. Each life-cycle is again nearly circular, and gives rise to new animals, the descendent generation. . . . The accumulation of such changes gives rise to the slow process of evolution, which is on a still larger scale than either the physiological or developmental ones. (Waddington 1952, quote from Griffiths and Stotz 2018, 227–228)

In this context, it again becomes clear that genes alone do not control development. Genes themselves are controlled in many ways, some by modifications of DNA sequences, some through regulation by the products of other genes and by the intracellular and extracellular context. The regulated expression of the coding regions of the genome depends on mechanisms that differentially activate and select the information in the coding sequences, depending on context and time. Biological information is distributed between the coding regions in the genome and regulatory processes. Griffiths and Stotz (2018) cite Waddington, who already argued that developmental outcomes are to be explained at the level of the whole system and not by single causes that "encode" or "instruct" that outcome, a view which has been reconfirmed by Noble (2006) and others. Developmental outcomes must also be explained dynamically, as trajectories in a space of possible states of genome expression. The role of epigenetic marks in development is to successively differentiate cells as a result of earlier stages of development, making genome expression in one tissue at a time a function of the history of these cells. In this sense DST is in accordance with the modern view of development, which takes account of epigenetic regulation.

Development Becomes Organismic

In face of these topics, which are currently being discussed in developmental biology, it is quite clear that the field is constantly moving toward an organismic perspective concerning the embryo. This includes the major properties that are discussed in the present book. There are *interdependencies* rather than single causes, and this includes interdependencies between different *system levels*. At every stage, development is a function of the whole system with its different levels of organization, from the form and shape of the whole embryo to the different organs and body areas, down to the behavior of tissues, cells, and genes. There is no system level that can be privileged to find a first cause of events (Noble 2017b). Gilbert (2014, 630) states: "Modern systems theory can be said to have had its start in developmental biology, which was one of the first sciences to apply these principles of causation, integration, and context dependency" (see also Gilbert and Sarkar 2000). Processes within certain sequences establish a *time autonomy* of events and step-by-step establish the *morphological* and *functional autonomy* of the growing organism, including the formation of its "gestalt" as the result of *processes of shape*. The various processes involved are active ones, so that *agency* is a characteristic of development as soon as growth and differentiation have begun after fertilization. Finally, all these sequences can themselves be changed and modulated, so that the modern field of evo-devo is able to detect sources of variability within the *evolutionary process* in the long run (Arthur 2004; Carroll 2005; Müller and Newman 2003). Overall, this picture corresponds well with the formulation of a *spatiotemporal organization of matter, information, and energy*.

4.14   Environmental Interrelationship

Organisms live in a complex and multilayered relation to their environment. They exhibit a certain relative self-determination toward its factors and fluctuations, have an extensive exchange with it, choose the scope of factors in their environment that are relevant for them, and construct and change their environment throughout their lifetime and across generations. Therefore, a central property of all organisms is their interrelationship with their environment.

Adaptation

In biology the term *adaptation* has been broadly used. Mahner and Bunge (1997) identified numerous meanings, which they summarized in eight categories. For the present topic, it is enough to regard adaptation as the ability of organisms to adapt to factors and changes within their environment. In the first place, this says nothing about the evolutionary process, which might have produced this ability. This understanding of adaptation is often called *adaptedness*. The ability of organisms to adapt is again a specific property of life, which is not found in the nonliving realm.

Adaptedness is a relational property of an organism or rather a property of the coupled organism-environment system. Then, autonomy and adaptation become a central pair in this system. Both are dependent on each other: On the one hand there is the organism, and on the other hand there is the environment. The organism—even in its simplest form—always establishes its life functions together with the generation of a boundary, and thus produces its "being different" from the surrounding environment. To maintain this state, the organism not only needs regulatory and stabilizing functions but also needs to react appropriately to cope successfully with the environmental influences. Self-assertion (relative autonomy) needs adaptation, and adaptation presupposes autonomy (see section 4.3; Di Paolo 2005; Rosslenbroich 2014).

The polar bear possesses great autonomy in regard to endothermy, flexible movement capacities, and flexible behavior, but to survive in the extreme climate, it needs adaptations such as the thick white fur and many others.

The dolphin (figure 4.29) has the extensive regulatory capacities and behavioral flexibility of mammals. These functions of autonomy are not prerequisites for life in water, because they are not present in fish, and most of them can be traced back to the dolphin's phylogenetic history. At the same time, the dolphin exhibits many adaptations to the aquatic environment, such as blubber for insulation, secondary homodont dentition, fins and fluke, streamlined shape, nasal openings that have migrated to the top of the head, and many more. This means that two elements are involved: first, the individual biological integrity, and second, maintaining its autonomy while contending with the factors of the environment. Autonomy theory describes how both elements change during evolution.

Niche Construction

As already depicted, in evolutionary biology it has become increasingly clear that the phrase "adaptation to a given environment" is one-sided. Organisms do not just adapt to a given environment but also can themselves strongly influence the environment. This has been discussed by means of the notion of "niche construction" (Laland and Sterelny 2006; Laland, Odling-Smee, and Endler 2017; Laland, Odling-Smee, and Feldman 2019; Lewontin 2000; Odling-Smee 2010; Odling-Smee et al. 2003). The concept describes the dynamic interaction between the organism and its environment and regards environment or niche not just as a given factor with which the organism has to deal. Niche construction describes the feedback processes between the organism and its environment.

The picture now becomes complicated: organisms that maintain their autonomy are intertwined with given and with constructed factors of their environment. These seemingly opposing features exhibit once more the principle of *concurrency*. The organism can adapt to some factors and change others as well to maintain its autonomy. Again, these feedback dynamics represent a system context. The complexity and interconnectedness of this relation is another crucial aspect of an organismic view of life (figure 4.30).

When Richard Lewontin wrote his first influential essays on niche construction (1982, 1983), the dominating conception in evolutionary biology was that organisms are adapted to a given inorganic and organic environment through a selection process. Although it was well known that organisms can profoundly change their environment, this activity of organisms was not part of major theories, even in ecology. Nevertheless, there were pioneers especially in ecological disciplines (Laland, Odling-Smee, and Feldman 2019; Looijen 2000).

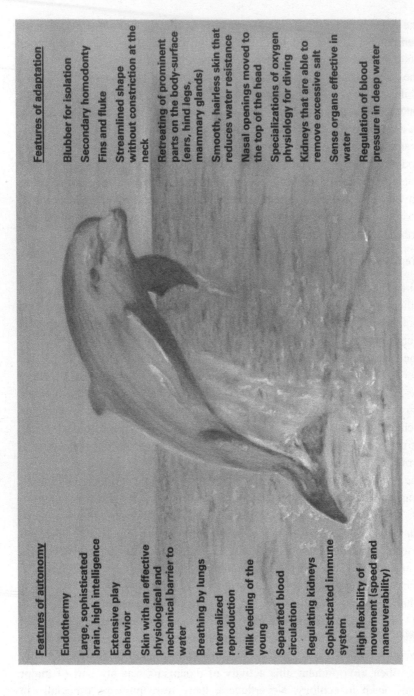

Features of autonomy

Endothermy

Large, sophisticated brain, high intelligence

Extensive play behavior

Skin with an effective physiological and mechanical barrier to water

Breathing by lungs

Internalized reproduction

Milk feeding of the young

Separated blood circulation

Regulating kidneys

Sophisticated immune system

High flexibility of movement (speed and maneuverability)

Features of adaptation

Blubber for isolation

Secondary homodonty

Fins and fluke

Streamlined shape without constriction at the neck

Retreating of prominent parts on the body-surface (ears, hind legs, mammary glands)

Smooth, hairless skin that reduces water resistance

Nasal openings moved to the top of the head

Specializations of oxygen physiology for diving

Kidneys that are able to remove excessive salt

Sense organs effective in water

Regulation of blood pressure in deep water

Figure 4.29
Features of autonomy and of adaptation in the dolphin. *Source:* Drawing courtesy of Marita Rosslenbroich.

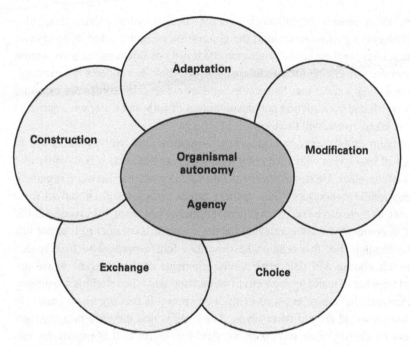

Figure 4.30
Some relations between organism and environment.

Laland et al. describe that this situation began to change around the 1990s, when the study of ecosystem engineering, a concept that overlaps with niche construction, began to develop within ecosystem ecology. Prior to that, the prevailing approach within population ecology had been to treat ecosystems as food webs dominated by trophic and competitive interactions, with little consideration of how organisms create habitat or resources for themselves and other organisms. It was not until the twenty-first century that the investigation of eco-evolutionary feedbacks and dynamics became a vigorous branch of evolutionary ecology.

Then increasingly the concept of niche construction was studied, demonstrating that the activities of organisms could exert a widespread and general influence on the flows of energy and matter in ecosystems, and that these activities could accumulate across individuals and over time to significantly modify evolutionary processes. The concept, however, is still being debated by conventional evolutionary biologists, as Laland et al. (2019) describe.

From a phenomenological point of view, it is obvious that not only animals construct components of their environments with artifacts such as nests, burrows, dams, and webs. In fact, the same reasoning applies to other organisms such as plants, which create shade, influence wind speeds, alter hydrological cycles, influence the cycling of nutrients and the generation of humus, and even choose environments for their seeds by altering the timing of flowering or germination and thereby influencing the conditions to which their descendants are exposed. Also, it is well known that bacteria as well as eukaryotic protists are essential for the profound construction and development of soils and water environments such as lakes, rivers, and the like.

Lewontin (2000) stresses how genes, organisms, and environments are in reciprocal interaction with each other in such a way that each is both cause and effect of the other. He describes the strong interdependence between organism and environment concisely: "Just as there can be no organism without an environment, so there can be no environment without an organism." (Lewontin 2000, 48). Environment is what is relevant for the organism's conduct of life, not just its surrounding. For this relation he describes four aspects: The first is that organisms choose and determine which elements of the external world are joined together to make up their environment and what the relations are among the elements that are relevant to them. The second is that organisms actively construct a world around themselves. The third is that they are in a constant process of altering their environment. And the fourth is that organisms can smooth out effects of variation and fluctuations in the environment, in the short term as well as over their lifetimes. This last point implies what has been described earlier as autonomy.

Laland et al. (2019) emphasize that the concept of niche construction presupposes organismal agency as a fundamental property of life, so that the present discussion has broad overlaps with section 4.4:

[W]e view agency to be an essential and inescapable aspect of nature. However, for us, agency is a property of individual organisms, not of sub-organismal components such as tissues or organs. We emphasize that living organisms are not just passively pushed around by external forces, but rather they act on their world according to intrinsically generated but historically informed capabilities. Organisms are self-building, self-regulating, highly integrated, functioning, and (crucially) "purposive" wholes, which through wholly natural processes exert a distinctive influence and a degree of control over their own activities, outputs, and local environments. Indeed, organisms must have these properties in order to be alive. (132)

When they assert that organisms are "purposive," Laland et al. mean that organisms exhibit goal-directed activities, such as foraging, courtship, or phototaxis, which are entirely natural tendencies with short-term local objectives and that have themselves evolved. The "goals" and "purposes" to which they refer can be defined with respect to general aspects of biological function, such as resource acquisition, stress avoidance, and reproduction. This corresponds to what has been characterized as "directed agency" in section 4.4.

Laland et al. summarize that Lewontin's writings, central to the conception of niche construction, have always been an appraisal of developmental processes as open and constructive through self-assembly and a corresponding rejection of the idea that organisms and their activities are fully specified by genetic programs. Organisms are regarded as influenced, but not determined, by their genes, and their activities are regarded as shaped by developmental information-gaining processes.

Laland (2004) illustrates niche construction with the example of earthworms, which have been brought to the attention of biologists by Charles Darwin. By means of their burrowing activities, their dragging organic material into the soil, their mixing it up with inorganic material and their casting, which serves as the basis for microbial activity, earthworms dramatically change the structure and chemistry of the soils in which they live. As a result, earthworms affect ecosystems by contributing to soil genesis, to the stability of soil aggregates, and to soil porosity, aeration, and drainage. Because their casting contains more organic carbon, nitrogen, and polysaccharides than the parent soil, earthworms can affect plant growth by ensuring the rapid recycling of many plant nutrients. In return, the earthworms benefit from the extra plant growth they induce by gaining an enhanced supply of plant litter (Lee 1985).

Many of these effects of earthworm niche construction typically depend on multiple generations, leading only gradually to cumulative changes in the soil. Thus, most contemporary earthworms inhabit a local selective environment, which has been radically altered by many generations of ancestors. It is likely that some earthworm phenotypes, such as epidermal structure, or the amount of mucus secreted, coevolved with earthworm niche construction. Accordingly, organisms not only acquire genes from their ancestors but also an ecological inheritance, which has been modified by the niche construction of their ancestors.

Originally, earthworms were aquatic and are now able to solve their water- and salt-balance requirements by means of tunneling, exuding mucus, eliminating calcite, and dragging leaf litter below ground, that is, through their niche construction. In this way, they retained the ancestral freshwater kidneys (nephridia) and have evolved few of the structural adaptations one would expect to see in an animal living on land. For instance, earthworms produce the high volumes of urine characteristic of freshwater animals in comparison to terrestrial animals. Consequently, earthworms are poorly adapted to cope with physiological challenges such as water- and salt-balance on land (Turner 2000). They can only survive in a terrestrial environment by co-opting the soils that they inhabit and by the tunnels they build to serve as accessory kidneys, which compensate for their poor adaptation. Also, they change the soil in a manner that makes it easier for them to draw water into their bodies. However, in the process, earthworms dramatically change their environment.

All of this earthworm activity, Laland (2004) explains, highlights a problem with the concept of "adaptation." In this case, it is the soil that is changed rather than the worm, to meet the demands of the worm's freshwater physiology. So, what is adapting to what? Standard evolutionary theory shortchanges the active role of organisms in constructing their environment.

In the present context, the important point is that there is a complementary match between organisms and their environments. This complementarity is an essential part of organismal properties, so that the organism-environment system and the context in which organisms exists are part of being alive (Holdrege 2021). An organism is not understandable without regarding its environment, and— metaphorically speaking—the organism does not end at its individual boundary layer.

4.15   Reproduction and Death

[Nature's] spectacle is always new, because she always creates a new audience. Life is her most beautiful invention, and death is her artifice to have abundant life.
—Johann Wolfgang von Goethe, 1783 (from Goethe 1981, 46, translation by author)

Reproduction, development, and death continuously compose all living entities, so that in principle organisms only exist as processes in this sense. This includes the fact that the processes of building and dismantling and of construction and degeneration act on different organizational levels and are permanently intertwined. Death and renewal concurrently generate the processuality of all forms of life.

Life Cycles in Multicellular Organisms

It is undisputed that reproduction exists only in living organisms. However, this process is as fascinating as mysterious, although it appears to be so natural. Again and again, a complex organism is constructed from simpler intermediate stages. The individual adult form itself only has a limited lifetime, but the species is preserved by means of the sequence of generations. Speculatively, a long-term continuation of individual life might be conceivable. The normal case, however, is that organisms go through a continuous process of reproduction, self-renewal, and death.

Self-renewal also takes place within the adult organism itself. Each cell sends its macromolecules and its organelles through a permanent process of degradation and reconstruction. The same is the case for many cells of a whole organism. In humans, there are tissues with higher and with lower rates of regeneration, using permanent processes of mitosis and apoptosis (programmed cell death), being part of normal healthy physiology (Hug 2000). Apoptosis also contributes essentially to embryonic development (Jacobson et al. 1997; Vaux and Korsmeyer 1999).

Reproduction, growth, development, and death are intimately intertwined. The same holds true for entire ecosystems, in which continuous assimilation must be in balance with degradation. How essential death processes are becomes especially obvious when they are missing. In ecosystems, dangerous overpopulations would occur, and in a single organism a constantly growing tumor threatens the life of the whole organism. Nevertheless, aging and death are among the least understood biological processes.

These phenomena also take into account that lifetimes are species specific and vary only within certain limits. The potential lifetime of an annual plant differs from that of an oak tree, and that of a mouse differs from that of an elephant. A horse is born, has a childhood, becomes adult and then old, and finally it dies. This sequence cannot be reversed or stopped. There seems to be a characteristic order of time, possibly intertwined with the physiological time structure, which has been described in section 4.9.

In many animal species there is also a strong interrelation between reproduction, aging, and death, for example, with salmon and eels. Atlantic salmon (*Salmo salar*) migrate back to the rivers in which they were born to spawn. Afterward, most of the adult individuals die. The European eel (*Anguilla anguilla*), on the other hand, migrates to a specific marine region, spawns there, and then dies. The imago of many insects dies when it has completed its reproductive cycle.

One can even say that only a momentary section is always visible on an organism. The whole of the organism presents itself completely only in time, not in space. Youth, maturity, and age phases are not present at the same time. Thus the organism is always only partially present as a sensually experienceable object, while the predominant part exists only in its time shape.

Reproduction, aging, and death expand the possibilities of evolution, as these processes increase the opportunity to produce variants. The phenotype is not just passed on from one generation to the next but must be constructed anew each and every time (Oyama 2000). It must run through its own process of shape each time, and in each generation the growing organism must integrate itself into the interdependence of individual autonomy and the conditions of the environment.

Penzlin (2014) described how an organism appeared so puzzling to the physicist Erwin Schrödinger, because it evades the rapid decay into a static state of equilibrium. This question also preoccupied Werner Heisenberg, according to Penzlin. Schrödinger noted that living organisms exhibit a degree of stability that complex nonliving structures could not possess simply based on physical and chemical laws. Other physicists have also pondered how the extremely improbable state of living systems far from thermodynamic equilibrium can be self-sustained. According to Penzlin, the organism achieves this not by preventing "decay" but by balancing the degrading and constructing (synthesizing) processes. Every organism, every single cell, exists in a state of permanent decay and reconstruction, in an uninterrupted process of self-renewal, even when it is not growing or is in an apparent state of rest.

Thus, life is only understandable within the continuous polarity between synthesis and degradation, between production and reduction, between life and death. In this sense, death is an inclusive element of life itself.

We are constantly in some danger of interpreting death in nature too strongly from an anthropomorphic perspective. Such a notion of death processes can obscure the balance between the processes that build up and those that break down. Therefore, it has been so difficult initially to understand the processes of apoptosis. Yet, an organismic understanding of life will have to consider death processes as much as life processes themselves. For individual human life, death is a catastrophic event, which we usually try to avoid and prevent at all costs. This is certainly true for most other organisms as well, and is well perceivable, at least in higher animals. Survival is a central motive in the realm of life and is also a major factor in evolution. But here the survival of the species is more important, and in many cases the survival of single individuals is subordinated to the survival of the species.

Multicellularity

In the early days of evolution, there was obviously a transition from eukaryotic unicellular organisms to multicellular organisms. The origin of multicellularity is the subject of many investigations and considerations in evolutionary research (Bich et al. 2019; Brunet and King 2017; Richter and King 2013).

It is particularly interesting that the eukaryotic single cell can exist autonomously as a complete organism. It is an individual. This changes in multicellularity, in which the context of many cells forms the organism and then appears as the individual. The cells involved can specialize and are subordinated to the overall function. Then regarding death processes, a truly remarkable procedure arises. First, starting from a germ bud (in case of sexual reproduction of a zygote), a multicellular organism is built up by continuous mitoses and by differentiation and formation processes. However, this organism has only a limited lifetime and then dies. What potentially survives are only the germ cells. That is, death is systematically and actively built into the reproductive cycle and becomes a regular constituent of the organism's existence. Death is thus internalized into the organization and life history of the organism.

Due to the development of somatic cells, the final death of organisms and the sequence of the succeeding generations with consecutive renewal by means of individual developments becomes the normal case.

This principle can be studied with an example. In Chlorophyceae, a class of green algae (Chlorophyta), a series can be established from the unicellular forms in *Chlamydomonas* through colonies of different cell numbers to the genus *Volvox* (figure 4.10). This series of Volvocales has often been used as a model for the origin of multicellularity in evolution (Kirk 1998, 2000; Ueki et al. 2010). The model character must be emphasized, because this series has undergone an independent development and has no connection with the evolution of multicellularity in plants and animals.

The cells of *Chlamydomonas* have two flagella at their front, with which they beat breaststroke-like to move in the water. Inside the cell, a large chloroplast exists that conducts photosynthesis. An eyespot appears as a red spot under the microscope.

In the further series of Volvocales, *Chlamydomonas*-type cells now remain connected in colonies. As the number of cells increases, a common gelatinous extracellular matrix is formed, from which the cells extend their flagella outward into the water environment and move the colony with a common flagellar beat. This has already been presented as an example of the integration of single cells into a higher-level system in section 4.2.

In colonies with low cell numbers, all cells are able to divide and form new colonies; in genera with higher cell numbers, a loss of the reproductive function of individual cells occurs to varying degrees. As somatic cells, these cells participate only in the nutrition and locomotion of the colonies, while specialized cells additionally carry out divisions and ensure reproduction.

In the genus *Volvox*, depending on the species, approximately 1,000–50,000 cells reside in a common matrix. The reproductive cells, the gonidia, are sunk into the matrix and are entirely specialized for reproduction. Only these cells form daughter colonies, while the somatic cells die by programmed cell death after release of the daughter colonies.

Figure 4.31 shows the vegetative life cycle of *Volvox carteri* (Kirk 1998, 2000). An asexual life cycle is completed in about 48 hours. When the gonidia inside the sphere are mature, they begin to divide synchronously and, with eleven or twelve rounds of division, form all the cells that will later be present in the adult sphere. Eventually, the mother sphere contains a group of young spheres that look like miniature adults. When they have grown to about half their final size, the daughter spheres are released by breaking open the mother sphere. The mother organism remains alive for about two more days before the somatic cells die synchronously and the sphere disintegrates. In sexual reproduction, there is specialized maturation of the gonidia into oocytes and sperm packets, through which fertilization then occurs, during which the mother sphere then also dies. Thus, if one observes this series in its entirety, it becomes clear that death processes become an integral part of the life cycle and reproduction in these organisms. Transition to multicellularity seems to internalize death processes.

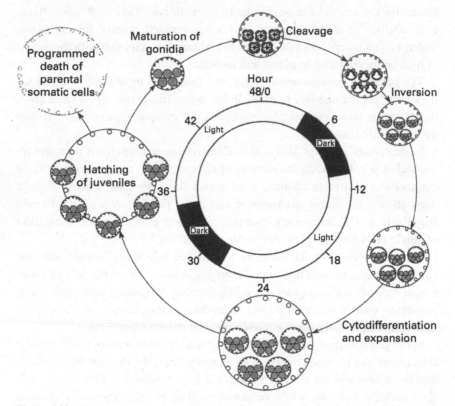

Figure 4.31
Life cycle of *Volvox carteri*. *Source*: From Kirk (1998, 125), reproduced with permission of Cambridge University Press through PLSclear.

Apoptosis

In many cases, death comes either from external events or as a result of aged and damaged organisms, compartments, and molecules. What has been surprising is that an active way of dying evolved. It is organized by genetic regulation and obviously has been present from the time the first single-celled life existed. An investigation of the origin and evolution of life usually focuses on the processes that promote life, the generation of traits that stabilized the first living systems and allowed them to evolve. For this reason, a study of the evolution of death seems, at a first glance, a peculiar endeavor. The fact that death can be a self-generated activity is a further aspect of the property of agency.

Morange (2008) depicts that much of the early observational evidence regarding apoptosis (programmed cell death) is more than a hundred years old, but for a long time it was neglected. In the nineteenth century, it was virtually inconceivable that death could be a normal part of the life of an organism, an indispensable component of embryonic development. Even in its smallest forms, death seemed obviously—almost by definition—to be confined to the end of life. Morange, however, explains that the evolution of multicellular organisms can now be seen to have introduced a top-down system of control, in which the fate of the organism dictates the fate of individual cells, rather than the other way around: it is the vitality of the organism as a whole that determines whether its constituent parts are alive or dead. Life, by its very nature, is a holistic phenomenon, Morange formulates.

Cells in a multicellular organism, which have been damaged, for example, by a pathogenic infection or have reached the end of their natural life, can initiate apoptosis. This is triggered by a complex network of different signal transduction pathways (Ameisen 2002; Huettenbrenner et al. 2003; Krammer 2000). During apoptosis, a cell's organelles, including the nucleus and its DNA, fragment, as do cytoplasmic structures. The cell shrinks and forms vesicles known as "blebs," which are eventually taken up and disposed of as so-called apoptotic bodies by specialized macrophages. Because nothing is released or left behind by the cell, neighboring cells are protected. Without apoptosis, the dying cell would leak and release its contents into the environment, leading to inflammatory reactions.

Signals to trigger apoptosis can come both from outside the cell and from inside the cell itself. In the case of external signals, corresponding molecules are released from other cells and trigger a signaling process in the target cell, which activates specific proteins and gene expression to initiate programmed cell death. If the signal comes from inside the cell itself, it is passed on through a series of protein–protein interactions until it finally triggers apoptosis. Both pathways of external and internal triggering of apoptosis are, again, also often interconnected.

Every second, several millions of cells in the human body undergo apoptosis, that is, in conditions of homeostasis each mitosis is compensated by one event of apoptosis. In the blood system, for example, programmed cell death must be as abundant as cell replication. If this fine-tuned balance is disturbed (even when occurring at a subdetectable extent), sooner or later life-threatening diseases such as lymphomas and leukemia will become manifest (Huettenbrenner et al. 2003). Also, in this sense the human body can be regarded as a process.

Apoptosis also plays a major role in embryonic development. Once cells and parts of tissue have been created, they can then be degraded at a later stage of development. These processes were studied using the model organism *Caenorhabditis elegans* as an example. The adult worm, about 1 mm long, consists of approximately 1,000 cells with exactly the same cell number in all individuals. During its development, precisely 131 of the originally formed cells die by apoptosis.

Genetic studies in *C. elegans* led to the discovery of two genes important for apoptosis, designated *ced-3* and *ced-4*. The products of these genes (Ced-3 and Ced-4) and most of the other proteins involved in apoptosis are constantly present in cells. However, they are permanently inhibited by the protein Ced-9 of the outer mitochondrial membrane. It acts as a molecular brake as long as apoptosis is not triggered by a signal. When the cell detects such an apoptotic signal, this brake is released, and the corresponding signaling pathway is activated by certain proteases and nucleases. The most important proteases involved in apoptosis are the so-called caspases.

In humans and mammals, there are several distinct signaling pathways of apoptosis involving a total of about fifteen different caspases. Mitochondria play a special role in apoptosis, involving proteins related to the Ced proteins of *C. elegans*. At key points in the apoptotic program, proteins integrate signals from several different sources and, if the signals are suitably constellated, can trigger apoptosis.

Ameisen (2002) explains that the identification of a genetic regulation of physiological cell death, and of its central role, not only in development but also in adult tissue homeostasis, has led to the acceptance of the idea that all cells from all multicellular animals may be intrinsically programmed to self-destruct, and that cell survival continuously depends on the repression of this self-destruction program by other cells. In other words, cells may survive only as long as they are signaled by other cells to suppress the induction of a default pathway leading to cell suicide. Hence, in a counterintuitive manner, a positive event—life—seems to proceed due to the continuous repression of a negative event—self-destruction. The coupling of the fate of each cell to the nature of the interactions it can establish with other cells has led to the concept of "social control" of cell survival and cell death, allowing a stringent regulation of cell numbers, of their localization, and a constant adjustment of the different cell types that constitute our organs and tissues. This principle, Ameisen concludes, might be one of the bases of our versatility and plasticity, allowing our bodies to build themselves, to constantly reconstruct, and to adapt to ever changing environments.

In the same sense, Durand (2020) formulates that the viability of any eukaryotic cell rests on a delicate balance between the biochemical activators and inhibitors, pro- and anti-apoptotic factors of programmed cell death.

Ameisen (2002) states that etymologically, the term *program* means "prewritten." However, the very concept of program in biology is ambiguous, suggesting a framework of design and finality, and favoring a confusion between the existence of prewritten genetic information and the multiple ways it can be used by the cells and the body. Accordingly, it is not the individual fate of each cell, its survival or its death, that is programmed (prewritten). Rather, it is the capacity of each cell to induce or repress its self-destruction as a reaction to the respective external or internal situation.

Regarding the physiological and molecular events during apoptosis, the process is not so much a "programmed" event, but rather regulated within the context of the whole organism. Thus, a better term would be "genetically regulated cell death." This term is occasionally used in research articles and is more organismic.

Regulated Cell Death in Unicellular Organisms

Before the turn of the last century, there was the widespread belief that single-celled organisms were in principle immortal, and that loss of viability regularly occurs through external factors such as nutrient depletion, physicochemical damage, predation, or the accumulation of metabolic waste products. Even in these cases, death processes belong essentially to their life, because populations of single cells would destroy their ecosystem and themselves, if they would live and reproduce without limits. Again, the balance between production and withdrawal is essential. But a genetically regulated active form of cell death was considered a hallmark of multicellular life.

However, around the turn to the twenty-first century, comparable forms of regulated death have also been observed in diverse unicellular organisms. This includes eukaryotic as well as prokaryotic cells (Durand 2020; Durand et al. 2016; Huettenbrenner et al. 2003; Lewis 2000). What started as a curiosity soon became a fundamental problem in biology. The existence of an active form of death in unicells was described as "enigmatic," "counterintuitive," "confusing," or simply an "anomaly" (Durand 2020).

This phenomenon was especially puzzling from an evolutionary point of view. Why would unicellular organisms harbor a suicidelike genetic program, for surely natural selection would not favor such an obviously lethal trait? In unicellular organisms, death of the cell equates to death of the organism. Because survival is expected to be the central incentive of all organisms, the phenomenon appears to be illogical and unreasonable.

Many of the genetic and protein elements involved in regulated cell death of multicellular organisms have now been identified in diverse unicells, and an argument can be made that regulated cell death is almost as old as cellular life itself (Durand 2020; Durand et al. 2016).

Durand (2020) describes that the study of processes in regulated cell death in unicellular organisms, while still rudimentary, has revealed a range of genetic components involved. Within a single taxon (especially with eukaryotic taxa), there is usually more than one molecular pathway leading to death, although there is frequently cross talk between them. The resultant phenotype may be morphologically and biochemically distinct or may include features that are associated with more than one morphotype of death. In addition to the variation in the death phenotype itself, it has also become clear that the molecular pathways overlap with other outcomes such as cell cycle arrest, dormancy, senescence, aging, spore formation, and sexual reproduction. The complexities of the molecular mechanisms have made dissecting their component parts and attributing functions to them, at least in a reductionist way, very challenging, Durand summarizes. Nevertheless, significant progress has been made, and in some instances, the molecular basis for a particular death phenotype has been dissected. In other cases, only a few of the key molecules have been identified.

In procaryotes, active cell death comes about somewhat differently (Durand 2020; Durand et al. 2016; Lewis 2000; Häcker 2013). Microscopic observations reveal the complexity of the cellular and subcellular changes, indicating that multiple molecular pathways must be involved. Even though there may be different phenotypic manifestations, the ultrastructural changes reveal that the mode of death is not passive cell lysis. This has subsequently been documented in diverse microbes under a range of conditions.

Beyond these observations, however, the impact of the phenomenon is not well understood. Regulated cell death in unicells is thought to have significance in the coexistence of colonies and populations. Today, it is becoming increasingly apparent that bacteria live and die in complex communities, which in many ways resemble a multicellular organism. The release of pheromones induces bacteria in a population to respond in concert by changing patterns of gene expression, a phenomenon called quorum sensing (Lewis 2000). Most bacterial species actually do not live as planktonic suspensions in vivo but form complex biofilms, tightly knit communities of cells. From this perspective, regulated death of damaged cells may be beneficial to a multicellular bacterial community. For example, apoptosis could limit the spread of a viral infection. In the case of damage by toxic factors, cells will donate their nutrients to their neighbors. Also, elimination of cells with damaged DNA would contribute to the maintenance of a low mutation rate.

Death as an Integral Part of Life

The important point in our context here is to emphasize that death is an integral part of life.

At first sight, life and death seem to be contrary and mutually exclusive. The one exists in the absence of the other. A living entity, whether a single organism or a complex biological ecosystem, can be alive only if it is not dead and vice versa. However, when the subject is explored further, it becomes clear that life and death processes are not always oppositional. They exhibit some coevolutionary features. In multicellular organisms, cellular death may serve to promote the life of the organism of which they are a part. A similar function seems to be involved in single cells as well.

Interestingly, Durand (2020) states that in many cases the molecular mechanisms have been difficult to pin down, and the reason is not particularly surprising. Complex traits, according to Durand, are difficult to disentangle using the reductionist approach in molecular biology: "This is not a problem unique to PCD [programmed cell death] biology. Organisms are integrated entities and reducing complex phenomena, of which PCD is certainly one, to atomized traits is problematic. Certainly, at the molecular level most traits are polygenic and difficult to dissect. For PCD, there may be key molecules, but the crosstalk between pathways, the variety of environmental factors associated with PCD, the phenotypic plasticity, and the context-dependent realization of the phenomenon all mean that uncovering the molecular basis is always going to be challenging" (Durand 2020, 94).

5 Synthesis

Taking Phenomena Seriously

In the previous chapters, fifteen specific properties of life have been described that are typically exhibited in organisms. An unbiased observation of what is directly manifested by living organisms will necessarily come up with such properties. Here "unbiased" means that theoretical presuppositions are avoided as far as possible. One such presupposition, for example, would be to assume that an organism is no different from a machine, or that all properties must be attributed to molecular processes. Another would be to assume primarily some form of wholeness in every case, or even to postulate some vital force as the origin of phenomena. All such reductionism can obscure the view of the phenomena.

Nevertheless, it is also clear that with the described properties some theoretical influences flow in. A pure perception of phenomena is hardly possible. And finally, it is not even desirable, because we want to understand the inter-relationships. A strict phenomenology would remain entirely with the description of what is perceptible. However, science consists not only of the purest possible observation but also of orienting ourselves in these processes, in penetrating them with the mind, and in finding explanations. We search for the concepts associated to the phenomena.

The philosopher Rudolf Steiner (1996) made this the starting point of his theory of science and described how important it is to work out both aspects carefully: observation on the one hand, which takes the primary phenomena seriously, and thinking about them on the other. Thinking and forming ideas about the phenomena, however, is proposed to be directly oriented to the phenomena. It can be problematic to take a starting point from a presupposed paradigm in the sense of Kuhn (1962).

Taking the phenomena seriously also includes that certain properties, which are actually observable, are not declared to be epiphenomena. In the foregoing, such properties have repeatedly been discussed. Ignoring autonomous system properties and focusing only on the underlying subprocesses is an example of this. Science has increasingly learned in recent years that this is biased. However, research that regards such properties is growing only slowly: too deeply ingrained is the focus on the constituents, from which any higher order is supposed to emerge. The opposite mistake, of course, would be to look only at the system components and neglect the constituents. Several times it has been described throughout the previous text that the *interdependence* itself is the decisive point. Only by taking this interdependence seriously as a phenomenon by itself, in its *concurrency*, and by bringing it into the center of consideration as a positive phenomenon, we actually find ourselves on the path to an organismic thinking.

Also, the universally perceptible intrinsic activity of organisms was repeatedly relegated to the background. In the historical chapter 2, the illuminating work of Riskin (2016) has been quoted. She shows how world views since the seventeenth century led to neglecting agency as a primary phenomenon. One view was that everything was to be regarded as driven only by a Creator. Another, that physical forces in the background were assumed to trigger and cause all events. The primordial formation of drives, impulses, and processes by the whole organism itself was not recognized.

The latter view has clearly left its mark on contemporary science, so that attempts are still being made to attribute all activities to underlying causes. Sometimes it is the genes that are supposed to drive everything, sometimes it is the molecular interactions, sometimes it is the environmental influences that force organisms to behave and adapt in certain ways. The very fact that organisms are active on their own, that every metabolism is primarily based on an intrinsic activity of the whole organism and does not occur in its entirety without the integrity of the system, is tacitly swept under the carpet. Or even that organisms can be actively involved in their evolution was unthinkable until recently. It is work like that of Walsh (2015) that brings the intrinsic activity of organisms back into the center of scientific consideration.

Of course, this displacement to an epiphenomenon is particularly evident in a property such as consciousness. Only recent investigations suggest that consciousness is a primary property of living organisms, the traces of which can be followed in evolution in the same way as the development of limbs or lungs. Steiner's view, quoted earlier, that the actual, positively observable phenomena should be made the primary starting point of science, takes on a special significance here.

However, we have also learned in the previous chapters that there have always been individual scientists who took such phenomena more seriously.

Yet, they were often more or less ignored by mainstream science. In chapter 2 many names of such personalities are compiled. It has been precisely such pioneers, who have helped to develop organismic aspects for the view on nature. Today there is so much experience, so many observations and facts of this kind that one can bring the aspects together to try to develop a coherent concept of the phenomenon of life.

We have now moved more onto the side of the phenomena. However, as mentioned previously, science also emphasizes the formation of concepts. According to this, it is not just a matter of orienting oneself to some extent regarding the phenomena, but only by thinking about them, by forming concepts and ideas, and by developing theories does one actually approach reality. Several times in the preceding chapters we have seen that a certain order underlies the organic. We can find this order in science, the underlying organization and information, through our theories and ideas. Only the tracking of this order, the recognition of the connections and structures, makes the scientific process complete. Therefore, every science consists of the careful observation of the phenomena on the one hand, and the recognition of the underlying order, the regularities, on the other hand. We are never satisfied with the mere collection and listing of phenomena, we want to understand "what holds the world together at its core." Thus, in essence, the merging of perception and phenomena on the one hand, and thought, ideas, and understanding on the other hand is the essential process of scientific work. By thinking about phenomena, we not only orient ourselves in them but we can replicate in our thinking the internal order that organisms present to us. It is not an exaggeration to say that in this way we can participate in the organization and processes of nature, as Johann Wolfgang von Goethe expected as an ideal of scientific work (Goethe 1981; Steiner 1985a).

However, theory building is subject to many influences and presuppositions. It is also influenced by certain traditions, as we saw at the beginning of chapter 2, thanks to the work of Kuhn and Lakatos. And it is subject to change over time. These processes are not to be viewed critically because they are simply the course of science. This kind of learning ability is, after all, what is healthy about this way of gradually understanding more of the world. Maybe it is sometimes too tenacious, too slow, and maybe science sometimes gets lost in various one-sidednesses. This need not be discussed further here. However, it has been made clear that a more complete, more comprehensive, more real properties-oriented way of thinking about organisms is needed. Many scientists cited in the course of the present inquiry are thinking along such lines and have developed proposals to that end. What has been presented in the preceding pages is merely an amalgamation of such aspects. It is an attempt to establish the intrinsic relationships between those aspects and to summarize the properties discovered in each case, which can be found and observed in living organisms.

As already emphasized in the formulation of the hypothesis (chapter 3), the properties described can certainly be compiled and understood differently. Perhaps even more properties can be found that have been overseen here, or that have not been detected by science so far. Or they may be summarized differently among themselves, so that there are fewer in the end. Therefore, the proposed overview is changeable and flexible and can be adapted to further knowledge. The key point is to consider the primary properties of the organic and to take them seriously as a starting point for their scientific treatment. This is called the organismic concept, and it is proposed that this is central to the further development of object-appropriate biology.

What should be wrong with taking seriously the primary phenomena of living beings, their original characteristics in their vitality, and learning to understand them as such, out of themselves?

Synthesis of Opposing Views

In chapter 2 it has been postulated that the history of science repeatedly showed that opposing alternative research programs can exist side by side, competing with each other for long periods of time, and that major progress is often made when a synthesis of both becomes realizable. After long argumentation, it may turn out that both sides are right from their particular perspective, and that there is a solution possible by merging aspects from both factions. Several examples from the history of science have been mentioned. Then the question was developed whether a similar synthesis would be possible regarding the dichotomy between mechanistic thinking on the one hand and approaches that search for special qualities and characteristics of organisms on the other hand.

Biology today has a rich knowledge of many details of the cell, of physiological processes, of anatomical features, of developmental processes, of evolution, and much more. Most—but surely not all—of this knowledge has been gained under mechanistic premises, which enabled researchers to focus on well-circumscribed subjects and topics. However, what often—but surely not always—has been lost during research on such details is the view from the context and the question of how the processes of the parts are successfully integrated within an organism as a whole. This not only provided a wealth of information about the species in question but also revealed how cells and biomolecules work in general.

The effectively collected results contradict thereby blatantly the assertion that one can somehow extrapolate the whole of the organism from the sum of the partial functions. The expected mechanistic resolution into cause-effect chains did not occur. Yet, the many individual findings can be integrated without contradiction, if one starts from integrative functions, from classical regulation, and from resilience (autonomy) of physiological functions and their highly temporal order: all elements that have been the basis of organismic thinking since the beginning of the twentieth century. Nowhere have these thoughts been contradicted by empirical research. Rather, the results of this work themselves indicate more and more clearly that even the early organismic thinkers of the first and second phases had developed accurate concepts with which to grasp and investigate context. It is often downright amazing how clearly the actual results point to this contextuality, even though the studies tended to have been conducted under mechanistic and particularistic premises. The organism often gives other responses than expected in one-sided presuppositions.

In the fifteen numbered sections of chapter 4, which constitute the main argument of the present book, an attempt has been made to place the knowledge and facts available today in this context. As stated at the beginning, the time was never so favorable and the knowledge so extensive as today to attempt such a synthesis and to get new answers to the question of life. The extensive knowledge of details in structures, functions, and genetic processes provides a new opportunity to understand integrative and systemic functions.

In this sense, a synthesis of both schools of thought is not only possible today but is virtually self-evident. All the molecular and genetic details are only understandable in context and in their time-autonomous order within the activity of the cell and the organism. Thus, an organismic approach does not contradict today's common empirical research but is at best a logical conceptual extension. It requires us, however, to leave the widespread mechanistic doctrine of life. Living organisms must be described primarily as animate.

I do not see what objections could be raised to such a picture of the living organism, because it does not relinquish the scientific basis nor use any mysterious and unprovable factors. No special metaphysics are necessary, but only consistent empiricism, which however takes its own results seriously. In this sense, organismic biology is a synthesis of the old opposites, but is now on a new level.

At the same time, it is also necessary to conduct further research directly according to such an organismic concept. If one starts primarily from the organismic qualities of living organisms and places them at the center of investigations from the outset, the results will also be closer to the actual processes. Some disciplines would change if an organismic concept were taken as a basis from the outset. This would be a great advancement for biology.

Occasionally there is talk of a crisis in biology (molecular biology) and major changes are predicted (e.g., Mesarovic et al. 2004; Strohman 1997). One can see this also somewhat differently: The opposition between micromechanistic and organismic ideas is old, and the entrenched atomism of some approaches in biology will not be overcome any time soon. This has to do with the fact that it is also a particular worldview and not merely a scientific paradigm. What is new is that organicism can now be given a more concrete program if some of its concepts are brought together in a synthesis. To work on such a synthesis, on such a program, is important for the future. This will not trigger the great paradigm shift, which will be a long time coming. But to develop a viable basic concept for organicism, on the basis of which one can carry out concrete practical research, is essential for the future. It is particularly interesting that quite a few scientists have been thinking organismically for a long time.

Synthesis of Properties

In chapter 2 the distinction between reduction and reductionism was made. Reduction is the scientifically necessary isolation of a certain area of phenomena or a certain process, which is to be examined more exactly. The separation of the intrinsically unitary organism into fifteen different properties, as described in the preceding chapters, is such a reduction.

However, the individual properties that have been described in this way are interconnected. In other words, they exist in the simultaneity and identity of the organism. They are merely different perspectives on the overall process. So, also here, globally seen, the principle of concurrency prevails. Therefore, an organismic approach cannot stop at the decomposition into the described properties but will have to develop a synthesis. One can perhaps envision imaginatively such a synthesis, but it (necessarily?) will remain vague and indefinite.

An attempt will be made here to identify and describe some obvious overlaps. Perhaps these are the first steps toward an overall picture, a synthesis, which itself must remain open.

One starting point for the description of organismal properties was that organisms are *processes* (chapter 3). There are always dynamics, changes, developments, and functions that constitute the organism. Therefore, Dupré and Nicholson (2018) strongly advocate a general processual view of life. A first synthesis has been proposed in section 4.7. Having discussed more properties, which are exhibited by organisms, it is now possible to reformulate this proposal somewhat more biologically. From the view of empirical biology, at least five characteristics need to be added to a view of general processuality. They are intrinsic to Dupré and Nicholson's draft, but need to be addressed more explicitly:

First, it is necessary to indicate what this process is processing. A process does not exist as such, it needs to substantiate in some form. Sections 4.5–4.7 have described that molecules as well as energy and information are processed.

Second, the process is mainly an activity. Rather than being some form of event or flow, it is a continuous agency as described in section 4.4.

Third, all processes have a detailed time order as described in section 4.9. Every process of an organism is organized within a specific regulated time structure, and most of this time regime is actively generated by the organism itself. Therefore, time autonomy is a decisive characteristic of all organismic processes.

Fourth, these active processes generate a certain autonomy toward the nonliving surrounding as described in section 4.3. The living organism is in some special way different from its immediate vicinity. It actively generates a disequilibrium.

Fifth, the process generates a shape, which has been called processes of shape in section 4.8.

Taking these features into account, we come to a process ontology, replenished by further properties, which can be observed in organisms. The order that constrains the process along these features has been called *organization* (chapter 3). Organization describes the typical structures, form, time order, and molecular, energetical, and informational regime that constitute a living organism.

Under this aspect it is neither enough to describe just the parts and molecules of an organism (substance ontology) nor only to describe the permanent changes of parts and components as such (process ontology). The essential questions are: How are these parts and these processes organized? How are the components arranged? How are the changes constrained in a particular way?

Another starting point in chapter 3 was the principle of *concurrency*. Throughout the text many examples of this principle were encountered. The principle of concurrency is typical for organic entities and makes it so difficult—but at the same time so fascinating—to understand life itself.

In chapter 3 it was explained that there is concurrency between process and organization, and the relation to substances has been called substance-process. Further examples of concurrency are the interdependencies and reciprocal relationships in functions and networks (section 4.1) and between different system levels (section 4.2). Organisms are relatively autonomous and yet concurrently open to their environment (section 4.3). They concurrently process molecules, energy, and information (sections 4.5–4.7); they are concurrently process and shape (section 4.8); they concurrently generate time sequences of very different frequencies (section 4. 9); and so on.

Another initial motif from chapter 3 was that it should be possible to find connections and overlaps between the various properties. Ultimately, they are each just one particular perspective on the overall process. Thus, in the end they are each related to one another. Nonetheless some overlaps may be especially obvious or more extensive. Many such overlaps have already become apparent in the discussion of individual properties. Here, some of these more obvious overlaps will be summarized. Although it is immediately clear that other characteristics would also have to be named in each case, only the particularly obvious ones will be addressed here.

A start has already been made with the preceding discussion of the process character of the living, with special importance given to agency. If agency is for a moment put into the center of consideration, basically all other properties need to be mentioned that are related to this process, or more concrete, that are generated by the activity of the organism. The activity of processes generates the different system levels and the relative autonomy. It processes molecules, energy, and information, and drives development. Obviously, it is also involved in evolutionary transitions, and much more. As agents, organisms are extended and differentiated not only in space but also in time. Even subjective experience is presumably not possible without internal activity. Therefore, it might be possible to reorganize some elements of figure 3.2 as proposed in figure 5.1.

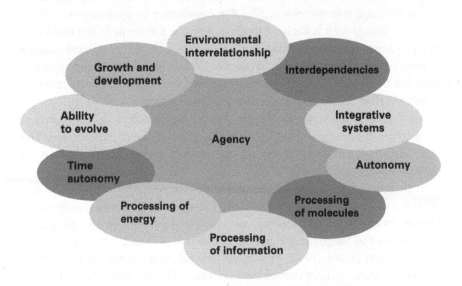

Figure 5.1
Synthesis of properties from the perspective of agency.

From a different perspective, relative autonomy can be brought into focus. Autonomy is actively generated by the life process and its activities. It is constructed by different system levels that contribute to the overall autonomy of the organism. Highly regulated time processes as well as the active handling of molecules, energy, and information establish a self-determination and establish the described disequilibrium. Simultaneously with expanded autonomy capabilities, the capacity for subjective experience expands, obviously leading to the appearance of consciousness (figure 5.2).

This complex can also be regarded primarily from the perspective of subjective experience and consciousness, collecting the most important properties that contribute to it (figure 5.3). Sensitivity, affectability, and processing of information may be the most important ones. However, to experience the environment, to have a certain degree of autonomy as well as an internal activity of some sort may also be involved.

As discussed in section 4.8. the shape of any organism is of fundamental importance. Several, if not all, properties contribute to the appearance of a particular spatial manifestation (figure 5.4). Shape is generated during the process of growth and development, is a manifestation of the system, and is supported by processing molecules, energy, and information. Especially information is essential to the generation of specific forms. In section 4.13, it has been characterized that the information for the structure of an organism is presumably not in the genome but is generated again and again during the developmental process.

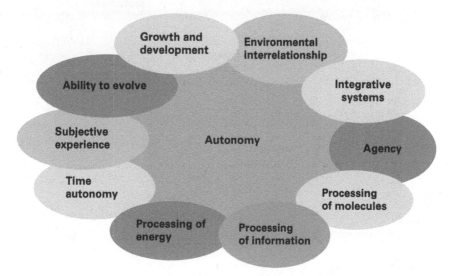

Figure 5.2
Synthesis of properties from the perspective of autonomy.

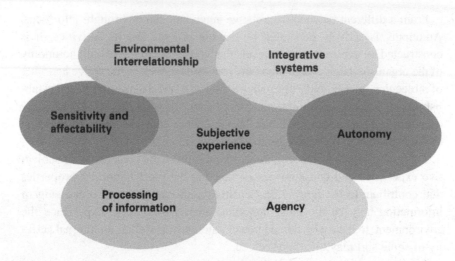

Figure 5.3
Synthesis of properties from the perspective of subjective experience.

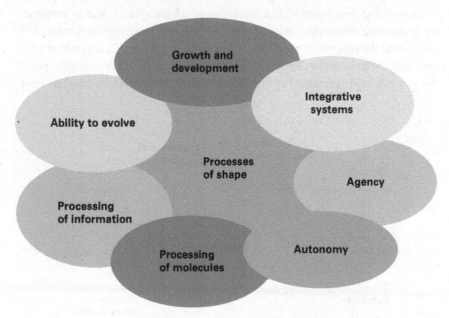

Figure 5.4
Synthesis of properties from the perspective of shape.

However, shape is also related to autonomy. Hans Jonas (1966) concluded that in an organism, form is emancipated from matter. Organismic form exhibits a degree of independence that enables the shape to continue to exist despite incessant material exchange. For Jonas, organisms invert the ontological relation between matter and form found in inanimate objects. In inanimate objects, form results from specifications of the material, and in organisms, matter is subordinated to form (Nicholson 2018). "The constituents of an organism at any particular instant are only the temporary realization of the self-producing organizational unity of the whole." The processual view of the organism "accurately reflects the fact that the matter of an organism is necessarily and continuously exchanged while its form is actively maintained. Bringing together the autonomous maintenance of form with the causal continuity of process that makes up a living entity over time," we can draw the ontological lesson, "which is that, as far as organisms are concerned, persistence is grounded in the continuous self-maintenance of form" (Nicholson 2018, 158).

A final example focuses on the ability of organisms to organize and manage information (figure 5.5). Information is obviously necessary for growth and development, something which modern developmental biology is describing extensively. Changes in these processes contribute to evolution. Also, information is actively organized, something which today molecular biology is determining in rich detail (reduplication of DNA, transcription, RNA-splicing, translation, RNA-processing, are some keywords), so that agency is one prerequisite.

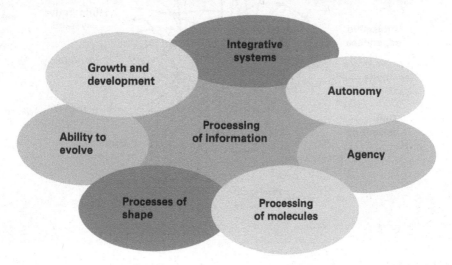

Figure 5.5
Synthesis of properties from the perspective of information.

It is not necessary to continue this description here, although some overlaps have not yet been mentioned. But the principle, the basic statement, should be clear: In each case the description is only about specific chosen segments out of the whole, which in its unified existence becomes scientifically comprehensible in this way. The vision is to think this unity in toto. Figure 5.6 may serve as a preliminary illustration.

Depending on the research question, one will usually only be able to examine some of these properties when studying an organism, or even just a single property, and even that can be complex. This is routine in research and does not need to be explained further. It is the process of reduction that makes the appropriate focus possible. When one studies an organism, say a particular specimen of an adult bird, some of its current characteristics are accessible, others are not (figure 5.7). In the case of this bird, for example, one cannot say something about its evolution. This would require data on its relationship with other species and their paleontological history, comparative morphological and

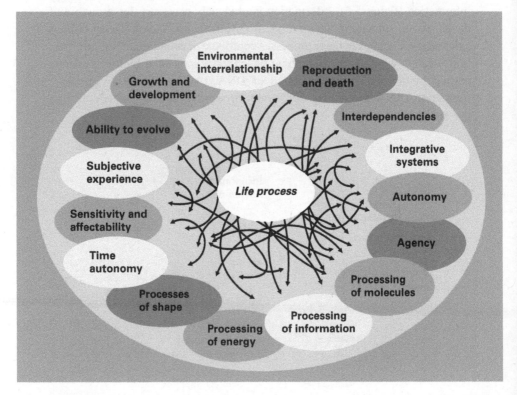

Figure 5.6
Preliminary illustration of a synthesis.

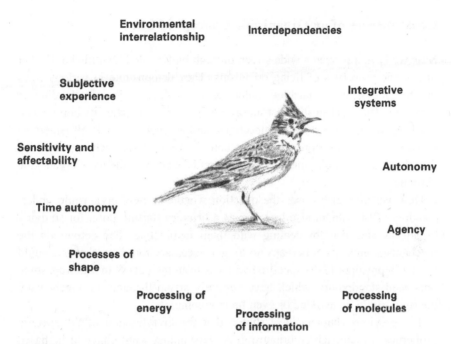

**Environmental
interrelationship**   **Interdependencies**

**Subjective
experience**   **Integrative
systems**

**Sensitivity and
affectability**   **Autonomy**

Time autonomy

Agency

**Processes of
shape**

**Processing of
energy**   **Processing
of information**   **Processing
of molecules**

Figure 5.7
Some features of an adult bird that can be studied. *Source*: Drawing courtesy of Angela Rosslenbroich.

genetic information, and other data, that is, information beyond that which can be obtained from that particular specimen. Likewise, one cannot make statements about its individual development. For that, one would have to study the embryonic processes that lead to this adult form. In any case it will be necessary to have the whole complex of properties in mind and to take them into account in the models and interpretations in order not to fall into the trap of reductionism.

It is important to understand that a synthesis is not the merging of different scientific disciplines. It is not just about the often demanded "interdisciplinarity." Different scientific disciplines may have different emphases in which they work on aspects of these properties, and an exchange between them may lead closer to a synthesis. But the actual synthesis is an ontological problem, a problem of the unity of the organism, whose properties we understand so far only under a certain reduction. To go beyond this reduction means to approach the wholeness of the organism, to build up an idea of how the unity of a living active organism is really shaped. This may be possible only very provisionally and incompletely, and it is rather complex. But it is obviously necessary to face this complexity.

Consequences of an Organismic Concept of Life

Now we have traversed a wide sweep through biology and identified a number of specific properties of living organisms. They demonstrate in various ways that the often-strained machine analogy is not sufficient for understanding an organism. The mechanistic doctrine in biology always reaches its limits when it comes to context, network connections, and autonomous and self-generated processes, and even more so when it comes to properties such as subjective experience and consciousness. Yet, these are the central elements of organismic existence.

However, this also raises the question whether a new organismic understanding of life cannot also be taken as a *primary* starting point for studying organisms and also for dealing with them in practice. The detour via the mechanistic analysis is perhaps no longer necessary today, so that one could primarily presuppose the specific liveliness from the outset. In this way, some one-sided standpoints, which have certainly arisen through the mechanistic doctrine, could be avoided or even be overcome.

The previous chapters have shown that the investigation of the specific properties, which such an organismic understanding would have to be based on, does not lead to any unscientific assumptions. All the specific properties mentioned can be investigated with scientific methods, even if essential developments are still necessary in many cases.

The further development of research on what, for example, constitutes an integrative system, what significance the self-activity of organisms or their specific autonomy have, is currently still severely impeded. In the standard research programs—and consequently in research funding—these properties are hardly addressed. Catchwords such as *molecular mechanisms*, *genetic determinants*, or *genetic engineering* largely dominate the funding business.

An organismic vision of life has at least two main implications. First it changes our general understanding, our idea of living beings. It is an important intellectual, scientific, and philosophic endeavor to come to an appropriate idea of life in general. Idea here means our way of thinking about the organism, the image, and the mental concept we have about living entities (Noble 2006, 2017b).

Second, an organismic vision of life has consequences for our practical dealings with the living, touching almost every field of human endeavor. This should not come as a surprise since most phenomena we deal with in our professional and everyday lives have to do with living organisms. Whether we talk about economics, the environment, education, health care, or agriculture and food, we are dealing with living organisms and ecosystems. Consequently, a fundamental shift of perception from the mechanistic to the organismic view of life would be relevant to all of these areas.

Scarfe (2013) makes clear that the full thrust of mechanistic thinking of the reductionist variety does not stop at the scientific explanation of natural phenomena. Research based on the mechanistic framework is, in many cases, committed to technical thinking and applications. By uncovering the set of "mechanisms" that are deemed to underlie natural phenomena, it is presumed that such mechanisms can then be manipulated to bring about a specific result of interest. The mechanistic lens presents nature as a series of levers, switches, and cogwheels, which can be manipulated through technological or biotechnological applications. In such a context the mechanistic explanation appears to be a necessary condition to heighten human power and control over the natural world by way of applications that they enable (see also Lewontin 1991).

Today the mechanistic framework is marred by its contribution to the ecological dysfunctions that pose threats to the well-being, the survival, and the evolutionary future of contemporary forms of life on this planet, Scarfe (2013) further explains. Being concerned for the integrity of ecosystems as well as for the humane treatment of non-human organisms, he believes that there are drastic consequences for most life-forms on this planet, including ourselves, if we do not de-emphasize the mechanistic framework and, at the very least, cultivate the value of an engagement in a phase of holistic reflection in scientific research that lays bare the abstractions and shortcomings that are created by mechanistically oriented research. Scarfe explains that there is a "blind spot" in our understanding of the natural world when looked at solely through the mechanistic lens. And it is largely because of this blind spot that human beings continue to develop and use technologies that may seem to heighten their power, render their lives easier, and improve their well-being in the short term, but that unwittingly do substantial harm to the biosphere's life-support systems and to the diverse life-forms that help to compose it.

As Scarfe (2013) further describes, because nature is such a complex interdependent system, many technological and biotechnological manipulations risk dispersing "unintended," "unexpected," "unpredicted," or "nontargeted" effects. Then, he asks, what does it mean when crops, genetically modified for the purpose of rendering them more resistant to herbicides and pesticides, have unintended effects on the host organism and unexpectedly cross-pollinate with other strains? Systemic pesticides and genetically modified organisms are also chief suspects in relation to the unintended collapse of insect populations, such as honeybee colonies, on which we depend for pollination. "In this light, the ecological degradation being perpetuated by the continual, indefinite scattering of 'non-targeted' effects throughout the biosphere, as a result of the employment of technologies and biotechnologies that originate directly from mechanistically-oriented biological research of the reductionist variety, provides proof of the fact that the mechanistic lens has a great blind spot in terms of its ability to provide a comprehensive understanding of organic nature" (40).

Mechanistically oriented researchers, in whom this metaphysics is very deeply entrenched, may respond to the preceding claim by suggesting that all we need to do in these instances is to continue to inquire and to discover the mechanisms behind such unintended effects, in order to "fix" them, without having to reflect on the adequacy of the underlying mechanistic metaphysics to provide a comprehensive understanding of life. They may claim that with time and progress, such problems will inevitably be resolved and overcome. They may think that if we somehow create an "error" in manipulating the mechanisms controlling the operations of the natural world, then through human technological-rational ingenuity, it will be assumed that we can just "pull a different lever" or "flip a different switch" to fix that error.

The lesson from the previous chapters is that we need to realize that nature is much more complex, interdependent, and subtle. Natural systems are not just made up of a network of efficient causal mechanisms that can be manipulated without ecological consequences. The message of organismic thinking is that, rather than merely amassing "biological knowledge," which will lead to an ever-increasing "technological control of nature," we need to cultivate what Conrad Hal Waddington called "biological wisdom" (Waddington 1960, 23, 30)

All of this is not to criticize scientists for creating ecological problems. On the contrary, many scientists contribute to the knowledge we have about these problems and are constantly warning about them. We saw in the previous chapters that the results of all this research, including those that have been gained under a mechanistic standpoint, are a prerequisite to formulate an organismic concept. However, the time has now come to draw the consequences from all that knowledge, and this includes a transition in the biological worldview.

Godfrey-Smith (2020, 203) provides a good example from praxis: across many countries, bee colonies quite suddenly started collapsing and consequently failing to pollinate all sorts of crops that rely on them. Given the economic importance of bees as pollinators, the cause of the collapses has been intensely studied. It has been assumed that it must be something worldwide, not local. Yet, the collapses occurred fairly quickly and unexpectedly. Was it a parasite? A fungus? Chemical toxins? Godfrey-Smith then reports on a conversation with an expert in that field who said that they are starting to get a handle on what is going on. So, what is the factor that is causing it? The expert replied that as far as they can tell there is no single factor. Instead, over many years, more and more small stresses have appeared in the lives of bees: more pollutants, more pesticides, more new microorganisms, less habitat. For a long while, as these stresses accumulated, bees were still able to cope. Colonies absorbed the stress by working harder. Although they were not obviously and visibly suffering, the capacity of the bees to buffer these problems was slowly being worn out. Eventually a critical point was reached, and honeybee colonies just started to collapse. They failed dramatically, not because some sudden pest had swept through, but because their capacity to absorb the stresses had run out. Now, fruit farmers desperately truck bee colonies thousands of miles from orchard to orchard, trying to get their crops pollinated with the bees that are still healthy enough to do the job.

Assuredly, that is also the case for insect populations worldwide, which are presently experiencing a veritable breakdown (Cardoso et al. 2020; Hallmann et al. 2017; Wagner et al. 2021). The same is true for the oceans (Landrigan et al. 2020). This sphere of biological creativity is so vast, that for centuries we have been able to do with them whatever we wanted. It absorbed the stresses, not invisibly, but often it was hard to observe and easy to ignore when money was involved. Now, our capacity to stress oceanic systems is much greater and the number of influences increases, so that the buffer capacities are overcharged.

Ecology

Ecology has a central importance. During evolution there have been at least five mass extinctions since the evolution of plants and animals. Presently we are experiencing the sixth mass extinction. However, this extinction is special insofar as it is caused by human activities, and it has a dimension that exceeds that of all other ones. This is not only beyond all responsibility we have for this planet and life on it but also threatens the very survival of humanity. It was therefore high time that this finally arrived in the social and political debates. In addition to many scientists, it is primarily a young generation that has begun to vehemently put this on the agenda. It is very much to be hoped that they will persevere and assert it against the many resistances.

But these alternatives must be based on an appropriate concept of organism and ecosystem. Science itself does not guarantee that there will be solutions to these problems. Especially the machine analogy is unsuitable to bring about such a turnaround.

Therefore, organismic approaches to ecology are crucial. Ecosystems must be understood as integrative systems with their interaction relationships, their time structures, and the intrinsic activity of the living beings within them. This has long been included in many concepts and formulations of ecology (for an overview, see Schwarz and Jax 2011) as well as in the concepts of sustainability, and it can be significantly strengthened by organismic approaches (Miller and Spoolman 2020; Steffen et al. 2005). However, there has been a long-standing dispute in ecology between "individualistic-reductionist" and "organicist-holistic" approaches.

Even the simplest examples from ecology show how interconnected organisms are in ecosystems and how much their activities create an organic community in which each species and individual forms an essential component. Understanding ecological interaction networks means understanding relationships and multiple feedback loops. Linear chains of cause and effect exist very rarely in ecosystems. Often a disturbance will not be limited to a single effect but is likely to spread out in multiple consequences. It can even be extended by related functions, which may obscure the original source of the disturbance. Solutions to problems are to be sought in the restoration of dynamic equilibria rather than in the manipulation of individual factors whose side effects within the complex system are usually impossible to assess.

It may also be necessary to correct the overemphasis on competition between organisms. There is much evidence today that there is just as much cooperation, exchange, and symbiosis between organisms. The cyclical exchanges of energy and resources in an ecosystem are sustained by pervasive cooperation. Partnership—the tendency to associate, establish links, live inside one another, and cooperate—is one of the hallmarks of life. Margulis and Sagan formulated (1997, 29): "Life did not take over the globe by combat, but by networking."

If we humans intervene in these networks, we should do so from a profound understanding of these relationships and try to maintain or promote the self-activity and stability of these systems. The processual relative autonomy that ecosystems can establish is discussed today under the term resilience and gains importance in face of manifold disturbances and challenges. Solutions to problems or the use of an ecosystem must always be sought from within the system rather than trying to manipulate it by technical means. It is to be feared that the knowledge and proficiency to act in such a system-context are still underdeveloped today.

Profound ecological knowledge of the complex interactions is a prerequisite for this and is increasingly available today. For example, it is becoming clear what a complex multidimensional living system the soil is, with its humus formation and the myriad of organisms in it. Accordingly, soil scientists are raising the alarm about the state of many soils on which primary production depends in view of the technology-oriented handling of them.

The term sustainability has become important in these developments as the task is to develop a sustainable society that satisfies its needs without spoiling the context in which it is living. This context needs to be preserved and fostered (Brown 1981, 2011). The breakdown of many ecosystems today shows that this ideal has hardly been approached so far. Inventing some new techniques is not enough to develop toward such a goal. Rather, we need a profound shift in our relation to nature, ecosystems, and life in general (Kather 2012).

Agriculture

Modern industrial agriculture relies heavily on fertilization and pesticides, which were developed according to monocausal principles. Where a substance is lacking, it is added in large quantities; where a pest appears, it is combated with a chemical specifically developed for this purpose. In recent years, the grave problems of this method of operating have become overwhelmingly clear: overfertilization, destruction of the soil, groundwater contamination, and pesticide exposures, all of which extend far beyond the immediate scope of application and are in the process of destroying the ecological balance of entire landscapes (Gowdy and Baveye 2018; Horrigan et al. 2002; Zaller 2020).

Some aspects are being discussed and put into practice today to get a grip on these problems, even though a powerful lobby that does its business with this mechanized form of agriculture resists. Yet, what is still poorly understood is the fundamental shift in our thinking about nature that is needed. Individual measures and some more sustainable techniques, as important as they may be, are not enough to bring about these changes.

For more than 100 years now, concepts of agriculture exist that successfully work without artificial fertilizers and without pesticides. They have increasingly learned to promote and care for the ecological context—that in which plants are produced and animals are kept—in such a way that the growing organisms remain healthy due to their own resilience within these systems. Soil, for example, is cared for and encouraged in its own development in such a manner that healthy humus forms the basis for healthy plants. Fertilizers are produced organically. The greatest attention is paid to the care and promotion of microorganisms and soil fauna. It was recognized early on that plants can mobilize their nutrients on their own, if the conditions are created for this agency. However, this only works in context.

In addition, there is the care of the entire environment, such as the planting of hedges and strips with diverse wild herbs and flowers. These attract insects and birds that ingest pests and improve the microclimate. So instead of poisoning insects, they are encouraged here. With such approaches, organic food of high quality is produced. This is practically applied organismic thinking or applied "biological wisdom." This is not a theory; it is successfully carried out today on thousands of organic farms worldwide. This practically applied wisdom will be able to feed the world (Freyer 2016; Halberg and Muller 2013; Horrigan et al. 2002; Muller et al. 2017; Reganold and Wachter 2016).

A wealth of questions and practical problems need to be addressed scientifically. The reflection on the basic theoretical concept in these approaches is also still in its infancy. If this research were performed at the outset from an organismic point of view, it could support practice considerably. Here and in similar areas, a large field of research is opening that needs to be addressed in the near future. At present, this development is proceeding far too slowly in view of the serious problems in our ecosystems.

The same situation applies to the management of animals. Only those who see animals as machines will confine them in tight cages and let them produce. If one acknowledges them as autonomous living beings with consciousness, one will be able to develop a sense of animal welfare.

Given the potentially drastic consequences of all these problems today, it is in our own interest to be flexible in respect to our metaphysical frameworks through which we interpret the world, rather than using mainly a mechanistic one as a fixed habit.

I do not think that the view I have developed in the preceding chapters delivers by itself solutions to the problems posed by mechanistic understanding, but it perhaps provides a starting point or a contribution. Biology as well as applied fields today should be able to move into a more organismic conception of research. Modern science needs to overcome concepts and ways of thinking of the nineteenth and twentieth century and take its own results seriously.

Biological, medical, and ecological methods of research in general would not have to be fundamentally changed. After all, many techniques and working methods are highly developed and sophisticated. Who would want to give this up? However, if the questions, hypotheses, and interpretations were geared to organismic concepts from the outset, the organisms' own performances, the systemic interrelationships, the time conditions, and all those factors discussed in the previous chapters could be given primary attention. One would not first take the circuitous route of mechanistic presuppositions, only to realize long afterward that the results actually point to an organismic connection.

There is research today in which such approaches are developed and being applied, so that a system concept, or the context of networks, for example, are assumed from the outset for the research question. Yet, these are only the first beginnings.

Medicine

Throughout history, the development of biology has gone hand in hand with that of medicine, so that the mechanistic view of life often influenced the attitudes of physicians toward health and illness. Medicine, however, is practically oriented and must deal with the living organism as well as with the psychological and personal characteristics of the patient. It is all too obvious that these different layers cannot be resolved mechanistically. This is, after all, the reality of practical medicine. It is primarily a broad body of experiences in dealing with the vitality of the organism and the personal characteristics and situation of the patient.

Science provides essential contributions and guidance for medical procedures, and modern medicine has integrated many of them very successfully into its work. However, medical science today is dominated to a considerable extend by the attempt to understand the biological "mechanisms" involved in illness. These mechanisms are studied from the point of view of cellular, molecular, and genetic biology, often ignoring many other influences and conditions, resulting in the so-called biomedical model. Out of the large network of phenomena that influence health and illness, the biomedical approach studies only a few physiological and pathophysiological aspects. Knowledge of these aspects is, of course, particularly useful, but at the same time may exhibit a profound one-sidedness (Joyner 2011a, 2011b; Soto and Sonnenschein 2018), so that Joyner et al. (2016, 1355) ask for a "wholesale reevaluation of the way forward in biomedical research."

It is even more modern to talk about "molecular medicine." Truly, this does not exist, because medicine is always about entire organisms or the whole human being. There are indeed molecular aspects of health and disease, but no medical system can be built on the molecular level alone. The so-called "precision medicine," which mainly tries to pin down medical problems to the genetic level or to some specific molecular defects exhibits an extreme one-sidedness (Joyner and Paneth 2019; Paneth and Joyner 2018).

Although the biomedical approach immerses deeper and deeper into molecular and genetic "mechanisms," major problems in medicine remain unresolved (Joyner et al. 2016; Van Regenmortel 2004). This will not change until medical science relates its studies on the various system levels that an organism represents; the time processes of life and illness; and the whole of the physical, emotional, and mental organization of the patient.

Along with the mechanical model of health and illness, the attention of physicians has moved away from the patient as a whole person. By concentrating on smaller and smaller fragments of the body, shifting its perspective from the study of organs and their function to that of cells, molecules, and genetic elements, modern medicine often loses sight of the context and of the autonomous self-activity during the healing process. Over the past decades, dissatisfaction with this form of medicine grew constantly, and many aspects of the problem have been discussed. However, a tremendous supremacy of the biomedical industry persists, so that alternative approaches are regularly being pushed into the background. Medical approaches that do not fit into this form of thinking are displaced into what is called "complementary medicine." This field is quite diverse, but it contains approaches that are grounded more or less explicitly in organismic thinking, do not neglect psychological and mental conditions of the patient, and attempt to specifically promote self-regulation. According to these aspects, some of these approaches are now being increasingly called "integrative medicine." Only recently research on such approaches has expanded, and in many cases has shown their potential for medicine.

Demarcations between "scientific medicine" and "complementary medicine" are not really helpful in realizing the best health care. Basically, it is possible to investigate scientifically many more things than are usually accepted. In recent years, more and more research has been done in the field of integrative medicine, so that earlier contradictions have been partially dissolved, and syntheses have become possible.

Nonetheless, much more could be learned from an organismic understanding. In this sense, there is an unexploited potential for medicine (Boyd and Noble 1993; Heusser 2016; Schaefer et al. 1977; Sonnenschein and Soto 2016; Soto and Sonnenschein 2018). Far too often, the one factor, the one gene, the one pathogen that could be the cause of a disease is being searched for. For example, the importance of genes for many diseases has long been overestimated (Krimsky and Gruber 2013; Noble 2008b, 2017a; Strohman 1993, 2003). Often, however, it is probably a matter of multilayered multidimensional situations, of complex interdependencies and simultaneously existing, mutually reinforcing imbalances, and of the context. Accordingly, multidimensional therapies may be more promising in many cases, not just by trying several possibilities but by developing a structured concept regarding these complex conditions. Especially in many chronic diseases this might be the case.

An organismic approach would broaden the scope from the levels of cells, genes, and molecules to the system levels involved and their integrative properties. The self-activity of organic functions would have to be taken into account as well as their time autonomy, their extensive interdependencies within networks, and their autonomy capacity and resilience. Since the organismic concept proposed here assumes a priori the property of consciousness in all living beings (section 4.11), its consideration in medicine is no longer a particular obstacle. As these different levels also include the molecular level, this would again be a synthesis.

Health and illness are multidimensional processes. From the organismic point of view, illness results from patterns of disorder, which reduce capacities of autonomy. They may become manifest at various levels of the organism. These include biological as well as psychological levels and the various interactions between the organism and the larger systems in which it is embedded. This means that the biological, cognitive, social, and ecological dimensions of life correspond to similar dimensions of health. The insight into these interrelationships virtually calls for an integrative medicine.

The Need for a New Conception of Life

Since the beginning of the twenty-first century at the latest, it has become increasingly clear what problems we are heading for if our technology, economy, and indeed our entire understanding of the world remain as they are. Especially the sphere of life has come under massive pressure due to our previous ways of thinking and of doing business. Species extinction, climate change, pollution of the oceans, and forest dieback are the keywords that are increasingly being discussed (see the concept of "planetary boundaries"; Persson et al. 2022; Rockström et al. 2009). Today, the problems are widely seen and extensively studied. Having reliable data of them is a first prerequisite for being able to overcome the problems one day. Yet, what is still being given too little attention is that these problems are also related to our way of thinking about nature.

Albert Einstein is attributed the sentence: "Problems cannot be solved with the same mind-set that created them." So far, many proposals and techniques for solutions to these problems are still thought of in a one-dimensional mechanistic way. This is primitive cause-and-effect thinking and stuck in the nineteenth century.

The image of the organism as a molecularly and genetically determined survival machine urgently needs to be revised. Such a new understanding of life is now emerging. Elements of systemic thinking or organismic thinking are being developed in laboratories around the world. Many pioneers have been quoted on the previous pages. Yet, this still has to gain much more momentum. Organismic thinking involves thinking in terms of relationships, patterns, and context. It takes the autonomy capacity of organisms and their activities seriously and respects them or skillfully uses them in a way that encourages and supports them.

Our whole attitude toward nature and living beings must change if we will understand life as living. Interventions in nature should then take place out of such an understanding of nature. The continuation of life on earth, and thus the survival of humanity, will depend on whether we learn to respond to the peculiarities and the systemic conditions of the planet and of life. We must develop a partnership with nature (Kather 2012; Suchantke 1993). This requires that we understand nature as nature and we do not try to understand and manipulate it from concepts that are alien and inappropriate to it.

The web of life, of which we ourselves are a part, needs thinking and acting according to its own laws. It needs "biological wisdom."

References

Agazzi, E., ed. 1991. *The Problem of Reductionism in Science*. Dordrecht: Kluwer.

Agutter, P. S., Malone, P. C., and Wheatley, D. N. 2000. Diffusion theory in biology: A relic of mechanistic materialism. *Journal of the History of Biology* 33 (1): 71–111.

Alberts, B., Johnson, A., Lewis, J., Morgan, D., Raff, M., Roberts, K., and Walter, P. 2017. *Molekularbiologie der Zelle*. Weinheim: Wiley-VCH.

Allen, C., and Bekoff, M. 1997. *Species of Mind: The Philosophy and Biology of Cognitive Ethology*. Cambridge, MA: MIT Press.

Allen, G. E. 2005. Mechanism, vitalism and organicism in late nineteenth and twentieth-century biology: The importance of historical context. *Studies in History and Philosophy of Science Part C: Studies in History and Philosophy of Biological and Biomedical Sciences* 36: 261–283.

Ameisen, J. C. 2002. On the origin, evolution, and nature of programmed cell death: A timeline of four billion years. *Cell Death & Differentiation* 9: 367–393.

Ames, P., Studdert, C. A., Reiser, R. H., and Parkinson, J. S. 2002. Collaborative signaling by mixed chemoreceptor teams in *Escherichia coli*: *Proceedings of the National Academy of Sciences of the United States of America* 99 (10): 7060–7065.

Amrine, F., Zucker, F., and Wheeler, H., eds. 1987. *Goethe and the Sciences: A Reappraisal*. Boston: Reidel.

Appert-Rollanda, C., Ebbinghaus, M., and Santen, L. 2015. Intracellular transport driven by cytoskeletal motors: General mechanisms and defects. *Physics Reports* 593: 1–59.

Arber, A. 1964. *The Mind and the Eye*. Cambridge: Cambridge University Press.

Arnellos, A. 2016. Biological autonomy: Can a universal and gradable conception be operationalized? *Biological Theory* 11: 11–24.

Arnellos, A., Spyrou, T., and Darzentas, J. 2010. Towards the naturalization of agency based on an interactivist account of autonomy. *New Ideas in Psychology* 28 (3): 296–311.

Arthur, W. 2004. *Biased Embryos and Evolution*. Cambridge: Cambridge University Press.

Arthur, W. 2011. *Evolution: A Developmental Approach*. Chichester: Wiley-Blackwell.

Auletta, G. 2010. A paradigm shift in biology? *Information* 1: 28–59. https://doi.org/10.3390/info1010028.

Auletta, G., Colage, I., and Jeannerod, M., eds. 2013. *Brains Top Down: Is Top-Down Causation Challenging Neuroscience?* Singapore: World Scientific.

Baedke, J. 2019. O organism, where art thou? Old and new challenges for organism-centered biology. *Journal of the History of Biology* 52 (2): 293–324.

Baeyer, H. C. 2003. *Information: The New Language of Science*. London: Weidenfeld & Nicolson.

Baquero, F. 2005. Evolution and the nature of time. *International Microbiology* 8 (2): 81–91.

Barabási, A. L., and Oltvai, Z. N. 2004. Network biology: Understanding the cell's functional organization. *Nature Reviews Genetics* 5: 101–113.

Barandiaran, X. E., Di Paolo, E., and Rohde, M. 2009. Defining agency: Individuality, normativity, asymmetry and spatio-temporality in action. *Journal of Adaptive Behavior* 17 (5): 367–386.

Barlan, K., Rossow, M. J., and Gelfand, V. I. 2013. The journey of the organelle: Teamwork and regulation in intracellular transport. *Current Opinion in Cell Biology* 25 (4): 483–488. https://doi .org/10.1016/j.ceb.2013.02.018.

Bateson, P. 2004. The active role of behaviour in evolution. *Biology and Philosophy* 19: 283–298.

Bateson, P. 2005. The return of the whole organism. *Journal of Biosciences* 30 (1): 31–39.

Bateson, P. 2017. Adaptability and evolution. *Interface Focus* 7: 20160126. https://doi.org/10 .1098/rsfs.2016.0126.

Bateson, P., Cartwright, N., Dupré, J., Laland, K., and Noble, D. 2017. New trends in evolutionary biology: Biological, philosophical and social science perspectives. *Interface Focus* 7: 20170051. https://doi.org/10.1098/rsfs.2017.0051.

Bawden, D., and Robinson, L. 2020. Still minding the gap? Reflecting on transitions between concepts of information in varied domains. *Information* 11 (2): 71. https://doi.org/10.3390/info11020071.

Bechtel, W. 2007. Biological mechanisms: Organized to maintain autonomy. In *Systems Biology: Philosophical Foundations*, ed. Boogerd, F., Bruggeman, F. J., Hofmeyr, J.-H. S., and Westerhoff, H. V., 269–302. Amsterdam: Elsevier.

Bechtel, W. 2010a. The cell: Locus or object of inquiry? *Studies in History and Philosophy of Biological and Biomedical Sciences* 41 (3): 172–182.

Bechtel, W. 2010b. The downs and ups of mechanistic research: Circadian rhythm research as an exemplar. *Erkenntnis* 73: 313–328. https://doi.org/10.1007/s10670-010-9234-2.

Bechtel, W. 2020. Hierarchy and levels: Analysing networks to study mechanisms in molecular biology. *Philosophic Transactions of the Royal Society B* 375: 20190320. https://doi.org/10.1098 /rstb.2019.0320.

Bechtel, W., and Richardson, R. C. 2010. *Discovering Complexity: Decomposition and Localization as Strategies in Scientific Research*. Cambridge, MA: MIT Press.

Bekoff, M. 2007. *The Emotional Lives of Animals*. Novato, CA: New World Library.

Bekoff, M., Allen, C., and Burghardt, G., eds. 2002. *The Cognitive Animal: Empirical and Theoretical Perspectives on Animal Cognition*. Cambridge, MA: MIT Press.

Bertalanffy, L. 1933. *Modern Theories of Development*. London: Oxford University Press.

Bertalanffy, L. 1952. *Problems of Life: An Evaluation of Modern Biological and Scientific Thought*. New York: Harper & Brothers.

Bertalanffy, L. 1973. *General System Theory: Foundations, Development, Applications*. Harmondsworth: Penguin University Books.

Bertolaso, M., Caianiello, S., and Serrelli, E. 2018. *Biological Robustness: Emerging Perspectives from Within the Life Sciences*. Cham: Springer,

Bhaskar, L., Krishnan, V. S., and Thampan, R. V. 2007. Cytoskeletal elements and intracellular transport. *Journal of Cellular Biochemistry* 101: 1097–1108.

Bich, L., Pradeu, T., and Moreau, J. F. 2019. Understanding multicellularity: The functional organization of the intercellular space. *Frontiers in Physiology* 10: 1170. https://doi.org/10.3389 /fphys.2019.01170.

Birch, J., Ginsburg, S., and Jablonka, E. 2020. Unlimited Associative Learning and the origins of consciousness: A primer and some predictions. *Biology & Philosophy* 35: 56. https://doi.org/10 .1007/s10539-020-09772-0.

Bissell, M. J., and Radisky, D. 2001. Putting tumors in context. *Nature Reviews Cancer* 1 (1): 46–54.

Bizzarri, M., Brash, D. E., Briscoe, J., Grieneisen, V. A., Stern, C. D., and Levin, M. 2019. A call for a better understanding of causation in cell biology. *Nature Reviews Molecular Cell Biology* 20: 261–262.

Bizzarri, M., Palombo, A., and Cucina, A. 2013. Theoretical aspects of systems biology. *Progress in Biophysics and Molecular Biology* 112: 33–43.

Bock, G., and Goode, J., eds. 1998. *The Limits of Reductionism in Biology: Papers From the Symposium Held at the Novartis Foundation, London 1997.* Chichester, UK: Wiley.

Boi, L. 2017. The interlacing of upward and downward causation in complex living systems: On interactions, self-organization, emergence and wholeness. In *Philosophical and Scientific Perspectives on Downward Causation*, ed. Paoletti, M. P., and Orilia, F., 180–202. New York: Routledge.

Bongard, J., and Levin, M. 2021. Living things are not (20th century) machines: Updating mechanism metaphors in light of the modern science of machine behavior. *Frontiers in Ecology and Evolution* 9:650726. https://doi.org/10.3389/fevo.2021.650726.

Boogerd, F., Bruggeman, F. J., Hofmeyr, J.-H. S., and Westerhoff, H. V., eds. 2007a. *Systems Biology: Philosophical Foundations.* Amsterdam: Elsevier.

Boogerd, F., Bruggeman, F. J., Hofmeyr, J.-H. S., and Westerhoff, H. V. 2007b. Towards philosophical foundations of systems biology: Introduction. In *Systems Biology: Philosophical Foundations*, ed. Boogerd, F., Bruggeman, F. J., Hofmeyr, J.-H. S., and Westerhoff, H. V., 3–19. Amsterdam: Elsevier.

Bortoft, H. 1996. *The Wholeness of Nature: Goethe's Way toward a Science of Conscious Participation in Nature.* New York: Lindisfarne Books.

Bortoft, H. 2012. *Taking Appearance Seriously: The Dynamic Way of Seeing in Goethe and European Thought.* Edinburgh: Floris.

Boto, L. 2010. Horizontal gene transfer in evolution: Facts and challenges. *Proceedings of the Royal Society B: Biological Sciences* 277 (1683): 819–827.

Boto, L., Doadrio, I., and Diogo, R. 2009. Prebiotic world, macroevolution, and Darwin's theory: A new insight. *Biology and Philosophy* 24: 119–128.

Boyd, C. A. R., and Noble, D. 1993. *The Logic of Life: The Challenge of Integrative Physiology.* Oxford: Oxford University Press.

Brandon, R. N. 1981. Biological teleology: Questions and explanations. *Studies in the History and Philosophy of Science* 12: 91–105.

Breidbach, O., and Jost, J. 2006. On the gestalt concept. *Theory in Biosciences* 125: 19–36.

Brigandt, I., and Love, A. 2017. Reductionism in biology. *The Stanford Encyclopedia of Philosophy* (Spring 2017 ed.), ed. Zalta, E. N. https://plato.stanford.edu/archives/spr2017/entries/reduction-biology/.

Briggs, D. E. G. 2014. Adolf Seilacher (1925–2014). *Nature* 509: 428.

Briggs, D. E. G. 2017. Seilacher, Konstruktions-Morphologie, morphodynamics, and the evolution of form. *Journal of Experimental Zoology Part B: Molecular and Developmental Evolution* 328B: 197–206.

Brown, D. 2017. The discovery of water channels (aquaporins). *Annals of Nutrition and Metabolism* 70 (suppl. 1): 37–42. https://doi.org/10.1159/000463061.

Brown, L. 1981. *Building a Sustainable Society.* New York: Norton.

Brown, L. 2011. *World on the Edge: How to Prevent Environmental and Economic Collapse.* New York: Norton.

Brumley, D. R., Polin, M., Pedley, T. J., and Goldstein, R. E. 2015. Metachronal waves in the flagellar beating of *Volvox* and their hydrodynamic origin. *Journal of the Royal Society: Interface* 12: 20141358. https://doi.org/10.1098/rsif.2014.1358.

Brumley, D. R., Wan, K. Y., Polin, M., and Goldstein, R. E. 2014. Flagellar synchronization through direct hydrodynamic interactions. *eLife* 3: 1–15. https://doi.org/10.7554/eLife.02750.

Brunet, F., and King, N. 2017. The origin of animal multicellularity and cell differentiation. *Developmental Cell* 43: 124–140.

Burute, M., and Kapitein, L. C. 2019. Cellular logistics: Unraveling the interplay between microtubule organization and intracellular transport. *Annual Review of Cell and Developmental Biology* 35: 29–54. https://doi.org/10.1146/annurev-cellbio-100818-125149.

Buzsáki, G. 2006. *Rhythms of the Brain*. Oxford: Oxford University Press.

Byrne, R. W. 1995. *The Thinking Ape: Evolutionary Origins of Intelligence*. Oxford: Oxford University Press.

Cannon, W. B. 1932. *The Wisdom of the Body*. New York: Norton.

Cantley, L., Hunter, T., Sever, R., and Thorner, J., eds. 2014. *Signal Transduction: Principles, Pathways, and Processes*. New York: Cold Spring Harbor Laboratory Press.

Capra, F., and Luisi, P. L. 2014. *The Systems View of Life: A Unifying Vision*. Cambridge: Cambridge University Press.

Cardoso, P., Barton, P. S., Birkhofer, K., Chichorro, F., Deacon, C., Fartmann, T., Fukushima, C. S., et al. 2020. Scientist's warning to humanity on insect extinctions. *Biological Conservation* 242: 108–426. https://doi.org/10.1016/j.biocon.2020.108426.

Carroll, R. L. 1997. *Patterns and Processes of Vertebrate Evolution*. Cambridge: Cambridge University Press.

Carroll, S. B. 2005. *Endless Forms Most Beautiful: The New Science of Evo Devo and the Making of the Animal Kingdom*. New York: Norton.

Carroll, S. B., Grenier, J. K., and Weatherbee, S. D. 2005. *From DNA to Diversity: Molecular Genetics and the Evolution of Animal Design*. Malden: Blackwell.

Cassirer, E. 1950. The argument over vitalism and the autonomy of living organisms. In *The Problem of Knowledge: Philosophy, Science, and History since Hegel*, trans., Woglom, W. H., and Hendel, C. W., 188–216. New Haven, CT: Yale University Press.

Cavalier-Smith, T. 2004. The membranome and membrane heredity in development and evolution. In *Organelles, Genomes and Eukaryote Phylogeny: An Evolutionary Synthesis in the Age of Genomics*, ed. Horner, D. S., and Hirt, R. P., 335–351. Boca Raton, FL: CRC Press.

Clack, J. A. 2012. *Gaining Ground: The Origin and Early Evolution of Tetrapods*. Bloomington: Indiana University Press.

Cleland, C. E. 2013. Is a general theory of life possible? Seeking the nature of life in the context of a single example. *Biological Theory* 7 (4): 368–379.

Cleland, C. E., and Chyba, C. F. 2002. Defining "Life." *Origins of Life and Evolution of Biospheres* 32: 387–393.

Comfort, N. 2018. Genetic determinism redux. *Nature* 561: 461–463.

Conway Morris, S. 2003. *Life's Solution: Inevitable Humans in a Lonely Universe*. Cambridge: Cambridge University Press.

Corning, P. 2020. Beyond the modern synthesis: A framework for a more inclusive biological synthesis. *Progress in Biophysics and Molecular Biology* 153: 5–12. https://doi.org/10.1016/j.pbiomolbio.2020.02.002.

Cornish-Bowden, A. 2006. Putting the systems back into systems biology. *Perspectives in Biology and Medicine* 49: 475–489.

Cornish-Bowden, A., and Cardenas, M. L. 2005. Systems biology may work when we learn to understand the parts in terms of the whole. *Biochemical Society Transactions* 33: 516–519.

Cornish-Bowden, A., Cárdenas, M. L., Letelier, J. C., and Soto-Andrade, J. 2007. Beyond reductionism: Metabolic circularity as a guiding vision for a real biology of systems. *Proteomics* 7 (6): 839–845. https://doi.org/10.1002/pmic.200600431.

Crick, F. 1966. *Of Molecules and Men*. Seattle: University of Washington Press.

Crick, F. 1994. *The Astonishing Hypothesis: The Scientific Search for the Soul*. New York: Macmillan.

Cubo, J. 2004. Pattern and process in constructional morphology. *Evolution and Development* 6 (3): 131–133.

Davies, P. 2012. The epigenome and top-down causation. *Interface Focus* 2: 42–48.

Davies, P. 2019. *The Demon in the Machine: How Hidden Webs of Information Are Finally Solving the Mystery of Life*. London: Penguin.

Dawkins, R. 1976. *The Selfish Gene*. Oxford: Oxford University Press.

Day, R. E., Kitchen, P., Owen, D. S., Bland, C., Marshall, L., Conner, A. C., Bill, R. M., and Conner, M. T. 2014. Human aquaporins: Regulators of transcellular water flow. *Biochimica et Biophysica Acta* 1840: 1492–1506.

de Waal, F. 2009. *The Age of Empathy: Natures Lessons for a Kinder Society*. New York: Harmony.

Dennett, D. 1995. *Darwin's Dangerous Idea: Evolution and the Meanings of Life*. New York: Touchstone.

Denton, M. J., Kumaramanickavel, G., and Legge, M. 2013. Cells as irreducible wholes: The failure of mechanism and the possibility of an organicist revival. *Biology and Philosophy* 28 (1): 31–52. https://doi.org/10.1007/s10539-011-9285-z.

Deppert, W., Kliemt, H., Lohff, B., and Schaefer, J., eds. 1992. *Wissenschaftstheorien in der Medizin. Ein Symposium*. Berlin: de Gruyter.

Di Paolo, E. A. 2005. Autopoiesis, adaptivity, teleology, agency. *Phenomenology and the Cognitive Sciences* 4: 429–452.

Doolittle, W. F. 1999. Phylogenetic classification and the universal tree. *Science* 284 (5423): 2124.

Dorn, G. W., and Mochly-Rosen, D. 2002. Intracellular transport mechanisms of signal transducers. *Annual Reviews in Physiology* 64: 407–429. https://doi.org/10.1146/annurev.physiol.64.08150 .155903.

Downes, S. M. 2001. The ontogeny of information. *Perspectives in Biology and Medicine* 44 (3): 464–469.

Drack, M., and Apftaler, W. 2007. Is Paul Weiss' and Ludwig von Bertalanffy's system thinking still valid today? *Systems Research and Behavioral Science* 24 (5): 537–546.

Drack, M., Apftaler, W., and Pouvreau, D. 2007. On the making of a system theory of life: Paul A Weiss and Ludwig von Bertalanffy's conceptual connection. *Quarterly Review of Biology* 82: 349–373.

Drack, M., and Wolkenhauer, O. 2011. System approaches of Weiss and Bertalanffy and their relevance for systems biology today. *Seminars in Cancer Biology* 21: 150–155.

Du Bois-Reymond, E. H. 1872. *Über die Grenzen des Naturerkennens*. Leipzig: Veit.

Duboule, D. 2003. Time for chronomics? *Science* 301: 277.

Dullemeijer, P. 1980. Functional morphology and evolutionary biology. *Acta Biotheoretica* 29: 151–250. https://doi.org/10.1007/BF00051368.

Dullemeijer, P., and Zweers, G. A. 1997. The variety of explanations of living forms and structures. *European Journal of Morphology* 35 (5): 354–364.

Dunlap, J. C., Loros, J. J., and DeCoursey, P. J. 2004. *Chronobiology: Biological Timekeeping*. Sunderland, MA: Sinauer.

Dupré, J. 2003. *Human Nature and the Limits of Science*. Oxford: Oxford University Press.

Dupré, J. 2012. *Processes of Life: Essays in the Philosophy of Biology*. Oxford: Oxford University Press.

Dupré, J., and Nicholson, D. J. 2018. A manifesto for a processual philosophy of biology. In *Everything Flows: Towards a Processual Philosophy of Biology*, ed. Nicholson, D. J., and Dupré, J., 3–45. Oxford: Oxford University Press.

Durand, P. M. 2020. *The Evolutionary Origins of Life and Death*. Chicago: University of Chicago Press.

Durand, P. M., Sym, S., and Michod, R. E. 2016. Programmed cell death and complexity in microbial systems. *Current Biology* 26: R587–R593.

Earnshaw, W. C., and Pluta, A. F. 1994. Mitosis. *Bioessays* 16 (9): 639–643. https://doi.org/10 .1002/bies.950160908.

Ebersbach, R., Weinzirl, J., and Heusser, P., eds. 2018. *Was ist Leben? Aktuelles zu Wirkursache und Erkenntnis des Lebendigen. Wittener Kolloquium Humanismus, Medizin und Philosophie, Band 5*, Würzburg: Königshausen & Neumann.

Eckert, R. 2000. *Tierphysiologie*. Stuttgart: Thieme.

Eigen, M. 1987. *Stufen zum Leben. Die frühe Evolution im Visier der Molekularbiologie*. Munich: Piper Verlag.

El-Hani, C. N., and Emmeche, C. 2000. On some theoretical grounds for an organism-centered biology: Property emergence, supervenience, and downward causation. *Theory in Bioscience* 119: 234–275.

Emmeche, C. 1997. Autopoietic systems, replicators, and the search for a meaningful biological definition of life. *Ultimate Reality and Meaning* 20: 244–264.

Endres, K. P., and Schad, W. 2002. *Moon Rhythms in Nature: How Lunar Cycles Affect Living Organisms*. Edinburgh: Floris.

Erwin, D. H. 2000. Macroevolution is more than repeated rounds of microevolution. *Evolution & Development* 2: 78–84.

Falkenburg, B. 2012. *Mythos Determinismus. Wieviel erklärt uns die Hirnforschung?* Berlin: Springer Spektrum.

Farias, S. T., Prosdocimi, F., and Caponi, G. 2021. Organic codes: A unifying concept for life. *Acta Biotheoretica* 69 (4): 769–782. https://doi.org/10.1007/s10441-021-09422-2.

Farnsworth, K. D. 2018. How organisms gained causal independence and how it might be quantified. *Biology* 7 (3): 38 https://doi.org/10.3390/biology7030038.

Farnsworth, K. D. 2022. How an information perspective helps overcome the challenge of biology to physics. *BioSystems* 217: 104683. https://doi.org/10.1016/j.biosystems.2022.104683.

Farnsworth, K. D., Nelson, J., and Gershenson, C. 2013. Living is information processing: From molecules to global systems. *Acta Biotheoretica* 61: 203–222. https://doi.org/10.1007/s10441-013-9179-3.

Feinberg, T. E., and Mallatt, J. M. 2016. *The Ancient Origins of Consciousness: How the Brain Created Experience*. Cambridge, MA: MIT Press.

Fleck, L. 1979. *Genesis and Development of a Scientific Fact*. Chicago: Chicago University Press.

Forgacs, G., and Newman, S. A. 2005. *Biological Physics of the Developing Embryo*. Cambridge: Cambridge University Press.

Fox Keller, E. 2000. *The Century of the Gene*. Cambridge, MA: Harvard University Press.

Fox Keller, E. 2011. Towards a science of informed matter: History and Philosophy. *Studies in History and Philosophy of Biological and Biomedical Sciences* 42 (2): 174–179.

Freyer, B., ed. 2016. *Ökologischer Landbau. Grundlagen, Wissensstand und Herausforderungen*. Bern: Haupt.

Fuchs, T. 2018. *Ecology of the Brain: The Phenomenology and Biology of the Embodied Mind*. Oxford: Oxford University Press.

Fuente, L., and Helms, J. A. 2005. Head, shoulders, knees, and toes. *Developmental Biology* 282: 294–306. https://doi.org/10.1016/j.ydbio.2005.03.036.

Fuentes, A. 2004. Cooperation or conflict? It's not all sex and violence: Integrated anthropology and the role of cooperation and social complexity in human evolution. *American Anthropologist* 106 (4): 710–718.

Fuentes, A. 2009. *Evolution of Human Behavior*. London: Oxford University Press.

Fuentes, A. 2017. *The Creative Spark: How Imagination Made Humans Exceptional*. New York: Dutton.

Furst, B. 2020. *The Heart and Circulation: An Integrative Model*. 2nd ed. Cham: Springer.

Gatlin, L. L. 1972. *Information Theory and the Living System*. New York: Columbia University Press.

Gawne, R., McKenna, K. Z., and Nijhout, H. F. 2018. Unmodern synthesis: Developmental hierarchies and the origin of phenotypes. *BioEssays* 40 (1): 1600265. https://doi.org/10.1002/bies.201600265.

Gayon, J. 2010. Defining life: Synthesis and conclusions. *Origins of Life and Evolution of Biospheres* 40 (2): 231–244.

Gerdes, S. Y., Scholle, M. D., Campbell, J. W., Balázsi, G., Ravasz, E., Daugherty, M. D., Somera, A. L., et al. 2003. Experimental determination and system level analysis of essential genes in *Escherichia coli* MG1655. *Journal of Bacteriology* 185 (19): 5673–5684. https://doi.org/10.1128/JB.185.19.5673-5684.2003.

Gerhart, J., and Kirschner, M. 1997. *Cells, Embryos, and Evolution: Toward a Cellular and Developmental Understanding of Phenotypic Variation and Evolutionary Adaptability*. Malden: Blackwell.

Giaever, G., Chu, A. M., Ni, L., Connelly, C., Riles, L., Véronneau S, Dow, S., et al. 2002. Functional profiling of the *Saccharomyces cerevisiae* genome. *Nature* 418: 387–391.

Gilbert, S. F. 2014. *Developmental Biology*, 10th ed. Sunderland, MA: Sinauer.

Gilbert, S. F., Opitz, J. M., and Raff, R. A. 1996. Resynthesizing evolutionary and developmental biology. *Developmental Biology* 173: 357–372.

Gilbert, S. F., and Sarkar, S. 2000. Embracing complexity: Organicism for the 21st century. *Developmental Dynamics* 219: 1–9.

Ginsburg, S., and Jablonka, E. 2019. *The Evolution of the Sensitive Soul: Learning and the Origins of Consciousness*. Cambridge, MA: MIT Press.

Ginsburg, S., and Jablonka, E. 2020. Consciousness as a mode of being. *Journal of Consciousness Studies* 27 (9–10): 148–162.

Godfrey-Smith, P. 2016. *Other Minds: The Octopus, the Sea, and the Deep Origins of Consciousness*. New York: Farrar, Strauss and Giroux.

Godfrey-Smith, P. 2020. *Metazoa: Animal Life and the Birth of the Mind*. New York: Farrar, Stauss and Giroux.

Goethe, J. W. 1981. *Naturwissenschaftliche Schriften, I. Hamburger Ausgabe, Band 13*. Munich: Beck Verlag.

Goodwin, B. 1994. *How the Leopard Changed Its Spots: The Evolution of Complexity*. New York: Scribner.

Goodwin, B. 2000. The life of form: Emergent patterns of morphological transformation. *Comptes Rendus de l'Academie des Sciences Serie III: Sciences de la Vie* 323: 15–21.

Goodwin, B. 2007. *Nature's Due: Healing Our Fragmented Culture*. Edinburgh: Floris.

Gould, J. L., and Gould, C. G. 1994. *Animal Mind*. New York: Freeman.

Gould, S. J. 1977. *Ontogeny and Phylogeny*. Cambridge, MA: Belknap Press.

Gould, S. J. 2002. *The Structure of Evolutionary Theory*. Cambridge, MA: Belknap Press.

Gould, S. J., and Lewontin, R. C. 1979. The spandrels of San Marco and the Panglossian paradigm. *Proceedings of the Royal Society London B* 205 (1161): 581–598.

Gowdy, J., and Baveye, P. 2018. An evolutionary perspective on industrial and sustainable agriculture. In *Agroecosystem Diversity: Reconciling Contemporary Agriculture and Environmental Quality*, ed. Lemaire, G., Carvalho, P., Kronberg, S., and Recous, S., 425–433. New York: Academic.

Green, S., Serban, M., Scholl, R., Jones, N., Brigandt, I., and Bechtel, W. 2018. Network analyses in systems biology: New strategies for dealing with biological complexity. *Synthese* 195 (4): 1751–1777. https://doi.org/10.1007/s11229-016-1307-6.

Greene, S. 2017. *Philosophy of Systems Biology: Perspectives from Scientists and Philosophers*. Springer: Cham.

Griffin, D. R. 1976. *The Question of Animal Awareness: Evolutionary Continuity of Mental Experience*. New York: Rockefeller University Press.

Griffin, D. R. 1984. *Animal Thinking*. Cambridge, MA: Harvard University Press.

Griffin, D. R. 2001. *Animal Minds: Beyond Cognition to Consciousness*. Chicago: University of Chicago Press.

Griffin, D. R., and Speck, G. B. 2004. New evidence of animal consciousness. *Animal Cognition* 7 (1): 5–18.

Griffiths, P., and Stotz, K. 2018. Developmental systems theory as a process theory. In *Everything Flows: Towards a Processual Philosophy of Biology*, ed. Nicholson, D. J., and Dupré, J., 225–245, Oxford: Oxford University Press.

Grisogono, A. M. 2017. (How) did information emerge? In *From Matter to Life: Information and Causality*, ed. Walker, S. I., Davies, P. C. W., and Ellis, G. F. R., 61–96. Cambridge: Cambridge University Press.

Grosberg, R. K., and Strathmann, R. R. 2007. The evolution of multicellularity: A minor major transition? *Annual Review of Ecology, Evolution, and Systematics* 38: 621–654. https://doi.org/10.1146/annurev.ecolsys.36.102403.114735.

Grunwald, A., Gutmann, M., and Neumann-Held, E., eds. 2002. *On Human Nature: Anthropological, Biological, and Philosophical Foundations*. Berlin: Springer.

Häcker, G. 2013. Is there, and should there be, apoptosis in bacteria? *Microbes and Infection* 15 (8/9): 640–644.

Haken, H. 1983. *Synergetics: An Introduction*. Berlin: Springer.

Halberg, N., and Muller, A. 2013. *Organic Agriculture for Sustainable Livelihoods*. Abingdon, UK: Routledge.

Haldane, J. B. S. 1931. *The Philosophical Basis of Biology*. London: Hodder and Stoughton.

Hallmann, C. A., Sorg, M., Jongejans, E., Siepel, H., Hofland, N., Schwan, H., Stenmans, W., et al. 2017. More than 75 percent decline over 27 years in total flying insect biomass in protected areas. *PLoS ONE* 12 (10): e0185809. https://doi.org/10.1371/journal.pone.0185809.

Harlan, V., ed. 2005. *Wert und Grenzen des Typus in der botanischen Morphologie*. Nümbrecht: Galunder.

Hartman, J., Garvik, B., and Hartwell, L. 2001. Principles of the buffering of genetic variation. *Science* 291: 1001–1004.

Hengeveld, R. 2011. Definitions of life are not only unnecessary, but they can do harm to understanding. *Foundations of Science* 16 (4): 323–325.

Henning, B. G., and Scarfe, A., eds. 2013. *Beyond Mechanism: Putting Life Back into Biology*. Lanham, MD: Lexington Books.

Herron, M. D. 2016. Origins of multicellular complexity: *Volvox* and the volvocine algae. *Molecular Ecology* 25 (6): 1213–1223.

Heusser, P. 2016. *Anthroposophy and Science: An Introduction*. Frankfurt: Peter Lang.

Hildebrandt, G. 1979. Rhythmical functional order and man's emancipation from the time factor. In *Basis of an Individual Physiology*, ed. Schaefer, K. E., Hildebrandt, G., and Macbeth, N., 15–43. Mount Kisco, NY: Futura.

Hildebrandt, G., Moog, R., and Raschke, F., eds. 1987. *Chronobiology and Chronomedicine*. Frankfurt: Verlag Peter Lang.

Hildebrandt, G., Moser, M., and Lehofer, M. 1998. *Chronobiologie und Chronomedizin: Biologische Rhythmen; Medizinische Konsequenzen*. Stuttgart: Hippokrates Verlag.

Hoffmeyer, J. 1996. *Signs of Meaning in the Universe*. Bloomington: Indiana University Press.

Hoffmeyer, J. 2008. *Biosemiotics: An Examination into the Signs of Life and the Life of Signs*. Scranton: University of Scranton Press.

Hoffmeyer, J. 2013. Why do we need a semiotic understanding of life? In *Beyond Mechanism: Putting Life Back into Biology*, ed. Henning, B. G., and Scarfe, A. C., 147–168. Lanham, MD: Lexington Books.

Hofmeyr, J.-H. S. 2007. The biochemical factory that autonomously fabricates itself: A systems biological view of the living cell. In *Systems Biology: Philosophical Foundations*, ed. Boogerd, F., Bruggeman, F. J., Hofmeyr, J.-H. S., and Westerhoff, H. V., 217–242. Amsterdam: Elsevier.

Hofmeyr, J.-H. S. 2017. Exploring the metabolic marketplace through the lens of systems biology. In *Philosophy of Systems Biology: Perspectives from Scientists and Philosophers*, ed. Greene, S., 117–124. Cham: Springer.

Hogenesch, J. B., and Herzog, E. D. 2011. Intracellular and intercellular processes determine robustness of the circadian clock. *FEBS Letters* 585 (10): 1427–1434.

Holdrege, C. 2005. Doing Goethean science. *Janus Head* 8: 27–52.

Holdrege, C. 2013. *Thinking Like a Plant: A Living Science of Life.* Great Barrington, MA: Lindisfarne.

Holdrege, C. 2017. *Do Frogs Come from Tadpoles? Rethinking Origins in Development and Evolution.* Ghent, NY: Evolving Science Association, The Nature Institute.

Holdrege, C. 2021. *Seeing the Animal Whole: And Why It Matters.* Great Barrington, MA: Lindisfarne.

Horrigan, L., Lawrence, R. S., and Walker, P. 2002. How sustainable agriculture can address the environmental and human health harms of industrial agriculture. *Environmental Health Perspectives* 110: 445–456.

Huang, S. 2011. Systems biology of stem cells: Three useful perspectives to help overcome the paradigm of linear pathways. *Philosophic Transactions of the Royal Society B* 366 (1575): 2247–2259. http://doi.org/10.1098/rstb.2011.0008.

Hubbart, R., and Wald, E. 1993. *Exploding the Gene Myth: How Genetic Information Is Produced and Manipulated by Scientists, Physicians, Employers, Insurance Companies, Educators, and Law Enforcers.* Boston: Beacon Press.

Huettenbrenner, S., Maier, S., Leisser, C., Polgar, D., Strasser, S., Grusch, M., and Krupitza, G. 2003. The evolution of cell death programs as prerequisites of multicellularity. *Mutation Research* 543: 235–249.

Hug, H. 2000. Apoptose: Die Selbstvernichtung der Zelle als Überlebensschutz. *Biologie in unserer Zeit* 30 (3): 128–135.

Jablonka, E., and Lamb, M. J. 1998. Bridges between development and evolution. *Biology and Philosophy* 13: 119–124.

Jablonka, E., and Lamb, M. J. 2005. *Evolution in Four Dimensions: Genetic, Epigenetic, Behavioral, and Symbolic Variation in the History of Life.* Cambridge, MA: MIT Press.

Jablonka, E., and Raz, G. 2009. Transgenerational epigenetic inheritance: Prevalence, mechanisms, and implications for the study of heredity and evolution. *Quarterly Review of Biology* 84 (2): 131–176.

Jablonski, D. 2020. Developmental bias, macroevolution, and the fossil record. *Evolution & Development* 22: 103–125.

Jacobson, M. D., Weil, M., and Raff, M. C. 1997. Programmed cell death in animal development. *Cell* 88: 347–354.

Jaeger, J., Irons, D., and Monk, N. 2008. Regulative feedback in pattern formation: Towards a general relativistic theory of positional information. *Development* 135 (19): 3175–3183. https://doi.org/10.1242/dev.018697.

Jagers op Akkerhuis, G. A. J. M. 2010. Towards a hierarchical definition of life, the organism, and death. *Foundations of Science* 15 (3): 245–262. https://doi.org/10.1007/s10699-010-9177-8.

Jahn, I. 2000. *Geschichte der Biologie—Theorien, Methoden, Institutionen, Kurzbiographien.* Heidelberg: Spektrum Verlag.

Jeong, H., Mason, S. P., Barabási, A. L., and Oltvai, Z. N. 2001. Lethality and centrality in protein networks. *Nature* 411: 41–42.

Jeong, H., Tombor, B., Albert, R., Oltvai, Z. N., and Barabási, A. L. 2000. The large-scale organization of metabolic networks. *Nature* 407: 651–654.

Jonas, H. 1966. *The Phenomenon of Life: Toward a Philosophical Biology.* New York: Harper and Row.

Jonas, H. 1992. *Philosophische Untersuchungen und metaphysische Vermutungen.* Frankfurt: Insel Verlag.

Jones, D. 2005. Niche construction. *Nature* 438: 14–16.

Joyner, M. J. 2011a. Giant sucking sound: Can physiology fill the intellectual void left by the reductionists? *Journal of Applied Physiology* 111 (2): 335–342. https://doi.org/10.1152/japplphysiol.00565.2011.

Joyner, M. J. 2011b. Why physiology matters in medicine. *Physiology* 26: 72–75.

Joyner, M. J. 2015. Has neo-Darwinism failed clinical medicine: Does systems biology have to? *Progress in Biophysics and Molecular Biology* 117: 107–112.

Joyner, M. J., and Paneth, N. 2019. Promises, promises, and precision medicine. *Journal of Clinical Investigations* 129 (3): 946–948. https://doi.org/10.1172/JCI126119.

Joyner, M. J., Paneth, N., and Ioannidis, J. P. 2016. What happens when underperforming big ideas in research become entrenched? *Journal of the American Medical Association* 316 (13): 1355–1356.

Joyner, M. J., and Pedersen, B. K. 2011. Ten questions about systems biology. *Journal of Physiology* 589 (5): 1017–1030.

Joyner, M. J., and Prendergast, F. G. 2014. Chasing Mendel: Five questions for personalized medicine. *Journal of Physiology* 592 (11): 2381–2388.

Juarrero, A. 2002. *Dynamics in Action: Intentional Behavior as a Complex System*. Cambridge, MA: MIT Press.

Kabachinski, G., and Schwartz, T. U. 2015. The nuclear pore complex—structure and function at a glance. *Journal of Cell Science* 128 (3): 423–429. https://doi.org/10.1242/jcs.083246.

Kaneko, K. 2006. *Life: An Introduction to Complex Systems Biology*. Berlin: Springer.

Kather, R. 2003. *Was ist Leben? Philosophische Positionen und Perspektiven*. Darmstadt: Wissenschaftliche Buchgesellschaft.

Kather, R. 2012. *Die Wiederentdeckung der Natur. Naturphilosophie im Zeichen der ökologischen Krise*. Darmstadt: Wissenschaftliche Buchgesellschaft.

Kauffman, S. A. 1993. *The Origins of Order: Self-Organisation and Selection in Evolution*. Oxford: Oxford University Press.

Kauffman, S. A., and Clayton, P. 2006. On emergence, agency, and organization. *Biology and Philosophy* 21: 501–521.

Kauffman, S. A., Logan, R. K., Este, R., Goebel, R., Hobill, D., and Shmulevich, I. 2008. Propagating organization: An enquiry. *Biology and Philosophy* 23: 27–45.

Kicheva, A., Cohen, M., and Briscoe, J. 2012. Developmental pattern formation: Insights from physics and biology. *Science* 338 (6104): 210–212. https://doi.org/10.1126/science.1225182.

King, N. 2004. The unicellular ancestry of animal development. *Developmental Cell* 7: 313–325.

Kirchhoff, T., and Voigt, A. 2010. Rekonstruktion der Geschichte der Synökologie. Konkurrierende Paradigmen, Transformationen, kulturelle Hintergründe. *Verhandlungen zur Geschichte und Theorie der Biologie. Band 15*, ed. Kaasch, M., and Kaasch, J., 181–196. Berlin: Verlag für Wissenschaft und Bildung.

Kirk, D. L. 1998. *Volvox: Molecular-Genetic Origins of Multicellularity and Cellular Differentiation*. Cambridge: Cambridge University Press.

Kirk, D. L. 2000. Volvox as a model system for studying the ontogeny and phylogeny of multicellularity and cellular differentiation. *Journal of Plant Growth Regulation* 19: 265–274.

Kirschner, M., Gerhart, J., and Mitchison, T. 2000. Molecular "vitalism." *Cell* 100 (1): 79–88.

Kirschner, M. W., and Gerhart, J. C. 2005. *The Plausibility of Life: Resolving Darwin's Dilemma*. New Haven, CT: Yale University Press.

Kitano, H. 2001. *Foundations of Systems Biology*. Cambridge, MA: MIT Press.

Kitano, H. 2004. Biological robustness. *Nature Reviews Genetics* 5: 826–837.

Kitano, H. 2007. Towards a theory of biological robustness. *Molecular Systems Biology* 3: 137. https://doi.org/10.1038/msb4100179.

Kitano, H., and Oda, K. 2006. Self-extending symbiosis: A mechanism for increasing robustness through evolution. *Biological Theory* 1 (1): 61–66.

Kleidon, A. 2002. Testing the effect of life on Earth's functioning: How Gaian is the Earth system? *Climatic Change* 66: 271–319.

Kleidon, A. 2010. Life, hierarchy, and the thermodynamic machinery of planet Earth. *Physics of Life Reviews* 7: 424–460.

Kleidon, A., Fraedrich, K., and Heimann, M. 2000. A green planet versus a desert world: Estimating the maximum effect of vegetation on land surface climate. *Climatic Change* 44: 471–493.

Klein, R. G. 2009. *The Human Career: Human Biological and Cultural Origins*. Chicago: University of Chicago Press.

Koestler, A., and Smythies, I. R., eds. 1969. *Beyond Reductionism: New Perspectives in the Life Sciences*. London: Hutchinson.

Kofler, W. 2014. "Information"—from an evolutionary point of view. *Information* 5 (2): 272–284. https://doi.org/10.3390/info5020272.

Kolb, V. M. 2007. On the applicability of the Aristotelian principles to the definition of life. *International Journal of Astrobiology* 6 (1): 51–57.

Koonin, E. V. 2009. Towards a postmodern synthesis of evolutionary biology. *Cell Cycle* 8 (6): 799–800.

Kostić, D., Hilgetag, C. C., and Tittgemeyer, M. 2020. Unifying the essential concepts of biological networks: Biological insights and philosophical foundations. *Philosophical Transactions of the Royal Society B* 375 (1796): 20190314. https://doi.org/10.1098/rstb.2019.0314.

Koukkari, W. L., and Sothern, R. B. 2006. *Introducing Biological Rhythms*. New York: Springer.

Kramer, I. M. 2015. *Signal Transduction*. New York: Academic.

Krammer, P. H. 2000. Apoptose. *Deutsches Ärzteblatt* 97 (25): 1752–1759.

Krimsky, S., and Gruber, J., eds. 2013. *Genetic Explanations: Sense and Nonsense*. Cambridge, MA: Harvard University Press.

Kritikou, E., Pulverer, B., and Heinrichs, A. 2006. All systems go! *Nature Reviews Molecular Cell Biology* 7: 801.

Kuhn, T. S. 1962. *The Structure of Scientific Revolutions*. Chicago: University of Chicago Press.

Kümmell, S. 2020. Autonomiezunahme und Autonomieverlust in der Evolution der Bewegungsfähigkeit der Säuger und ihrer Vorläufer. In *Perspektiven einer Biologie der Freiheit. Autonomieentwicklung in Natur, Kultur und Landschaft*, ed. Rosslenbroich, B., 157–280. Stuttgart: Freies Geistesleben.

Kümmell, S., and Frey, E. 2012. Digital arcade in the autopodia of Synapsida: Standard position of the digits and dorsoventral excursion angle of digital joints in the rays II–V. *Palaeobiodiversity and Palaeoenvironments* 92 (2): 171–196. https://doi.org/10.1007/s12549-012-0076-6.

Lakatos, I. 1980. *The Methodology of Scientific Research Programmes: Volume 1: Philosophical Papers*. Cambridge: Cambridge University Press.

Laland, K. N. 2004. Extending the extended phenotype. *Biology and Philosophy* 19: 313–325.

Laland, K. N., Odling-Smee, J., and Endler, J. 2017. Niche construction, sources of selection and trait coevolution. *Interface Focus* 7 (5): 20160147. https://doi.org/10.1098/rsfs.2016.0147.

Laland, K. N., Odling-Smee, J., and Feldman, M. W. 2019. Understanding niche construction as an evolutionary process. In *Evolutionary Causation: Biological and Philosophical Reflections*, ed. Uller, T., and Laland, K. N., 127–152., Cambridge, MA: MIT Press.

Laland, K. N., Odling-Smee, F. J., and Turner, S. 2014. The role of internal and external constructive processes in evolution. *Journal of Physiology* 592 (11): 2413–2422.

Laland, K. N., and Sterelny, K. 2006. Seven reasons (not) to neglect niche construction. *Evolution* 60: 1751–1762.

Laland, K. N., Uller, T., Feldman, M., Sterelny, K., Müller, G. B., Moczek, A., Jablonka, E., et al. 2014. Does evolutionary theory need a rethink? *Nature* 514: 161–164.

Laland, K. N., Uller, T., Feldman, M. W., Sterelny, K., Müller, G. B., Moczek, A., Jablonka, E., and Odling-Smee, F. J. 2015. The extended evolutionary synthesis: Its structure, assumptions and predictions. *Proceedings of the Royal Society B* 282 (1813): 20151019.

Landrigan, P., Stegeman, J., Fleming, L., Allemand, D., Anderson, D., Backer, L., Brucker-Davis, F., et al. 2020. Review: Human health and ocean pollution. *Annals of Global Health* 86 (1): 1–64.

Larhlimi, A., Blachon, S., Selbig, J., and Nikoloski, Z. 2011. Robustness of metabolic networks: A review of existing definitions. *BioSystems* 106 (1): 1–8.

Laubichler, M. D. 2017. The emergence of theoretical and general biology: The broader scientific context for the Biologische Versuchsanstalt. In *Vivarium: Experimental, Quantitative, and Theoretical Biology at Vienna's Biologische Versuchsanstalt*, ed. Müller, G. B., 95–114. Cambridge, MA: MIT Press.

Laubichler, M. D., and Renn, J. 2015. Extended evolution: A conceptual framework for integrating regulatory networks and niche construction. *Journal of Experimental Zoology Part B: Molecular and Developmental Evolution* 324: 565–577.

Laubichler, M. D., and Wagner, G. P. 2001. How molecular is molecular developmental biology? A reply to Alex Rosenberg's Reductionism redux: Computing the embryo. *Biology and Philosophy* 16: 53–68.

Lee, K. E. 1985. *Earthworms: Their Ecology and Relation with Soil and Land Use*. London: Academic.

Levin, M. 2012. Morphogenetic fields in embryogenesis, regeneration, and cancer: Non-local control of complex patterning. *BioSystems* 109 (3): 243–261.

Lewis, K. 2000. Programmed death in bacteria. *Microbiology and Molecular Biology Reviews* 64 (3): 503–514.

Lewontin, R. 1982. Organism and environment. In *Learning, Development and Culture*, ed. Plotin, H. C., 151–170. New York: Wiley.

Lewontin, R. 1983. Gene, organism and environment. In *Evolution: From Molecules to Men*, ed. Bendall, D. S., 273–285. Cambridge: Cambridge University Press.

Lewontin, R. 1991. *Biology as Ideology*. New York: Harper.

Lewontin, R. 1993. *The Doctrine of DNA: Biology as Ideology*. London: Penguin.

Lewontin, R. 2000. *The Triple Helix: Gene, Organism and Environment*. Cambridge: Harvard University Press.

Lewontin, R. 2009. Foreword: Carving nature at its joints? In *Mapping the Future of Biology*, ed. Barberousse, A., Morange, M., and Pradeu, T., v–vii. Boston: Springer.

Lewontin, R., Rose, S., and Kamin, L. J. 1984. *Biology, Ideology, and Human Nature: Not in Our Genes*. New York: Pantheon.

Liljenström, H. 2016. Multi-scale causation in brain dynamics. In *Cognitive Phase Transitions in the Cerebral Cortex: Enhancing the Neuron Doctrine by Modeling Neural Fields*, ed. Kozma, R., and Freeman, W., 177–186. New York: Springer.

Lindholm, M. 2015. DNA dispose, but subjects decide: Learning and the extended synthesis. *Biosemiotics* 8: 443–461. https://doi.org/10.1007/s12304-015-9242-3.

Lockley, M. 1999. *The Eternal Trail: A Tracker Looks at Evolution*. Reading, MA: Perseus.

Lockley, M. 2014. Bernd Rosslenbroich: On the origin of autonomy; A new look at the major transitions in evolution. *Acta Biotheoretica* 62: 537–541. https://doi.org/10.1007/s10441-014-9238-4.

Loeb, J. 1912. *The Mechanistic Conception of Life*. Chicago: University of Chicago Press.

Longo, G., Miquel, P. A., Sonnenschein, C., and Soto, A. M. 2012. Is information a proper observable for biological organization? *Progress in Biophysics and Molecular Biology* 109: 108–114.

Looijen, R. C. 2000. *Holism and Reductionism in Biology and Ecology: The Mutual Dependence of Higher and Lower Level Research Programmes. Episteme 23*. Dordrecht: Kluwer Academic.

Love, A. C. 2015. *Conceptual Change in Biology: Scientific and Philosophical Perspectives on Evolution and Development*. Dordrecht: Springer.

Lovelock, J. E., and Margulis, L. 1974. Atmospheric homeostasis by and for the biosphere: The Gaia hypothesis. *Tellus* 26: 2–10.

Low, Philip. 2012. The Cambridge declaration on consciousness. Presented at the Francis Crick Memorial Conference on Consciousness in Human and Non-Human Animals, Churchill College, University of Cambridge, July 7, 2012. https://fcmconference.org/img/CambridgeDeclarationOn Consciousness.pdf.

Luisi, P. L. 2003. Autopoiesis: A review and a reappraisal. *Die Naturwissenschaften* 90: 49–59.

Lyon, P., Keijzer, F., Arendt, D., and Levin, M. 2021. Introduction: Reframing cognition; Getting down to biological basics. *Philosophical Transactions of the Royal Society B: Biological Sciences* 376 (1820): 20190750. https://doi.org/10.1098/rstb.2019.0750.

Mahner, M., and Bunge, M. 1997. *Foundations of Biophilosophy*. Berlin: Springer.

Malik, K. 2002. *Man, Beast, and Zombie: What Science Can and Cannot Tell Us about Human Nature*. New Brunswick, NJ: Rutgers University Press.

Margulis, L. 1993. *Symbiosis in Cell Evolution*. San Francisco: Freeman.

Margulis, L. 1999. *Symbiotic Planet: A New Look at Evolution*. New York: Basic Books.

Margulis, L., and Sagan, D. 1997. *Microcosm: Four Billion Years of Microbial Evolution*. Berkeley: University of California Press.

Margulis, L., and Sagan, D. 2002. *Acquiring Genomes: A Theory of the Origins of Species*. New York: Basic Books.

Marijuan, P. C. 2004. Information and life: Towards a biological understanding of informational phenomena. *Triple-C* 2: 6–19.

Marshall, M. 2021. The pace of development. *Nature* 592: 682–684.

Masel, J., and Siegal, M. L. 2009. Robustness: Mechanisms and consequences. *Trends in Genetics* 25 (9): 395–403.

Masel, J., and Trotter, M. V. 2010. Robustness and evolvability. *Trends in Genetics* 26 (9): 406–414.

Maturana, H., and Varela, F. 1980. *Autopoiesis and Cognition: The Realization of the Living*. Dordrecht: Reidel.

Maturana, H. R., and Varela, F. J. 1987. *The Tree of Knowledge: The Biological Roots of Human Understanding*. Boston: Shambhala.

Mayr, E. 1988. *Toward a New Philosophy of Biology: Observations of an Evolutionist*. Cambridge, MA: Harvard University Press.

Mayr, E. 1996. The autonomy of biology: The position of biology among the sciences. *Quarterly Review of Biology* 71 (1): 97–106.

Mayr, E. 1997. *This Is Biology: The Science of the Living World*. Cambridge, MA: Belknap Press.

McIntosh, J. R. 2016. Mitosis. *Cold Spring Harbor Perspectives in Biology* 8 (9): a023218. https://doi.org/10.1101/cshperspect.a023218.

McKinney, M. L., and McNamara, K. J. 1991. *Heterochrony: The Evolution of Ontogeny*. New York: Plenum.

McNamara, K. J., ed. 1990. *Evolutionary Trends*. London: Belhaven.

McNamara, K. J. 2012. Heterochrony: The evolution of development. *Evolution: Education and Outreach* 5 (2): 203–218.

McShea, D. W. 2015. Bernd Rosslenbroich: On the origin of autonomy; A new look at the major transitions in evolution. *Biology and Philosophy* 30: 439–446. https://doi.org/10.1007/s10539-015-9474-2.

Melham, T., Bard, J., Werner, E., and Noble, D. 2013. Conceptual foundations of systems biology. *Progress in Biophysics and Molecular Biology* 111: 55–56.

Mesarovic, M. D., Sreenath, N., and Keene, J. 2004. Search for organizing principles: Understanding in systems biology. *Systems Biology* 1: 19–27.

Miller, G. T., and Spoolman, S. 2020. *Living in the Environment*. Belmont, CA: Brooks/Cole.

Minelli, A., and Pradeu, T., eds. 2014. *Towards a Theory of Development*. Oxford: Oxford University Press.

Mohawk, J. A., Green, C. B., and Takahashi, J. S. 2012. Central and peripheral circadian clocks in mammals. *Annual Review of Neuroscience* 35: 445–462.

Moore-Ede, M. C., Sulzman, F. M., and Fuller, C. A. 1982. *The Clocks That Time Us: Physiology of the Circadian Timing System*. Cambridge, MA: Harvard University Press.

Mor, A., White, M. A., and Fontoura, B. M. A. 2014. Nuclear trafficking in health and disease. *Current Opinion in Cell Biology* 28: 28–35. https://doi.org/10.1016/j.ceb.2014.01.007.

Morange, M. 2008. *Life Explained*. New Haven, CT: Yale University Press.

Moreno, A., Etxeberria, A., and Umerez, J. 2008. The autonomy of biological individuals and artificial models. *BioSystems* 91: 309–319.

Moreno, A., and Mossio, M. 2015. *Biological Autonomy: A Philosophical and Theoretical Enquiry*. Dordrecht: Springer.

Morris, K. V., and Mattick, J. S. 2014. The rise of regulatory RNA. *Nature Reviews Genetics* 15: 423–437.

Moser, M., Frühwirth, M., and Kenner, T. 2008. The symphony of life: Importance, interaction, and visualization of biological rhythms. *IEEE Engineering in Medicine and Biology Magazine* 27 (1): 29–37. https://doi.org/10.1109/MEMB.2007.907365.

Moss, L. 2003. *What Genes Can't Do*. Cambridge, MA: MIT Press.

Moss, L. 2006. The question of questions: What is a gene? *Theoretical Medicine and Bioethics* 27: 523–534.

Muller, A., Schader, C., Scialabba, N. E., Brüggemann, J., Isensee, A., Erb, K. H., Smith, P., et al. 2017. Strategies for feeding the world more sustainably with organic agriculture. *Nature Communications* 8: 1290. https://doi.org/10.1038/s41467-017-01410-w.

Müller, G. B. 2003. Homology: The evolution of morphological organization. In *Origination of Organismal Form: Beyond the Gene in Development and Evolutionary Biology*, ed. Müller, G. B., and Newman, S. A., 51–69, Cambridge, MA: MIT Press.

Müller, G. B. 2007. Evo-devo: Extending the evolutionary synthesis. *Nature Review Genetics* 8: 943–949.

Müller, G. B., ed. 2017a *Vivarium: Experimental, Quantitative, and Theoretical Biology at Vienna's Biologische Versuchsanstalt*. Cambridge, MA: MIT Press.

Müller, G. B. 2017b. Why an extended evolutionary synthesis is necessary. *Interface Focus* 7 (5): 20170015. https://doi.org/10.1098/rsfs.2017.0015.

Müller, G. B., and Newman, S. A., eds. 2003. *Origination of Organismal Form: Beyond the Gene in Development and Evolutionary Biology*. Cambridge, MA: MIT Press.

Müller, J., Bickelmann, C., and Sobral, G. 2018. The evolution and fossil history of sensory perception in amniote vertebrates. *Annual Review of Earth and Planetary Sciences* 46: 495–519. https://doi.org/10.1146/annurev-earth-082517-010120.

Müller, W. A., and Hassel, M. 2006. *Entwicklungsbiologie und Reproduktionsbiologie von Mensch und Tieren*. Berlin: Springer.

Nagel, T. 2012. *Mind and Cosmos: Why the Materialist Neo-Darwinian Conception of Nature Is Almost Certainly False*. Oxford: Oxford University Press.

Neumann-Held, E. 2002. Can we find human nature in the human genome? In *On Human Nature: Anthropological, Biological, and Philosophical Foundations*, ed. Grunwald, A., Gutmann, M., and Neumann-Held, E. M., 141–161. Berlin: Springer.

Newman, S. A. 2002. Developmental mechanisms: Putting genes in their place. *Journal of Biosciences* 27 (2): 97–104. https://doi.org/10.1007/BF02703765.

Newman, S. A. 2012. Physico-genetic determinants in the evolution of development. *Science* 338 (6104): 217–219. https://doi.org/10.1126/science.1222003.

Newman, S. A. 2014. Physico-genetics of morphogenesis: The hybrid nature of developmental mechanisms. In *Towards a Theory of Development,* ed. Minelli, A., and Pradeu, T., 95–113. Oxford: Oxford University Press.

Nicholson, D. J. 2014a. The machine conception of the organism in development and evolution: A critical analysis. *Studies in History and Philosophy of Biological and Biomedical Sciences* 48: 162–174.

Nicholson, D. J. 2014b. The return of the organism as a fundamental explanatory concept in biology. *Philosophy Compass* 9 (5): 347–359.

Nicholson, D. J. 2018. Reconceptualizing the organism: From complex machine to flowing stream. In *Everything Flows: Towards a Processual Philosophy of Biology,* ed. Nicholson, D. J., and Dupré, J., 139–166. Oxford: Oxford University Press.

Nicholson, D. J., and Dupré, D. J., eds. 2018. *Everything Flows: Towards a Processual Philosophy of Biology.* Oxford: Oxford University Press.

Nicholson, D. J., and Gawne, R. 2015. Neither logical empiricism nor vitalism, but organicism: What the philosophy of biology was. *History and Philosophy of the Life Sciences* 37: 345–381.

Nijhout, H. F. 1990. Metaphors and the roles of genes in development. *BioEssays* 12: 441–446.

Niklas, K. J., Bondos, S. E., Dunker, A. K., and Newman, S. A. 2015. Rethinking gene regulatory networks in light of alternative splicing, intrinsically disordered protein domains, and post-translational modifications. *Frontiers in Cell and Developmental Biology* 3: 8. https://doi.org/10.3389/fcell.2015.00008.

Noble, D. 2006. *The Music of Life: Biology beyond Genes.* Oxford: Oxford University Press.

Noble, D. 2008a. Claude Bernard, the first systems biologist, and the future of physiology. *Experimental Physiology* 93 (1): 16–26.

Noble, D. 2008b. Genes and causation. *Philosophical Transactions of the Royal Society A: Mathematical, Physical and Engineering Sciences* 366 (1878): 3001–3015.

Noble, D. 2011a. Neo-Darwinism, the modern synthesis and selfish genes: Are they of use in physiology? *Journal of Physiology* 589 (5): 1007–1015.

Noble, D. 2011b. Systems: What's in a name? *Physiology* 26: 126–128.

Noble, D. 2012. A theory of biological relativity: No privileged level of causation. *Interface Focus* 2 (1): 55–64. https://doi.org/10.1098/rsfs.2011.0067.

Noble, D. 2013a. A biological relativity view of the relationships between genomes and phenotypes. *Progress in Biophysics and Molecular Biology* 111: 59–65.

Noble, D. 2013b. Physiology is rocking the foundations of evolutionary biology. *Experimental Physiology* 98 (8): 1235–1243. https://doi.org/10.1113/expphysiol.2012.071134.

Noble, D. 2017a. *Dance to the Tune of Life: Biological Relativity.* Cambridge: Cambridge University Press.

Noble, D. 2017b. Evolution viewed from physics, physiology and medicine. *Interface Focus* 7 (5): 20160159. https://doi.org/10.1098/rsfs.2016.0159.

Noble, D., Jablonka, E., Joyner, M. J., Müller, G. B., and Omholt, S. W. 2014. Evolution evolves: Physiology returns to centre stage. *Journal of Physiology* 592 (11): 2237–2244.

Noble, R., and Noble, D. 2017. Was the watchmaker blind? Or was she one-eyed? *Biology* 6 (4): 47. https://doi.org/10.3390/biology6040047.

Noble, D., Noble, R., and Schwaber, J. 2013. What is it to be Conscious? In *The Claustrum: Structural, Functional, and Clinical Neuroscience,* ed. Smythies, J., Edelstein, L. and Ramachandran, V., 353–363. Amsterdam: Academic.

Normandin, S., and Wolfe, C. T., eds. 2013. *Vitalism and the Scientific Image in Post-Enlightenment Life Science, 1800–2010.* Dordrecht: Springer.

Nurse, P. 2008. Life, logic and information. *Nature* 454: 424–426.

Odling-Smee, F. J. 2010. Niche inheritance. In *Evolution: The Extended Synthesis*, ed. Pigliucci, M., and Müller, G. B., 175–207. Cambridge, MA: MIT Press.

Odling-Smee, F. J., Laland, K. N., and Feldman, M. W. 2003. *Niche Construction: The Neglected Process in Evolution*. Princeton, NJ: Princeton University Press.

Oliver, J. D., and Perry, R. S. 2006. Definitely life but not definitively. *Origins of Life and Evolution of the Biosphere* 36 (5–6): 515–521. https://doi.org/10.1007/s11084-006-9035-4.

O'Malley M. A., and Dupré, J. 2005. Fundamental issues in systems biology. *BioEssays* 27: 1270–1276.

Oyama, S. 2000. *The Ontogeny of Information: Developmental Systems and Evolution*. Durham: Duke University Press.

Oyama, S. 2002. The nurturing of natures. In *On Human Nature: Anthropological, Biological, and Philosophical Foundations*, ed. Grunwald, A., Gutmann, M., and Neumann-Held, E. M., 163–170. Berlin: Springer.

Oyama, S., Griffiths, P. E., and Gray, R. D. 2001. *Cycles of Contingency: Developmental Systems and Evolution*. Cambridge, MA: MIT Press.

Paneth, N., and Joyner, M. J. 2018. Editors' introduction to the special issue. *Perspectives in Biology and Medicine* 61 (4): 467–471. https://doi.org/10.1353/pbm.2018.0057.

Parrington, J. 2015. *The Deeper Genome: Why There Is More to the Human Genome Than Meets the Eye*. Oxford: Oxford University Press.

Penttonen, M., and Buzáki, G. 2003. Natural logarithmic relationship between brain oscillators. *Thalamus and Related Systems* 2 (2): 145–152. https://doi.org/10.1017/S1472928803000074.

Penzlin, H. 2009. Jakob von Uexküll legte die Grundlagen zu seiner "Umweltlehre." *Biologie in unserer Zeit* 39 (5): 349–352.

Penzlin, H. 2012. Was heißt "lebendig"? Der Selbst-Organisation auf der Spur. *Biologie in unserer Zeit* 42 (1): 56–63. https://doi.org/10.1002/biuz.201210470.

Penzlin, H. 2014. *Das Phänomen Leben. Grundfragen der Theoretischen Biologie*. Berlin Heidelberg: Springer Spektrum.

Persson, L., Carney Almroth, B. M., Collins, C. D., Cornell, S., de Wit, C. A., Diamond, M. L., Fantke, P., et al. 2022. Outside the safe operating space of the planetary boundary for novel entities. *Environmental Science & Technology*. 56 (3): 1510–1521. https://doi.org/10.1021/acs.est.1c04158.

Peterson, E. L. 2016. *The Life Organic: The Theoretical Biology Club and the Roots of Epigenetics*. Pittsburgh: University of Pittsburgh Press.

Piccolo, S. 2013. Mechanics in the embryo. *Nature* 504: 223–225.

Pigliucci, M. 2014. Between holism and reductionism: A philosophical primer on emergence. *Biological Journal of the Linnean Society* 112 (2): 261–267. https://doi.org/10.1111/bij.12060.

Pigliucci, M., and Müller, G. 2010. *Evolution: The Extended Synthesis*. Cambridge, MA: MIT Press.

Plessner, H. 1975. *Die Stufen des Organischen und der Mensch*. Berlin: De Gruyter.

Polanyi, M. 1968. Life's irreducible structure. *Science* 160: 1308–1312.

Popa, R. 2010. Necessity, futility and the possibility of defining life are all embedded in its origin as a punctuated-gradualism. *Origins of Life and Evolution of Biospheres* 40 (2): 183–190. https://doi.org/10.1007/s11084-010-9198-x.

Prigogine, I., and Stengers, I. 1984. *Order out of Chaos: Man's new dialogue with nature*. Toronto: Bantam.

Purnell, B. A. 2012. Forcefull thinking. *Science* 338: 209.

Radlanski, R. J., and Renz, H. 2006. Genes, forces, and forms: Mechanical aspects of prenatal craniofacial development. *Developmental Dynamics* 235 (5): 1219–1229. https://doi.org/10.1002/dvdy.20704.

Ravasz, E., Somera, L., Mongru, D. A., Oltvai, N., and Barabási, L. 2002. Hierarchical organization of modularity in metabolic networks. *Science* 297: 1551–1555.

Reber, A. S. 2019. *The First Minds: Caterpillars, Karyotes, and Consciousness.* Oxford: Oxford University Press.

Reganold, J. P., and Wachter, J. W. 2016. Organic agriculture in the twenty-first century. *Nature Plants* 2: 15221. https://doi.org/10.1038/nplants.2015.221.

Rehmann-Sutter, C. 2000. Biological organicism and the ethics of the human-nature relationship. *Theory in Biosciences* 119: 334–354.

Rehmann-Sutter, C. 2002. Genetics, embodiment and identity. In *On Human Nature: Anthropological, Biological, and Philosophical Foundations*, ed. Grunwald, A., Gutmann, M., and Neumann-Held, E. M., 23–50. Berlin: Springer.

Reilly, S. W., Wiley, E. O., and Meinhardt, D. J. 1997. An integrative approach to heterochrony: The distinction between interspecific and intraspecific phenomena. *Biological Journal of the Linnean Society* 60 (1): 119–143.

Rensing, L. 1973. *Biologische Rhythmen und Regulation.* Stuttgart: Fischer.

Richter, D. J., and King, N. 2013. The genomic and cellular foundations of animal origins. *Annual Review of Genetics* 47: 509–537.

Riedl, R. 1978. *Order in Living Organisms: A Systems Analysis of Evolution.* New York: Wiley.

Riedl, R. 1979. Über die Biologie des Ursachen-Denkens—ein evolutionistischer, systemtheoretischer Versuch. In *Mannheimer Forum 78/79*, ed. Ditfurth, H. Mannheim: Boehringer.

Riedl, R. 1984. *Die Strategie der Genesis.* Munich: Piper.

Riegner, M. 1985. Horns, hooves, spots, and stripes: Form and pattern in mammals. *Orion Nature Quarterly* 4 (4): 22–35.

Riegner, M. 2008. Parallel evolution of plumage pattern and coloration in birds: Implications for defining avian morphospace. *The Condor* 110 (4): 599–614.

Riegner, M. 2013. Ancestor of the new archetypal biology: Goethe's dynamic typology as a model for contemporary evolutionary developmental biology. *Studies in History and Philosophy of Biological and Biomedical Sciences* 44: 735–744.

Riskin, J. 2016. *The Restless Clock: A History of the Centuries-Long Argument over What Makes Living Things Tick.* Chicago: University of Chicago Press.

Ritter, W. 1919. *The Unity of the Organism, or the Organismal Conception of Life.* Boston: Gorham.

Ritter, W., and Bailey, E. W. 1928. The organismal conception: Its place in science and its bearing on philosophy. *University of California Publications in Zoology* 31: 307–358.

Rockström, J., Steffen, W., Noone, K., Persson, A., Chapin, F. S., Lambin, E., Lenton, T. M., et al. 2009. Planetary boundaries: Exploring the safe operating space for humanity. *Ecology and Society* 14 (2): 32.

Rodríguez-Pascual, F. 2019. How evolution made the matrix punch at the multicellularity party. *Journal of Biological Chemistry* 294 (3): 770–771.

Rose, H., and Rose, S. 2013. *Genes, Cells and Brains: The Promethean Promises of the New Biology.* London: Verso.

Rose, S. 1981. *Against Biological Determinism.* London: Alison and Busby.

Rose, S. 1988. Reflections on reductionism. *Trends in Biochemical Sciences* 13: 160–162.

Rose, S. 1997. *Lifelines: Biology, Freedom, Determinism.* Harmondsworth: Penguin.

Rosen, R. 1991. *Life Itself: A Comprehensive Inquiry into the Nature, Origin, and Fabrication of Life.* New York: Columbia University Press.

Rosslenbroich, B. 2011a. Outline of a concept for organismic systems biology. *Seminars in Cancer Biology* 21 (3): 156–164. https://doi.org/10.1016/j.semcancer.2011.06001.

Rosslenbroich, B. 2011b. Patterns and processes in macroevolution. *Annals of the History and Philosophy of Biology (Universitätsverlag Göttingen 2013)* 16: 171–184.

Rosslenbroich, B. 2014. *On the Origin of Autonomy: A New Look at the Major Transitions in Evolution.* Cham: Springer.

Rosslenbroich, B. 2016a. Gegensatz und Synthese von Denkstilen am Beispiel der Evolutionstheorie. In *Denkstile und Schulenbildung in der Biologie*, ed. Kaasch, M,, Kaasch, J., and Himmel, T. K. D., 139–149. Verhandlungen zur Geschichte und Theorie der Biologie, bd. 19. Berlin: VWB.

Rosslenbroich, B. 2016b. The significance of an enhanced concept of the organism for medicine. *Evidence-Based Complementary and Alternative Medicine (Hindawi)* 2016: 1587652. https://doi.org/10.1155/2016/1587652.

Rosslenbroich, B. 2023. Human success as a complex of autonomy, adaptation and niche construction. In *Human Success: Evolutionary Origins and Ethical Implications*, ed. Desmond, H., and Ramsey, G., 84–117. Oxford: Oxford University Press.

Ruiz-Mirazo, K., and Moreno, A. 2012. Autonomy in evolution: From minimal to complex life. *Synthese* 185: 21–52. https://doi.org/10.1007/s11229-011-9874-z.

Ruiz-Mirazo, K., Peretó, J., and Moreno, A. 2004. A universal definition of life: Autonomy and open-ended evolution. *Origins of Life and Evolution of the Biosphere* 34: 323–346.

Ruse, M. 2013. *The Gaia Hypothesis: Science on a Pagan Planet*. Chicago: University of Chicago Press.

Saetzler, K., Sonnenschein, C., and Soto, A. M. 2011. Systems biology beyond networks: Generating order from disorder through self-organization. *Seminars in Cancer Biology* 21: 165–174.

Salthe, S. N. 1985. *Evolving Hierarchical Systems: Their Structure and Representation*. New York: Columbia University Press.

Santolini, M., and Barabási, M. L. 2018. Predicting perturbation patterns from the topology of biological networks. *Proceedings of the National Academy of Sciences of the United States of America* 115 (27): E6375–E6383. www.pnas.org/cgi/doi/10.1073/pnas.1720589115.

Scarfe, A. 2013. On a "life-blind spot" in neo-Darwinism's mechanistic metaphysical lens. In *Beyond Mechanism: Putting Life Back into Biology*, ed. Henning, B. G., and Scarfe, A., 25–64. Lanham, MD: Lexington Books.

Schad, W. 1982. Biologisches Denken. In *Goetheanistische Naturwissenschaft*, bd. 1, *Allgemeine Biologie*, ed. Schad, W., 9–25. Stuttgart: Freies Geistesleben.

Schad, W. 1992. *Die Heterochronien in der Evolution der Wirbeltierklassen*. Dissertation, Faculty of Health, School of Medicine, Witten/Herdecke University.

Schad, W. 1993. Heterochronical patterns of evolution in the transitional stages of vertebrate classes. *Acta Biotheoretica* 41: 383–389.

Schad, W. 1997. *Die Zeitintegration als Evolutionsmodus*. Habilitation, Faculty of Health, School of Medicine, Witten/Herdecke University.

Schad, W. 2006. Der wissenschaftliche Zugang zum Ätherischen. *Jahrbuch für Goetheanismus*, 82–140. Niefern-Öschelbronn: Tycho de Brahe Verlag.

Schad, W. 2016. *Zeitbindung in Natur, Kultur und Geist*. Stuttgart: Freies Geistesleben.

Schad, W. 2020. *Understanding Mammals: Threefoldness and Diversity*. New York: Adonis.

Schaefer, K. E., Hensel, H., and Brady, R., eds. 1977. *Toward a Man-Centered Medical Science: A New Image of Man in Medicine*, vol. 1. Mt. Kisco, NY: Futura.

Schlosser, M. 2015. "Agency." In *The Stanford Encyclopedia of Philosophy*, ed. E. N. Zalta. Stanford University, 2015. Article published August 10, 2015. http://plato.stanford.edu/archives/fall2015/entries/agency.

Schrödinger, E. 1944. *What Is Life? The Physical Aspect of the Living Cell*. New York: Macmillan.

Schwarz, A., and Jax, K., eds. 2011. *Ecology Revisited: Reflecting on Concepts, Advancing Science*. Dordrecht: Springer.

Seamon, D., and Zajonc, A., eds. 1998. *Goethe's Way of Science: A Phenomenology of Nature*. New York: State University of New York Press.

Seilacher, A., and Gishlick, A. D. 2015. *Morphodynamics*. Boca Raton, FL: CRC Press.

Shapiro, J. A. 2009. Revisiting the central dogma in the 21st century. *Annals of the New York Academy of Sciences* 1178 (1): 6–28. https://doi.org/10.1111/j.1749-6632.2009.04990.x.

Shapiro, J. A. 2010. Mobile DNA and evolution in the 21st century. *Mobile DNA* 1: 4. https://doi.org/10.1186/1759-8753-1-4.

Shapiro, J. A. 2011. *Evolution: A View from the 21st Century*. Upper Saddle River, NJ: FT Press Science.

Shapiro, J. A. 2013a. How life changes itself: The Read–Write (RW) genome. *Physics of Life Reviews* 10 (3): 287–323. https://doi.org/10.1016/j.plrev.2013.07.001.

Shapiro J. A. 2013b. Rethinking the (im)possible in evolution. *Progress in Biophysics and Molecular Biology* 111: 92–96.

Shapiro, J. A. 2014. Physiology of the read–write genome. *Journal of Physiology* 592 (11): 2319–2341.

Shapiro, J. A. 2017. Biological action in Read–Write genome evolution. *Interface Focus* 7 (5): 20160115. https://doi.org/10.1098/rsfs.2016.0115.

Sharov, A., Maran, T., and Tønnessen, M. 2015a. Organisms reshape sign relations. *Biosemiotics* 8: 361–365.

Sharov, A., Maran, T., and Tønnessen, M. 2015b. Towards synthesis of biology and semiotics. *Biosemiotics* 8: 1–7.

Sharov, A. A. 1992. Biosemiotics: Functional-evolutionary approach to the problem of the sense of information. In *Biosemiotics: The Semiotic Web*, ed. Sebeok, T. A., and Umiker-Sebeok, J., 345–373. New York: Mouton de Gruyter.

Sharov, A., and Tønnessen, M. 2021. *Semiotic Agency: Science beyond Mechanism*. Cham: Springer.

Shubin, N. 2008. *Your Inner Fish: A Journey into the 3.5-Billion-Year History of the Human Body*. New York: Pantheon.

Shubin, N. H., and Marshall, C. R. 2000. Fossils, genes, and the origin of novelty. *Paleobiology* 26 (4 Supplement): 324–340.

Smith, C. U. M. 2010. Darwin's unsolved problem: The place of consciousness in an evolutionary world. *Journal of the History of the Neurosciences* 19 (2): 105–120. https://doi.org/10.1080/09647040903504781.

Sonnenschein, C., and Soto, A. M. 1999. *The Society of Cells: Cancer and Control of Cell Proliferation*. New York: Taylor & Francis.

Sonnenschein, C., and Soto, A. M. 2013. Cancer genes: The vestigial remains of a fallen theory. In *Genetic Explanations: Sense and Nonsense*, ed. Krimsky, S., and Gruber, J., 81–93. Cambridge, MA: Harvard University Press.

Sonnenschein, C., and Soto, A. M. 2016. Carcinogenesis explained within the context of a theory of organisms. *Progress in Biophysics and Molecular Biology* 122: 70–76.

Soto, A. M., and Sonnenschein, C. 2005a. Emergentism as a default: Cancer as a problem of tissue organization. *Journal of Biosciences* 30: 103–118.

Soto, A. M., and Sonnenschein, C. 2005b. The somatic mutation theory of cancer: Growing problems with the paradigm. *BioEssays* 26: 1097–1107.

Soto, A. M., and Sonnenschein, C. 2006. Emergentism by default: A view from the bench. *Synthese* 151 (3): 361–376.

Soto, A. M., and Sonnenschein, C. 2011a. The tissue organization field theory of cancer: A testable replacement for the somatic mutation theory. *Bioessays* 5: 332–340.

Soto, A. M., and Sonnenschein, C. 2011b. Why systems biology and cancer? *Seminars in Cancer Biology* 21: 147–149.

Soto, A. M., and Sonnenschein, C. 2012. Is systems biology a promising approach to resolve controversies in cancer research? *Cancer Cell International* 12: 12. https://doi.org/10.1186/1475-2867-12-12.

Soto, A. M., and Sonnenschein, C. 2018. Reductionism, organicism, and causality in the biomedical sciences: A critique. *Perspectives in Biology and Medicine* 61 (4): 489–502. https://doi.org/10.1353/pbm.2018.0059.

Soto, A. M., Sonnenschein, C., and Miquel, P. A. 2008. On physicalism and downward causation in development and cancer biology. *Acta Biotheoretica* 56 (4): 257–274.

Stamp Dawkins, M. 1993. *Through Our Eyes Only? The Search for Animal Consciousness.* Oxford: Freeman/Spektrum.

Steffen, W., Sanderson, A., Tyson, P., Jäger, J., Matson, P., Moore, B., Oldfield, F., et al. 2005. *Global Change and the Earth System: A Planet under Pressure.* Berlin: Springer.

Steiner, R. 1985a. *Goethe's World View.* New York: Spring Valley.

Steiner, R. 1985b. *The Origins of Natural Science.* New York: Spring Valley.

Steiner, R. 1996. *The Science of Knowing.* New York: Spring Valley.

Stelling, J., Sauer, U., Szallasi, Z., Doyle, I. F. J., and Doyle, J. 2004. Robustness of cellular functions. *Cell* 118: 675–685.

Sterelny, K. 2005. Made by each other: Organisms and their environment. *Biology and Philosophy* 20: 21–36.

Stewart, I. 2002. *Does God Play Dice?* Malden, MA: Blackwell.

Stonier, T. 1996. Information as a basic property of the universe. *BioSystems* 38: 135–140.

Stonier, T. 1997. *Information and Meaning: An Evolutionary Perspective.* Berlin: Springer.

Strohman, R. 1993. Ancient genomes, wise bodies, unhealthy people: Limits of a genetic paradigm in biology and medicine. *Perspectives in Biology and Medicine* 37: 112–145.

Strohman, R. 1997. The coming Kuhnian revolution in biology. *Nature Biotechnology* 15: 194–200.

Strohman, R. 2001. A new paradigm for life beyond genetic determinism. *California Monthly*, April 2001, 24–27.

Strohman, R. 2002. Maneuvering in the complex path from genotype to phenotype. *Science* 296: 701–703.

Strohman, R. 2003. Genetic determinism as a failing paradigm in biology and medicine: Implications for health and wellness. *Journal of Social Work Education* 39 (2): 169–191.

Suchantke, A. 1993. *Partnerschaft mit der Natur.* Stuttgart: Urachhaus.

Suchantke, A. 2010. *Metamorphosis: Evolution in Action.* Edinburgh: Floris.

Thompson, D. W. 1917. *On Growth and Form.* Cambridge: Cambridge University Press.

Thompson, E. 2007. *Mind in Life: Biology, Phenomenology, and the Sciences of Mind.* Cambridge, MA: Harvard University Press.

Travis, J. 2013. Mysteries of development. *Science* 340: 1156.

Tseng, Q., Duchemin-Pelletier, E., Deshiere, A., Balland, M., Guillou, H., Filhol, O., and Théry, M. 2012. Spatial organization of the extracellular matrix regulates cell–cell junction positioning. *Proceedings of the National Academy of Sciences of the United States of America* 109 (5): 1506–1511. https://doi.org/10.1073/pnas.1106377109.

Tsokolov, S. A. 2009. Why is the definition of life so elusive? Epistemological considerations. *Astrobiology* 9 (4): 401–412.

Turner, J. 2000. *The Extended Organism: The Physiology of Animal-Built Structures.* Cambridge, MA: Harvard University Press.

Turner, J. S. 2007. *The Tinkerer's Accomplice: How Design Emerges from Life Itself.* Cambridge, MA: Harvard University Press.

Turner, J. S. 2013. Homeostasis and the forgotten vitalist roots of adaptation. In *Vitalism and the Scientific Image in Post-Enlightenment Life Science, 1800–2010*, ed. Normandin, S., and Wolfe, C. T., 271–291. Dordrecht: Springer.

Turner, J. S. 2017. *Purpose and Desire: What Makes Something "Alive" and Why Modern Darwinism Has Failed to Explain It.* New York: Harper Collins.

Ueki, N., Matsunaga, S., Inouye, I., and Hallmann, A. 2010. How 5000 independent rowers coordinate their strokes in order to row into the sunlight: Phototaxis in the multicellular green alga *Volvox*. *BMC Biology* 8: 103. https://www.biomedcentral.com/1741-7007/8/103.

Uexküll, J. 1909. *Umwelt und Innenwelt der Tiere*. Berlin: Springer Verlag.

Uexküll, J. 1973. *Theoretische Biologie*. Frankfurt aM: Suhrkamp.

Uexküll, T., and Wesiak, W. 1998. *Theorie der Humanmedizin—Grundlagen ärztlichen Denkens und Handelns*. Munich: Urban & Schwarzenberg.

Vahle, H. C. 2007. *Die Pflanzendecke unserer Landschaften. Eine Vegetationskunde*. Stuttgart: Freies Geistesleben.

Vale, R. D. 2003. The molecular motor toolbox for intracellular transport. *Cell* 112: 467–480.

van der Steen, W. J. 1997. Limitations of general concepts: A comment on Emmeche's definition of "life." *Ultimate Reality and Meaning* 20: 317–320.

Van Regenmortel, M. H. V. 2004. Reductionism and complexity in molecular biology. *EMBO Reports* 5: 1016–1020.

Varela, F., Maturana, H., and Uribe, R. 1974. Autopoiesis: The organization of living systems, its characterization and a model. *BioSystems* 5: 187–195.

Varela, F. J. 1979. *Principles of Biological Autonomy*. New York: North Holland.

Vaux, D. L., and Korsmeyer, S. J. 1999. Cell death in development. *Cell* 96: 245–254.

Venter, C. 2007. *Life Decoded*. London: Penguin.

Verkman, A. S., and Mitra, A. K. 2000. Structure and function of aquaporin water channels. *American Journal of Physiology Renal Physiology* 278: F13–F28.

Vliet, C., Thomas, E. C., Merino-Trigoa, A., Teasdaleb, R. D., and Gleeson, P. A. 2003. Intracellular sorting and transport of proteins. *Progress in Biophysics & Molecular Biology* 83: 1–45.

Waddington, C. H. 1952. The evolution of developmental systems. In *Proceedings of the Twenty-Eighth Meeting of the Australian and New Zealand Association for the Advancement of Science*, ed. Herbert, D. A., 155–159. Brisbane: A. H. Tucker, Government Printer.

Waddington, C. H. 1957. *The Strategy of the Genes: A Discussion of Some Aspects of Theoretical Biology*. London: Allen and Unwin.

Waddington, C. H. 1960. *The Ethical Animal*. New York: Atheneum.

Wagner, A. 2007. Distributional robustness versus redundancy as causes of mutational robustness. *Bioessays* 27: 176–188.

Wagner, A. 2012. The role of robustness in phenotypic adaptation and innovation. *Proceedings of the Royal Society B* 279 (1732): 1249–1258. https://doi.org/10.1098/rspb.2011.2293.

Wagner, D. L., Grames, E. M., Forister, M. L., Berenbaum, M. R., and Stopak, D. 2021. Insect decline in the Anthropocene: Death by a thousand cuts. *Proceedings of the National Academy of Sciences USA* 18 (2): e2023989118. https://doi.org/10.1073/pnas.2023989118.

Wagner, G. P. and Laubichler, M. D. 2004. Rupert Riedl and the re-synthesis of evolutionary and developmental biology: Body plans and evolvability. *Journal of Experimental Zoology Part B: Molecular and Developmental Evolution* 302: 92–102.

Wake, D. 1982. Functional and evolutionary morphology. *Perspectives in Biology and Medicine* 25 (4): 603–620.

Walker, S. I. 2014. Top-down causation and the rise of information in the emergence of life. *Information* 5 (3): 424–439. https://doi.org/10.3390/info5030424.

Walker, S. I., Davies, P. C. W., and Ellis, G. F. R. 2017. *From Matter to Life: Information and Causality*. Cambridge: Cambridge University Press.

Walsh, D. 2015. *Organisms, Agency, and Evolution*. Cambridge: Cambridge University Press.

Weingarten, M. 1993. *Organismen—Objekte oder Subjekte der Evolution? Philosophische Studien zum Paradigmawechsel in der Evolutionsbiologie*. Darmstadt: Wissenschaftliche Buchgesellschaft.

Weiss, K. M. 2018. Genetic pointillism versus physiological form. *Perspectives in Biology and Medicine* 61 (4): 503–516.

Weiss, P. A. 1959. Cellular dynamics. *Reviews of Modern Physics* 31 (1): 11–20.

Weiss, P. A. 1962. From cell to molecule. In *The Molecular Control of Cellular Activity*, ed. Allen, J. M., 1–72. New York: McGraw-Hill.

Weiss, P. A. 1963. The cell as unit. *Journal of Theoretical Biology* 5: 389–397.

Weiss, P. A. 1968. *Dynamics of Development: Experiments and Inferences.* New York: Academic.

Weiss, P. A. 1969. The living system: Determinism stratified. In *Beyond Reductionism: New Perspectives in the Life Sciences*, ed. Koestler, A., and Smythies, J. R., 3–55. London: Hutchinson.

Weiss, P. A. 1971. The basic concept of hierarchic systems. In *Hierarchically Organized Systems in Theory and Practice*, ed. Weiss, P. A., 1–43. New York: Hafner.

Weiss, P. A. 1973. *The Science of Life: The Living System—A System for Living.* New York: Futura.

Weiss, P. A. 1977. The system of nature and the nature of systems: Empirical holism and practical reductionism harmonized. In *Toward a Man-Centered Medical Science: A New Image of Man in Medicine*, vol. 1., ed. Schaefer, K., Hensel, H., and Brady, R., 17–63. Mt. Kisko, NY: Futura.

West-Eberhard, M. J. 2003. *Developmental Plasticity and Evolution.* Oxford: Oxford University Press.

Wheatley, D. N. 1998. Diffusion theory, the cell and the synapse. *Biosystems* 45 (2): 151–163. https://doi.org/10.1016/s0303-2647(97)00073-7.

Willmer, P. 2003. Convergence and homoplasy in the evolution of organismal form. In *Origination of Organismal Form: Beyond the Gene in Development and Evolutionary Biology*, ed. Müller, G. B., and Newman, A., 33–49. Cambridge, MA: MIT Press.

Wilson, E. O. 1995. *Naturalist.* New York: Warner Books.

Winning, J., and Bechtel, W. 2018. Rethinking causality in biological and neural mechanisms: Constraints and control. *Minds and Machines* 28 (2): 287–310. https://doi.org/10.1007/s11023 -018-9458-5.

Winzeler, E. A., Shoemaker, D. D., Astromoff, A., Liang, H., Anderson, K., Andre, B., Bangham, R., et al. 1999. Functional characterization of the *S. cerevisiae* genome by gene deletion and parallel analysis. *Science* 285 (5429): 901–906.

Woese, C. 2004. A new biology for a new century. *Microbiology and Molecular Biology Reviews* 68 (2): 173–186.

Wolpert, L. 1991. *The Triumph of the Embryo.* Oxford: Oxford University Press.

Woodger, J. H. 1929. *Biological Principles: A Critical Study.* London: Routledge & Kegan Paul.

Wozniak, M. A., and Chen, C. S. 2009. Mechanotransduction in development: A growing role for contractility. *Nature Reviews in Molecular Cell Biology* 10: 34–43.

Wuketits, F. M. 1981. *Biologie und Kausalität: Biologische Ansätze zur Kausalität, Determination und Freiheit.* Berlin: Parey Verlag.

Wuketits, F. M. 2008. Kausalität. *Naturwissenschaftliche Rundschau* 61 (5): 269–270.

Yamashita S., Arakaki, Y., Kawai-Toyooka, H., Noga, A., Hirono, M., and Nozaki, H. 2016. Alternative evolution of a spheroidal colony in volvocine algae: Developmental analysis of embryogenesis in *Astrephomene* (Volvocales, Chlorophyta). *BMC Evolutionary Biology* 16 (1): 243. https://doi.org/10.1186/s12862-016-0794-x.

Yu, B. J., Sung, B. H., Koob, M. D., Lee, C. H., Lee, J. H., Lee, W. S., Kum, M. S., and Kim, S. C. 2002. Minimization of the *Escherichia coli* genome using a Tn5-targeted Cre/*loxP* excision system. *Nature Biotechnology* 20: 1018–1023.

Zaller, J. G. 2020. *Daily Poison: Pesticides—an Underestimated Danger.* Cham: Springer.

Index

Page numbers in italics refer to figures; page numbers in bold refer to tables.